Remote Sensing Techniques and GIS Applications in Earth and Environmental Studies

Abhisek Santra
Haldia Institute of Technology, India

Shreyashi Santra Mitra
Haldia Institute of Technology, India

A volume in the Advances in Geospatial Technologies (AGT) Book Series

IGI GLOBAL
DISSEMINATOR OF KNOWLEDGE

www.igi-global.com

Published in the United States of America by
IGI Global
Information Science Reference (an imprint of IGI Global)
701 E. Chocolate Avenue
Hershey PA 17033
Tel: 717-533-8845
Fax: 717-533-8661
E-mail: cust@igi-global.com
Web site: http://www.igi-global.com

Library of Congress Cataloging-in-Publication Data

Names: Santra, Abhisek, editor.
Title: Remote sensing techniques and GIS applications in earth and environmental studies / Abhisek Santra and Shreyashi Santra Mitra, editors.
Description: Hershey PA : Information Science Reference, [2017]
Identifiers: LCCN 2016045759| ISBN 9781522518143 (hardcover) | ISBN 9781522518150 (ebook)
Subjects: LCSH: Environmental monitoring--Remote sensing. | Environmental monitoring--Geographic information systems. | Earth sciences--Remote sensing. | Earth sciences--Geographic information systems.
Classification: LCC GE45.R44 R49 2017 | DDC 628.028/7--dc23 LC record available at https://lccn.loc.gov/2016045759

This book is published in the IGI Global book series Advances in Geospatial Technologies (AGT) (ISSN: 2327-5715; eISSN: 2327-5723)

British Cataloguing in Publication Data
A Cataloguing in Publication record for this book is available from the British Library.

Advances in Geospatial Technologies (AGT) Book Series

ISSN:2327-5715
EISSN:2327-5723

Mission

The geospatial technology field is a fast-paced, high growth industry that is involved in a variety of fields, including military planning, public health, land use, environmental protection, and Google Earth mapping. With such a diverse body of applications, the research in geospatial technologies is always evolving and new theories, methodologies, tools, and applications are being developed. **Advances in Geospatial Technologies (AGT) Book Series** is a reference source and outlet for research that discusses all aspects of geographic information, including areas such as geomatics, geodesy, GIS, cartography, remote sensing, and other areas. Because geospatial technologies are so pervasive in such a variety of areas, AGT also includes books that address interdisciplinary applications of the technologies.

Coverage

- Cartography
- Public Sector Use of Geospatial Technologies
- Geomatics
- Geospatial Analysis
- Geographical Information Systems
- Digital Terrain Modeling
- Global Positioning Systems
- Environmental Monitoring
- Surveying Technologies
- Spatial Reference Systems

IGI Global is currently accepting manuscripts for publication within this series. To submit a proposal for a volume in this series, please contact our Acquisition Editors at Acquisitions@igi-global.com or visit: http://www.igi-global.com/publish/.

The Advances in Geospatial Technologies (AGT) Book Series (ISSN 2327-5715) is published by IGI Global, 701 E. Chocolate Avenue, Hershey, PA 17033-1240, USA, www.igi-global.com. This series is composed of titles available for purchase individually; each title is edited to be contextually exclusive from any other title within the series. For pricing and ordering information please visit http://www.igi-global.com/book-series/advances-geospatial-technologies/73686. Postmaster: Send all address changes to above address.

Titles in this Series

For a list of additional titles in this series, please visit: www.igi-global.com

Handbook of Research on Geographic Information Systems Applications and Advancements
Sami Faiz (LTSIRS Laboratory, University of Tunis El Manar, Tunisia) and Khaoula Mahmoudi (LTSIRS Laboratory, University of Tunis El Manar, Tunisia)
Information Science Reference • copyright 2017 • 650pp • H/C (ISBN: 9781522509370) • US $315.00 (our price)

Geo-Intelligence and Visualization through Big Data Trends
Burçin Bozkaya (Sabanci University School of Management, Turkey) and Vivek Kumar Singh (Rutgers, The State University of New Jersey, USA & Massachusetts Institute of Technology, USA)
Information Science Reference • copyright 2015 • 348pp • H/C (ISBN: 9781466684652) • US $225.00 (our price)

Handbook of Research on Emerging Digital Tools for Architectural Surveying, Modeling, and Representation
Stefano Brusaporci (University of L'Aquila, Italy)
Engineering Science Reference • copyright 2015 • 829pp • H/C (ISBN: 9781466683792) • US $525.00 (our price)

Geographic Information Analysis for Sustainable Development and Economic Planning New Technologies
Giuseppe Borruso (University of Trieste, Italy) Stefania Bertazzon (University of Calgary, Canada) Andrea Favretto (University of Trieste, Italy) Beniamino Murgante (University of Basilicata, Italy) and Carmelo Maria Torre (Polytechnic of Bari, Italy)
Information Science Reference • copyright 2013 • 434pp • H/C (ISBN: 9781466619241) • US $180.00 (our price)

Qualitative Spatio-Temporal Representation and Reasoning Trends and Future Directions
Shyamanta M. Hazarika (Tezpur University, India)
Information Science Reference • copyright 2012 • 381pp • H/C (ISBN: 9781616928681) • US $195.00 (our price)

Universal Ontology of Geographic Space Semantic Enrichment for Spatial Data
Tomaž Podobnikar (Scientific Research Centre of the Slovenian Academy of Sciences and Arts, Slovenia & University of Ljubljana, Slovenia) and Marjan Čeh (University of Ljubljana, Slovenia)

Table of Contents

Detailed Table of Contents

Solar radiation on hitting a target surface may be transmitted, absorbed or reflected. Different materials reflect and absorb differently at different wavelengths. The reflectance spectrum of a material is a plot of the fraction of radiation reflected as a function of the incident wavelength and serves as a unique signature for the material. In principle, a material can be identified from its spectral reflectance signature if the sensing system has sufficient spectral resolution to distinguish its spectrum from those of other materials. This premise provides the basis for multispectral remote sensing. Nguyen Dinh Duong (1997) proposed a method for decomposition of multi-spectral image into several sub-images based on modulation (spectral pattern) of the spectral reflectance curve. The hypothesis roots from the fact that different ground objects have different spectral reflectance and absorption characteristics which are stable for a given sensor. This spectral pattern can be considered as invariant and be used as one of classification rules.

Earth's land surface temperature is considered to be very important for modeling the environment. Following the trend of increasing global population, urban areas are expanding in spatio-temporal domain. In this way it is affecting the urban climate and subsequently the global climate. Thus, scientific understanding is required to conceive the knowledge about interaction between urban land use/land cover and the atmospheric conditions prevailing in that area. In this chapter the land surface temperature estimation and urban heat island detection are perceived from remote sensing perspective. The chapter in this context highlights three major aspects, viz. the theoretical background, description about some of the common thermal sensors and widely used algorithms to retrieve surface temperature from these satellite sensors.

It is widely recognized that interpretations about the planet Mars must begin by using Earth as a reference, because Earth analogues can provide ground truth to constrain interpretations on the geological history of Mars. Fortunately, Deccan Basalt and its derivatives, exposed in Kutch area of Gujarat, can be considered as best candidate from Indian subcontinent to approximate this analogue in terms of petrology, stratigraphy and geochemical similarity. It is believed that Deccan Volcanic province with its extensive volcanic plains, cones and craters as well as abundant hydrous sulphates of secondary origin, approximates the geological, geomorphological and environmental conditions on the Mars.

Researchers worldwide use remotely sensed imagery in their projects, in both the social and natural sciences. However, users often encounter difficulties working with satellite images and aerial photographs, as image interpretation requires specific experience and skills. The best way to acquire these skills is to go into the field, identify your location in an overhead image, observe the landscape, and find corresponding features in the overhead image. In many cases, personal observations

The study aims to predict landslide hazard zones near Tehri Dam in Uttarakhand State located in the Western Himalayas in India. Four different models were analysed: Weight Factor Model (M1), Multiple Factor Model (M2), Statistical Bivariate model (M3) and Analytical Hierarchical Processes (AHP) model (M4). Five different combination of reference landslides were used for deriving weights of the classes in the factor maps: all the landslides from 1990, 2002 & 2010 (C1); landslides from 2010 (C2); landslides from 1990 & 2002 (C3); landslides located within 500m from roads (C4); landslides located beyond 500m from roads (C5). The accuracy resulted from each model in each combination was [Mn:C1, C2...Cn]: M1: 60,44,46,38,66%; M2:70,76,79,73,71%; M3:45,37,23,36,85%; M4:64,51,51,64,36%. Multiple Factor Model (M2) resulted in a consistently high accuracy in all the combinations. Finally, the 20 different model outputs were integrated to derive unified hazard zonation maps based on: (a) mean (85% accuracy), (b) penalisation (57% accuracy) and (c) k-means cluster (80% accuracy) approaches.

The Dulung River flows across West Bengal and Jharkhand in India. The geographical variables present within the basin area are categorised into groups like Physiographic, Morphometric and Land Use-Land Cover (LULC) attributes. These facets are mapped and overlain in a GIS environment and correlations drawn between them. Factor Scores obtained through Principal Component Analysis are further compared and correlated. The different variables are fused to obtain a comprehensive grouping of the above three facets that is reflective of the overall terrain attributes and its overlying LULC classes. Through this, within the Dulung River Basin, three broad Physiographic-Soil-Land Use Units (PSLUs) are identified, which comprise of the structural ridges and residual hills, piedmont plains and floodplains. For further insight into existing LULC-landform relations, select villages across the basin landscape are examined in detail. The relations derived help in suggesting possible land management practices in this region.

Remote sensing technologies offer the potential for contributing the security to human existence on arid zones in the country in variety of ways. Remote Sensing in agriculture particularly for natural resource management. It provides important coverage, mapping and classification of land cover features. The remote view of the sensor and the ability to store, analyze, and display the sensed data on field maps are make remote sensing a potentially important tool for agriculture. The aerial photography gives two main advantages viz., speedy survey in very large area or remote area and precise description and recording of resources status. Remotely sensed images provide a means to assess field conditions and gave valuable insights into agronomic management. It led to understanding of leaf reflectance and leaf emittance changes in response to leaf thickness, species, canopy shape, leaf age, nutrient status, and water status. Understanding of leaf reflectance has led to quantify various agronomic parameters, e.g., leaf area, crop cover, biomass, crop type, nutrient status, and yield.

Immediately following a natural disaster, it is imperative to accurately assess the damages caused by the disaster for effective rescue and relief operations. Passive remote sensing imageries have been analyzed and used for over four decades for such assessments. However, they do have their limitations including inability to collect data during violent weather conditions, medium to low spatial resolution, and assessing areas and pixels on a damages/no damage basis. Recent advances in active remote sensing data collection methods can resolve some of these limitations. In this chapter, the basic theories and processing techniques of active remote sensing data is first discussed. It then provides some of the advantages and limitations of using active remote sensing data for disaster damage assessments. Finally, the chapter concludes by discussing how data from active sensors are used to assess damages from various types of natural disasters.

Remote Sensing and Geographic Information Systems are a boon to the modern scientific domain as it provides a strong base for understanding the Earth's behavior in so many ways and help us to understand its response to certain conditions. With the exploitation of these two modules important spatial decisions can be made at ease with quite a satisfactory accuracy level. With the context of global warming water scarcity is being identified as sector demanding strategical implementation of plans for sustainable development. To ensure water supply water conservation is required. Runoff harvesting could be an option to that if the area under plan experience substantial runoff in a year. Lower Kasai Keleghai interfluve situated in West Bengal falls under monsoonal climatic regime. GIS based site-suitability analysis was adopted here based on LULC and Curve Number to identify potential zones for runoff harvesting. The result shows that 18.71% of the interfluve is suitable for providing a base for runoff harvesting.

Forecasting of Potential Fishing Zone (PFZ) is considered as economically and environmentally significant towards ensuring profitable base of economy and planning for sustenance of existing fishing pool. Changes in environmental conditions affect the distribution, abundance and availability of fish. The traditional sampling approaches for PFZ identification using boats and vessels are not only costly and time consuming but practically absurd considering the vastness of seas and oceans. In this chapter importance of alternative but effective methods of airborne and satellite remote sensing has been given. The chapter elucidates the factors for PFZ identification like thermal condition in sea controlling its thermal circulation, chlorophyll-a concentration estimated from ocean color dynamics, etc. Tools/system to prepare PFZ advisories and also the platforms for dissemination of the same, have been illustrated based on Indian scenario.

Before an individual can evaluate wildlife habitat and make management recommendations, some basic concepts about habitat and its relationships to different wildlife species should be understood. In this chapter, some of the basic concepts will be described; mainly analyzing of habitat alterations, landscape analysis, networking and creation of corridor between protected areas, wildlife habitat suitability analysis using Remote Sensing & GIS. Since most of the contest will be based on these concepts. Like other natural resource fields, wildlife management is both an art and science that deals with complex interactions in the environment. This means that management includes art or judgment based on experience as well as sound factual information based on scientific studies.

Ecological Footprint (EF) analysis is the spatial measurement of ecological load exerted by the humans on the earth that arises from the concept of sustainability and sustainable use of Earth's resources. A region-based EF study is conducted for Birla Institute of Technology, Mesra (India) campus to improve its sustainability. Highlight of the study is the explicitness of the methodology for determining the EF that incorporates analysis derived from conversion factors mentioned in the Ecological Footprint consultancy publications along with inputs from GIS domain. Questionnaire-based survey from the respondents regarding resource utilization and geospatial enumeration of land use land cover that harbors the population and their resources are the two integral parts of the analysis. Total EF of the institution campus is calculated to be 0.645 gha/ individual. This analysis provides a strong framework for combining efforts in a manner that can communicate the immediate priorities for improving the sustainability strategy of the campus area.

Foreword

It gives me pleasure to say few words regarding this book entitled "Remote Sensing Techniques and GIS Applications in Earth and Environmental Studies" and its Editors. Multifrequency techniques for identification purpose using delta – DOR which is essential for long distance remote sensing to get sky measurements of right ascension and declination are considered. VLBI based techniques with large baselines east-west and north-south direction used. Many important issues like ground truth validation, public health management, mapping of landslide, river basin studies using medium gain antenna operating in dual circular polarization with capability of handling high power on board satellites are well discussed. Circularly polarized Quadrifilar Helix Antenna provides a circularly polarized cadoids shaped radiation pattern with a high front to back ratio, better circular symmetry in the pattern and very good polarization discrimination over wide coverage are being used for studies. These types of antenna used for transmission of satellite telemetry data with omni-directional coverage in C-band for India's all upcoming Geo-stationary and Navigational satellite. In observation of very useful parameters, the microwave and millimeterewave radiometers like Total Power Radiometer, Fast fourier Transform detector, Dicke Radiometer, Ratiometric Dicke Radiometer, Meech Ratiometric Radiometer are utilised. Limb sounding techniques are used for atmospheric parameters detection. In short in this book all latest technological application outcomes are well covered. The application of SAR imaging is discussed in other easy to understand approach. Both the editors are widely experienced in this area with long research and teaching experiences. While forwarding this excellent book I recall my working experience with RCI, DRDO and GMRT, TIFR, Pune through University of Calcutta and feel extremely happy to see the topics of Applications of superbly efficient technological systems for the benefit of human civilisations. I hope the editors Dr. Abhisek Santra and Dr. Shreyashi Santra Mitra will be very successful in their great effort and students, teachers and professionals will find this book very useful.

Manab Kumar Sengupta
Haldia Institute of Technology, India

Preface

Remote Sensing may be conceptualized as the acquisition of information about an object or phenomenon by means of a sensor without being any physical contact between that object or phenomenon and the sensor. These sensors are capable of detecting and measuring the reflected or emitted radiations from the objects or phenomena under study. Measurements are made in different spectral regions of electromagnetic spectrum on interactions between the targets and the electromagnetic radiation (EMR). The data acquisition process is not dependent upon this interaction only, but also upon the interaction between the EMR and the atmosphere. After interacting with the targets and the atmosphere, the reflected or emitted EMR reaches to the airborne and/or space borne sensors. It was all started just after the World War-I with the aerial photography. The scope was enhanced with the subsequent development of radars in Germany, U.S.A. and U.K. The era of satellite remote sensing began in 1960s when the first satellite TIROS-1 was launched. It was designed for climatological and meteorological observations. During this phase, some of the remote sensing devices originally developed for defence purpose, were released for civilian use as more advanced devices became available for military reconnaissance. These devices enhanced the reach of aerial observation outside the visible spectrum into infrared and microwave regions. The term 'Remote Sensing' established its entity beyond the scope of 'Aerial Photography' in broader sense. To support remote sensing research at the institution level throughout the country, United States National Aeronautics and Space Administration (NASA) established a research programme in remote sensing in early 1960s. National Academy of Sciences (NAS) scrutinized the opportunities for remote sensing applications in the field of agriculture and forestry. As a result of the contributions from different directions, NASA launched Landsat-1, the first of many earth-orbiting satellites designed for monitoring earth's land surface. Since then, there has been a rapid growth in science, technology and applications of remote sensing. Today, we have a series of satellite and airborne sensors governed by different countries like U.S.A., Canada, India, Japan, China, France, Russia, Brazil, etc. For example, Landsat, MODIS, IRS, SPOT, Radarsat, ENVISAT, etc. provide earth observation data at various resolutions and at regular interval to monitor both the land and water bodies of the earth. Synoptic earth observation data is now being used as the most dependable and cost effective means of monitoring and analyzing our earth features and processes.

Geographical Information System (GIS) on the other hand, encircles the scientific study of geospatial patterns and processes. It takes into account the location and distribution of natural and anthropogenic phenomena on earth. GIS can be explained as a computer system for retrieving, storing, querying, analyzing and displaying geospatial data. However, the ability of handling and processing geospatial data distinguishes GIS from other information systems. The age of GIS is comparatively less than that of remote sensing. Since the late 1960s, computers have been used to store and process geospatial data. GIS works in 1960s and 70s included computer mapping of university premises, national land inventory, urban road network, etc. For several years, GIS was considered as a difficult and expensive means of analysis. The idea has been revolutionized in 1990s with the advent of powerful hardware and software packages and with the introduction of Graphical User Interface (GUI). A set of commercial software packages came to the market and made GIS, a mainstream mode of decision making process. In the early 21st century, the open-source and free GIS software packages brought millions of users in the GIS community. Together with the remote sensing, GIS is now even considered as one of the most advanced courses of science and technology at the major universities of the world.

Remote Sensing and GIS have revolutionized our perception about planet earth as an integrated system, giving us a growing capability to monitor the dynamism of earth features and processes. Remote sensing and GIS are considered as important tools in earth and environmental management, providing detailed and up-to-date information. The synergetic use of remote sensing and GIS offers a cost-effective periodic update of earth and environmental information at various scales starting from local to global. From the end-user perspective, development in planning & policy decision making has been triggered and encouraged by remote sensing and GIS. On the other hand, from technological point of view, these tools would not flourish without the advancement of research in pure and applied sectors of physics, space science, computer science & engineering and electronics & telecommunication. This dual nature of remote sensing and GIS brought us to the idea of dividing the book into two broad sections – techniques and applications. Our motivation for this book is to illustrate the potential for remote sensing and GIS in various aspects of earth and environmental studies. This book covers some of the latest development areas in terms of techniques and applications of remote sensing and GIS in earth and environmental studies by praiseworthy researchers, scholars and experts in this field. A total of fourteen chapters are presented by the contributors from different corners of the world in form of this book. Key ideas of these chapters are presented subsequently.

Chapters 1 to 5 illustrate some of the important technical as well as developmental aspects of remote sensing and GIS. Chapter 1 describes the role of spectral indices in identifying the geospatial objects. The author has applied NDVI, NDWI and MNDWI as spectral indices to identify stagnant water bodies and compared the results with their spectral pattern algorithm generated results. The author has identified the scope of Artificial Neural Network (ANN) technique to get more realistic outputs.

Chapter 2 provides the techniques of remote estimation of land surface temperature. The chapter reviews the theoretical background and common algorithms to retrieve land surface temperature. A brief account on the major thermal sensors is also given. Following the trend, the author showed the future direction of research in this field of remote sensing.

Chapter 3 outlines the spectroscopic vis-à-vis geochemical study of Jarosite & Goethite as an extent of suitability as terrestrial mineralogical / geochemical analogue to Mars from earth observation point of view. The methodology adopted by the author addresses the approximation of terrestrial and Martian Jarosites.

Chapter 4 discusses the significance of terrestrial photographs as a substitute for field observations and explains the issues of data collection. The authors present results of experimental estimation of the effectiveness of the use of terrestrial ground truth photographs for interpretation of remote sensing images. The chapter introduces the concept of GeoTruth – a web-based collaborative framework for collecting, storing and distributing ground truth terrestrial photographs and corresponding metadata.

Chapter 5 proposes a framework for an essential public health information visualization platform for Japanese Encephalitis disease. The author used Erdas Apollo 2010 to develop the architecture for web-GIS based public health information system. A GUI has been created in this regard using Java Server Pages (JSP) for its customization.

Chapters 6 to 14 show some of the application areas of remote sensing and GIS in various domains of earth and environmental studies. Chapter 6 illustrates the modeling aspects of debris flow and risk assessment of selected landslides occurred in Uttarakhand state of India. Rapid Mass Movements (RAMMS) numerical model was used to estimate the motion of geophysical mass movements. The author also shows the information about model input and frictional parameters of the model.

Chapter 7 aims to integrate spatial as well as temporal aspects of landslides in Western Himalayas. Four different models were analyzed to divide the study area into different landslide hazard zones. The authors adopted five different combinations of reference landslides.

Chapter 8 establishes the correlation between land use and terrain characteristics in the piedmont tract of Eastern India. The authors estimated various basin morphometric attributes along with lithology, soil and climate characteristics. Factor scores were obtained through principal component analysis and compared & correlated further. The outcomes of the correlations help in suggesting possible land management practices in the study area.

Chapter 9 offers the techniques and applications of remote sensing in arid environment of India. The author describes alternate land use system, assessment and improvement of degradation status of rangelands, moisture stress, rainwater management, weed and nutrient management, soil and water resources assessment, pest management and disease control, precision crop management, animal productivity, etc. under the light of arid environment.

Chapter 10 highlights the applications of active remote sensing for assessing damage caused by natural disaster. The chapter gives glimpses of basic theories and processing techniques of active remote sensing data. The author also provides the advantages and limitations of using such data for disaster damage assessment.

Chapter 11 underpins the idea of site suitability analysis to identify potential sites to construct rain-water harvesting systems. The author applied Modified SCS-CN method and identified the suitable areas for providing a base for runoff harvesting.

Chapter 12 identifies the role of remote sensing in forecasting potential fishing zones (PFZs). The authors elucidate the various factors for PFZ identification based on satellite and airborne remote sensing data. Tools / system to prepare PFZ advisories and dissemination platforms have also been illustrated in Indian scenario.

Chapter 13 evaluates wildlife habitat in relation with wildlife species. Habitat alterations, landscape analysis, networking and creation of corridor between protected areas, wildlife habitat suitability analysis using remote sensing and GIS are discussed in this chapter. The authors also make some management recommendations.

Finally, chapter 14 measures the dynamics of ecological footprint as an index of environmental sustainability at the regional level. The authors estimated the ecological capacity and ecological footprint of the study area based on geospatial model. The analysis also provides a strong framework for combining efforts in such a manner that can communicate the immediate priorities for improving the sustainability strategy of the study area.

On this foundation, many new tools, techniques and applications developed for remote sensing and GIS may be realized across these fourteen distinctive chapters. The book is expected to be a guide or handbook for earth and environmental planners and policy makers; researchers and scientists; students at the post graduate level and for those with some experience in the areas of synoptic earth and environmental monitoring. We hope the success of the book through contributing in the growing recognition of remote sensing and GIS as an important spatially based tool for earth and environmental studies.

Acknowledgment

We would like to acknowledge every one of those inspirations who have contributed towards the publication of the book. Many people are implicitly linked to the book, many more explicitly. I am indebted to IGI Global Publishing for taking this initiative to host this book to be published. We are indebted to all of the eminent Editorial Board Members, who have helped us a lot to ensure the qualitative parameters of the book. We would like to convey sincere thanks to Professor Soumitra Mukherjee of Jawaharlal Nehru University, Professor C. Jeganathan Chokalingham of BIT Mesra, Dr. D. Mitra of Indian Institute of Remote Sensing, Professor Asish Sarkar of Chandannagar Govt. College for their continuous support and invaluable suggestions for the betterment of this book. We were privileged to work under the guidance of eminent scientists like Prof. M.N. Bandyopadhyay, Director of Haldia Institute of Technology and Prof. Prabir Paul., Indian Institute of Education, Science and Technology. We thank all the respected authors for sharing their valuable researches with us in the form of chapters. The authors were the foundations of this project and the level and extent of their works gave us constant encouragement towards maintaining the qualitative aspect of the project, apart from the quantitative issues. All the reviewers put efforts to maintain the standard of the chapters by providing meticulous judgment and strict evaluation. They deserve special thanks from us. We would like to take this opportunity to acknowledge our colleagues and peers for sharing their academic ideas and professional tips for the betterment of the book. Lastly, we would like to express heartfelt acknowledgements to those audiences who have been the motivation of this project. This book will only be successful and effective if the students, researchers, academicians and other personnel getting benefitted from the vast uses of Remote Sensing and GIS can be benefitted from this book.

Section 1
Techniques of Remote Sensing and GIS for Environmental Applications

Chapter 1
Identification of Geospatial Objects Using Spectral Pattern

Subhabrata Barman
Haldia Institute of Technology, India

ABSTRACT

Solar radiation on hitting a target surface may be transmitted, absorbed or reflected. Different materials reflect and absorb differently at different wavelengths. The reflectance spectrum of a material is a plot of the fraction of radiation reflected as a function of the incident wavelength and serves as a unique signature for the material. In principle, a material can be identified from its spectral reflectance signature if the sensing system has sufficient spectral resolution to distinguish its spectrum from those of other materials. This premise provides the basis for multispectral remote sensing. Nguyen Dinh Duong (1997) proposed a method for decomposition of multispectral image into several sub-images based on modulation (spectral pattern) of the spectral reflectance curve. The hypothesis roots from the fact that different ground objects have different spectral reflectance and absorption characteristics which are stable for a given sensor. This spectral pattern can be considered as invariant and be used as one of classification rules.

DOI: 10.4018/978-1-5225-1814-3.ch001

INTRODUCTION

Through the use of satellites, we now have a continuing program of data acquisition for the entire world with time frames ranging from a couple of weeks to a matter of hours. Very importantly, we also now have access to remotely sensed images in digital form, allowing rapid integration of the results of remote sensing analysis into a GIS. The development of digital techniques for the restoration, enhancement and computer-assisted interpretation of remotely sensed images initially proceeded independently and somewhat ahead of GIS. However, the raster data structure and many of the procedures involved in these Image Processing Systems (IPS) were identical to those involved in raster GIS. As a result, it has become common to see IPS software packages add general capabilities for GIS, and GIS software systems add at least a fundamental suite of IPS tools.

DIGITAL IMAGE

A digital remotely sensed image is typically composed of picture elements (pixels) located at the intersection of each row i and column j in each K bands of imagery. Associated with each pixel is a number known as Digital Number (DN) or Brightness Value (BV) that depicts the average radiance of a relatively small area within a scene. A smaller number indicates low average radiance from the area and the high number is an indicator of high radiant properties of the area. The size of this area effects the reproduction of details within the scene. As pixel size is reduced more scene detail is presented in digital representation.

COLOUR COMPOSITES

While displaying the different bands of a multi spectral data set, images obtained in different bands is displayed in image planes (other than their own) the colour composite is regarded as False Colour Composite (FCC). High spectral resolution is important when producing colour components. For a true colour composite an image data used in red, green and blue spectral region must be assigned bits of red, green and blue image processor frame buffer memory. A colour infrared composite 'standard false colour composite' is displayed by placing the infrared, red and green in the red, green and blue frame buffer memory. In this healthy vegetation shows up in shades of red because vegetation absorbs most of green and red energy but reflects approximately half of incident Infrared energy. Urban areas reflect equal portions of NIR, R & G, and therefore they appear as steel grey.

There are three steps to generate colour composites from panchromatic images:

1. Read the panchromatic images using *geotiffread* function.
2. Channel the panchromatic images into a RGB image frame as required.
3. Visualise the RGB image using *imshow* function.

NORMALIZED DIFFERENCE GEO-SPATIAL INDICES

When dealing with images, the knowledge of the techniques of basic image processing always comes handy. Image processing, itself being a vast domain, we mentioned only those methods which are mainly related to our work, in the previous chapter. However, these methods can be applied to the multispectral satellite images with one band at a time. To enhance or extract features from satellite images which cannot be clearly detected in a single band, you have to use the spectral information of the object recorded in multiple bands. These images may be separate spectral bands from a single multispectral data set, or they may be individual bands from data sets that have been recorded at different dates or using different sensors. The operations of addition, subtraction, multiplication and division, are performed on two or more co-registered images of the same geographical area (Sreenivas & Chary, 2011). This section deals with multi-band operations.

The following operations will be treated:

- The use of ratio images to reduce topographic effects.
- Various indexes, some of which are more complex than ratio's only.

Various mathematical combinations of satellite bands, have been found to be sensitive indicators of the presence and condition of green vegetation. These band combinations are thus referred to as vegetation indices. Two such indices are the simple vegetation index (VI) and the normalized difference vegetation index (NDVI).

$$NDVI = (NIR - Red) / (NIR + Red)$$

Remotely sensed imagery has long been used in water resources assessment and coastal management. These applications have involved the delineation of open water using thematic information extraction techniques. There are various methods for the extraction of water information from remote sensing imagery, which, according to the number of bands used, are generally divided into two categories, i.e. single-band and multi-band methods. The single-band method usually involves choosing a band from a multispectral image to extract open water information (Rundquist, Lawson et al. 1987). A threshold is then determined for the band to discriminate water form land. However, the subjective selection of the threshold value may lead to an over- or under-estimation of open water area and the extracted water information is often mixed with shadow noise. The multi-band method takes advantage of reflective differences of each involved band. There are two ways to extract water information using the multi-band method. One is through analysing signature features of each ground target among different spectral bands, finding out the signature differences between water and other targets based on the analysis, and then using an if-then-else logic tree to delineate land from open water (Yu, Huang et al., 2001; Xu, 2002). The other one is a band-ratio approach using two multispectral bands. One is taken from visible wavelengths and is divided by the other usually from near infrared (NIR) wavelengths. As a result, vegetation and land presences are suppressed while water features are enhanced. However, the method can suppress non-water features but not remove them, and therefore the normalized difference water index (NDWI) was proposed by McFeeters (1996) to achieve this goal. The NDWI is expressed as follows (McFeeters, 1996):

$$NDWI = (Green - NIR) / (Green + NIR)$$

Where Green is a green band such as TM band 2, and NIR is a near infrared band such as TM band 4.

To remove the built-up land noise, its signature features were examined. The reflectance pattern of built-up land in the green band (TM 2) and NIR band (TM 4) is similar with that of water, i.e. they both reflect green light more than they reflect near Infrared light. As a result, the computation of the NDWI also produces a positive value for built-up land just as for water. This is why the enhanced water presence in the NDWI-image is often mixed with built-up land noise. However, detailed examination of the spectral signatures reveals that the average digital number of TM band 5, representing middle infrared (MIR) radiation, is much greater than that of TM band 2 (green band) for built up land. Therefore, if a MIR band is used instead of the NIR band in the NDWI, the built-up land should have negative values. Based on this assumption, the NDWI is modified by substituting the MIR band for the NIR band. The modified NDWI (MNDWI) is expressed as follows:

$$(MNDWI) = (Green-MIR) / (Green + MIR)$$

Where MIR is a middle infrared band such as TM band 5.

Spectral Pattern

When solar radiation hits a target surface, it may be transmitted, absorbed or reflected. Different materials reflect and absorb differently at different wavelengths. The reflectance spectrum of a material is a plot of the fraction of radiation reflected as a function of the incident wavelength and serves as a unique signature for the material. In principle, a material can be identified from its spectral reflectance signature if the sensing system has sufficient spectral resolution to distinguish its spectrum from those of other materials. This premise provides the basis for multi-spectral remote sensing.

Duong (1997) proposed a method for decomposition of multi-spectral image into several sub-images based on modulation (spectral pattern) of the spectral reflectance curve. The hypothesis roots from the fact that different ground objects have different spectral reflectance and absorption characteristics which are stable for a given sensor. This spectral pattern can be considered as invariant and be used as one of classification rules.

Spectral Pattern of Water

Spectral Pattern of Water body Given a dataset with four spectral bands: green, red, near infrared and short wave infrared *X(b1, b2, b3, b4)* where the bi stands for reflectance in band *i* converted from digital number DN of each pixel using the gain and bias coefficients provided along with image data. A spectral reflectance curve can be constructed by simple plotting of reflectance values in a graph as showed. If we use the *C* for value of relative positions of each vertex, then modulation or spectral pattern of the spectral reflectance curve could be expressed as: *C12C13C14C23C24C34*. The C value is of {0, 1, 2}. For example, for pixel vector (130, 100, 70, 76) the code will be 222220.

Because the reflectance of clear water is high in green band, low in red and near infrared and almost absorbed in short wave infrared band therefore it is always *b1>b3, b1>b4, b2>b3, b2>b4,* so the *C13, C14, C23, C24* are always 2. For turbid water it can happen that reflectance in red band due to sedimentation is higher than reflectance in green band. Based on experiments the author found that water body is involved mainly in pattern 222222, 022222, and 002222 (Figure 1).

Water Body Extraction Algorithm

- **Pre-Requisite:** Conversion of DN to reflectance is a two-step process. First you must convert DNs to radiance values. Then you need to convert these radiance values to reflectance values (USGS, 2012). From the said process of

Figure 1. Three main modulations of spectral reflectance curve

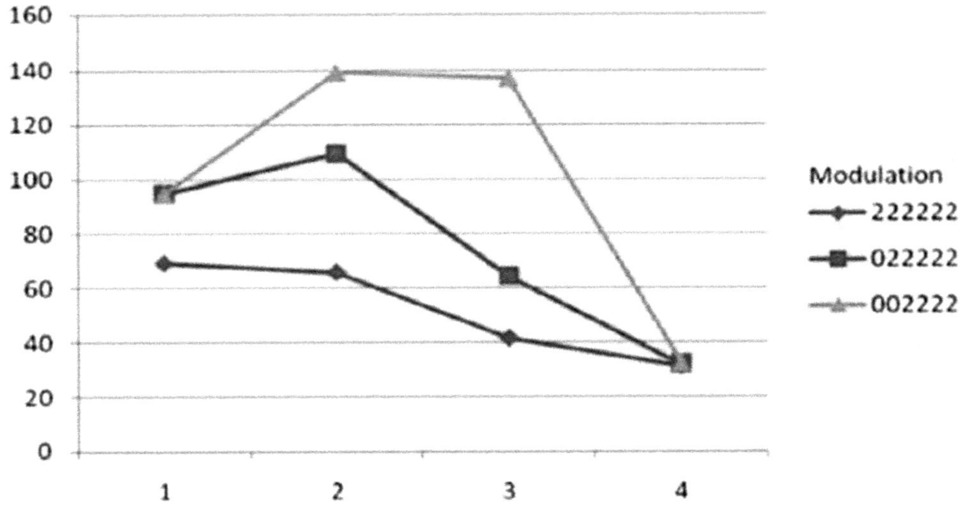

conversion, we can safely conclude that reflectance is directly proportional to DN. As such, we have tried to modify the algorithm (Duong, 2012) by substituting reflectance with DN.

Algorithm:

Step 1: Reading image data.

Step 2: Calculating *C12C13C14C23C24C34* for each pixel using 4 panchromatic images.

Step 3: Case 222222: applying different threshold values for SWIR band of TM sensors to extract water.

Step 4: Case 022222: applying different threshold values for SWIR band of TM sensors to extract water.

It is obvious that the algorithm is composed of two parts: decomposition of the image into sub images according to spectral reflectance patterns and level slicing into water and land using different threshold values for the SWIR band. If we apply a single threshold value for water extraction there might be under or over estimation in clear and turbid water area because reflectance of turbid water is always higher than clear water.

Figure 2. TrueColor composite of the region of interest

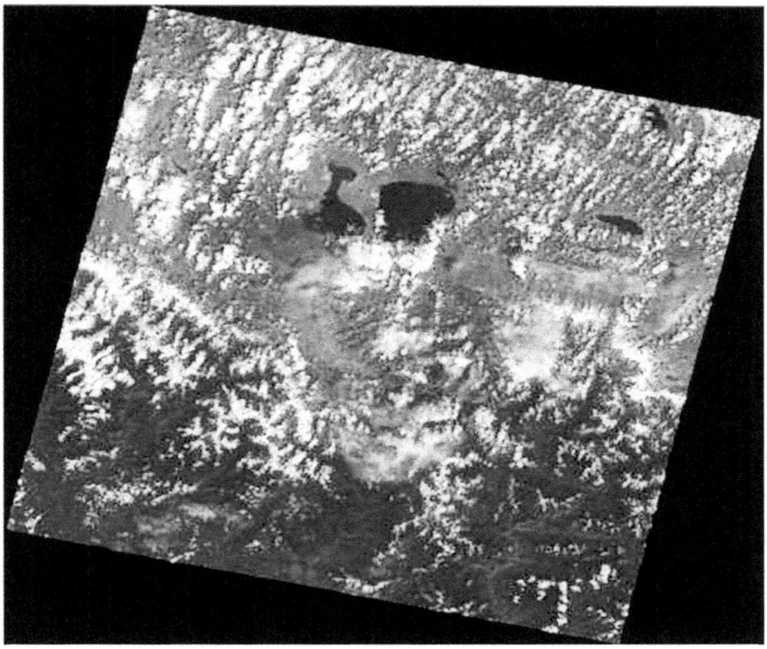

Figure 3. CIR composite of the region of interest
Image information: LANDSAT 8 imagery 2014, File no. LC81440392014250LGN00

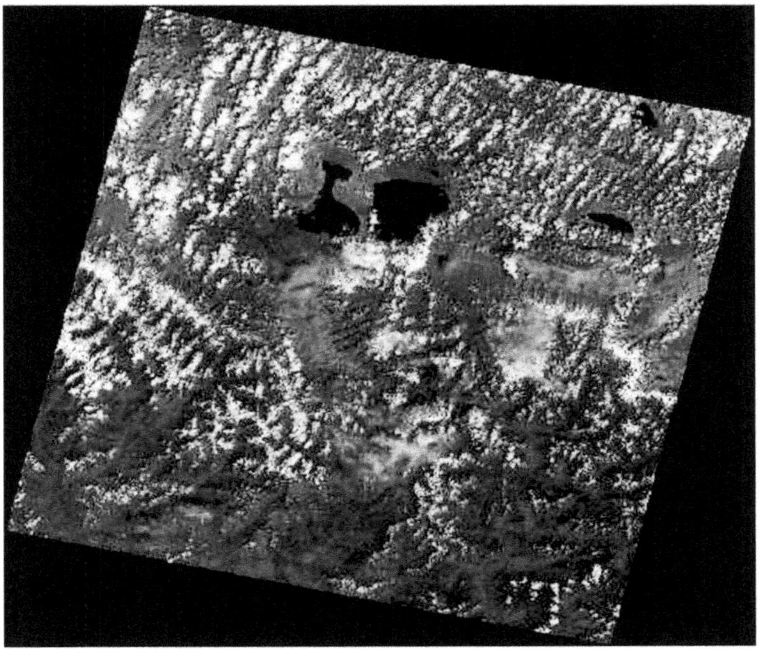

Implementation Overview

The test image (Figure 2 and Figure 3) that we have used in our project is a L1 processed compressed file (Landsat 8 image) which is available in a downloadable format from the website of Earth Explorer, i.e. http://earthexplorer.usgs.gov. On unzipping the compressed file and further extracting, we get the particular 11 different bands, which are panchromatic images. Along with these, a band quality file and Meta data is also given for use. The 11 different bands in which we get the image are mentioned there.

As we can see in the above image, few water bodies are present in the given area. Our main objective is focused on only that area where there is a chance of presence of water reserves. For this purpose, we chalk out a Region of Interest from the given image and applying our algorithms on the concerned area. On applying the algorithm, we got the output images as given below. Since this approach gives a distinct idea about the presence of water bodies, no threshold value was applied here. Furthermore, this method helps us to distinguish between the different forms of water as clear or turbid which can be interpreted from the resultant images (Figure 4 and Figure 5).

The output image so obtained gives, the area of water body as 18.2368% of the total area without turbid water and 19.2278% with turbid water. The difference in

Figure 4. Image showing the regions having clear water

Figure 5. Image showing the water contained regions (both turbid and clear)

the percentage is due to the presence of turbid water which has higher reflectance value than clear water. The Area of water body is calculated using numel() function in Matlab.

Water_body_cover = 100 × numel(R(R(:))./numel(R))

Advantages of Spectral Pattern Methodology over NDWI

The spectral pattern approach helps us in identifying not only water resources, but also distinguish between them as clear water and turbid water, whereas the first one simply helps in detection of water reserves. In the NDWI, a threshold value is needed to clearly identify the water bodies from the rest of the image. This threshold value is obtained from the Hit and Trial method and might not be same for different images. This disadvantage is overcome in the spectral pattern method since no threshold value is required there. The major problem that both the methods face is the presence of cloud pattern in the image. Due to the presence of cloud cover, it becomes difficult to correctly determine the water bodies.

PIXEL BASED CLASSIFICATION BASED ON ARTIFICIAL NEURAL NETWORK

Modelling an environmental phenomenon usually needs some spatial information about the distribution and the types of land cover and land use (LCLU) as well as soil types. Proper classification of LCLU is a very essential requirement for all modelling tasks in environmental problems. However, in remote arid areas this is difficult to obtain easily due to lack of information and inaccessibility of these areas. Therefore, utilizing automatic remote sensing techniques will provide a reasonable answer to this problem. Nevertheless, knowing the best classification method to perform this task is a very important aspect in order to utilize the right approach for classification (Al-Ahmadi & Hames, 2009). In the current work, the authors have tried to classify images from IRS-LISS III and LANDSAT 7 ETM+ using artificial neural networks. For the LISS image, we have classified the pixels into two classes that is water and non-water. As for LANDSAT 7 imagery, we have done a pixel based classification to classify the pixels into clear water, muddy water and non-water classes using ANN.

CLASSIFICATION OF SATELLITE IMAGE

Image classification refers to the task of extracting information classes from a multiband raster image. The resulting raster from image classification can be used to create thematic maps. The overall objective of image classification is to automatically categorize all pixels in an image into and cover classes or themes. Normally, multispectral data are used to perform the classification, and the spectral pattern present within the data for each pixel is used as numerical basis for categorization. That is, different feature types manifest different combination of DNs based on their inherent spectral reflectance and emission properties. Classification is a widely studied issue in remote sensing image processing. The ability to map open surface water is an integral part to many hydrologic and agricultural models, wildlife management programmes, and recreational and natural resource studies.

Pixel Based Classification

Pixel based image analysis means that the classic image classification method that classifies remote sensing images according to the spectral information in the image and the classification manner is pixel by pixel and one pixel only belongs to one class. In supervised classification, the image analysis supervises the pixel categorization process by specifying, to the computer algorithm, numerical descriptors of the various land-cover types present in an image. Training samples that describe the typical spectral pattern of the land cover classes are defined. Pixels in the image are compared numerically to the training samples and are labelled to the land cover class that has similar characteristics. There are three basic steps involved in the supervised classification method: training stage, classification stage and accuracy assessment stage, these three stages are applied in the classification process of ETM+ spectral bands. A pixel of unknown identity may be classified by computing the distance between the value of the unknown pixel and each of the class means, if the pixel were further than an analyst defined distance (distance threshold) from any class mean, it would be classified as "unknown". This distance threshold could vary for each class depending on the expected degree of compactness of that class. Compactness might be estimated from the standard deviation for each feature of the pixels making up the training sample for a given class (Qian & Zhoua et al. 2007).

A neural network can be trained to associate an output to a given input. This is obtained by giving a set of training inputs and the corresponding targets. During training, the weights change according to a specified rule until they reach an 'optimal' value. To perform the classification with Neural Networks, the DNs (0-255) values of 5 TM bands of LANDSAT 7 (all bands except the blue and thermal one for the image) for each pixel are presented as input. The target classes are represented in vector form *i.e.* they are represented by vectors (1 *x* number of classes) in which the activation of unit *i* is set to 1 when class *i* is being represented and all other units are set to 0.(For example, target for class 2 is [0 1 0 0 0 0 0…] and so on).The network scheme adopted is a 3-layers feed-forward (back-propagation)network, layers with appropriate transfer function. The back-propagation algorithm is the most used in multilayer Neural Networks. It defines two sweeps of the network: a forward one from the input layer to the output layer, and then a backward one when the errors are back-propagated to determine how the weights should be changed during training. Learning is achieved using a generalization of the delta rule (Callan, 1999). A full description of the back-propagation algorithm can be found in any basic Neural Network book like Callan (1999). Scale conjugate gradient back-propagation is used for training the network based on the given datasets. The Epoch vs MSE graph is given in Figure 8 and Figure 9. Once the training and validation is done for the given datasets, the neural network system is deployed to classify the entire image. The LISS-III and LANDSAT 7 classified output image are given in Figure 6 and Figure 7 respectively.

Figure 6. LISS-III image classified into water bodies and non-water bodies (white denotes the water regions)

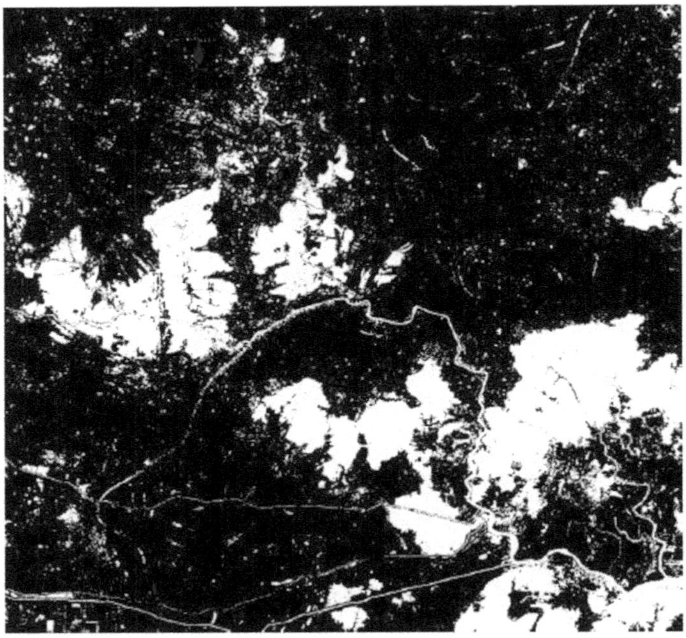

Figure 7. LANDSAT 7 image classified into water, muddy water, and non-water bodies (red denotes clear water bodies, green denotes muddy water, and blue denotes non water bodies)

Figure 8. Epoch vs. mean squared error for LANDSAT image

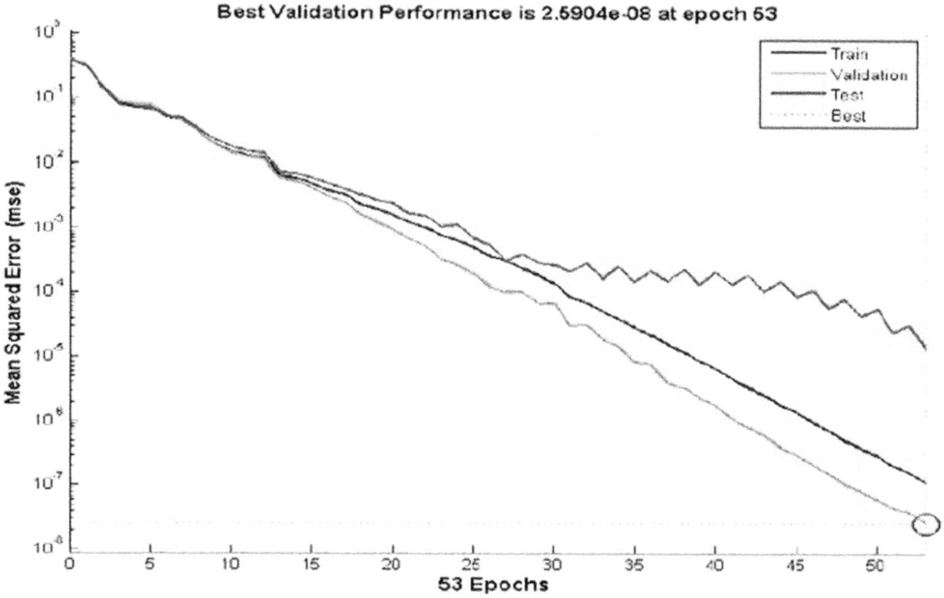

Figure 9. Epoch vs. mean squared for LISS-III image

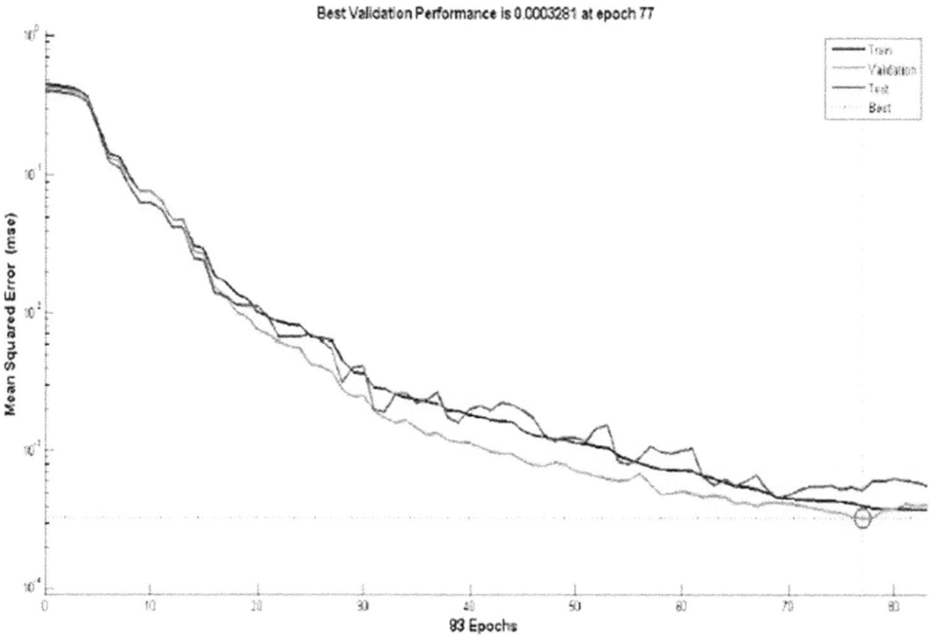

RESULTS AND DISCUSSION

Though visual interpretation of the classified image, we can easily verify that the pixels are efficiently classified their respective classes. However, the neural network method of classification suffers from an inherent drawback, which is the availability of accurate datasets for training purposes. Given the lack of ground data, we have tried to sample the training data by visually interpreting the panchromatic images of the specified location. The best validation performance is 2.6904e-08 at epoch 53 for LANDSAT 7 image and for LISS-III image it is 0.0003281 at epoch 77, and thereby we can say that it is highly efficient despite the inherent drawbacks. For the LANDSAT 7 image, 30.81% of the total area of the image is covered with water.

A comparative study of the algorithms that have implemented on the images obtained from IRS LISS-III satellite is presented. We begin with comparing the combined outputs of MNDWI and NDVI with that generated by the object extraction algorithm using spectral pattern. The two techniques are mainly used to identify the land objects from the imagery. However, they cannot be used to differentiate between clear water sources and turbid water sources. Also, they are highly prone to errors due to the presence of clouds, which alter the spectral pattern for a given pixel. Thus, a classification algorithm, namely Artificial Neural Network was being used to give a better idea of the images. The supervised classification algorithm depends mainly on the accuracy of the training data sets. Through visual interpretation of the individual band images, we were able to gather the training sets and thereby classify the entire image.

CONCLUSION

Further modification can be done in the procedure using ANN technique to get more realistic outputs. In order to eradicate the erroneous values, more accurate ground data is required. More accurate results can be obtained by making atmospheric corrections and using additional noise removal techniques. The techniques implemented in this project can also be used to develop a standalone software that would be capable of identifying the various spectral signature from each pixels and classify them into various geo-spatial objects.

REFERENCES

Al-Ahmadi, F. S., & Hames, A. A. (2009). Comparison of Four Classification Methods to Extract Land Use and Land Cover from Raw Satellite Images forSome Remote Arid Areas, Kingdom of Saudi Arabia. *Earth Science*, *20*(1), 167–191.

Callan, R. (1999). The essence of Neural Networks: Prentice Hall Europe.

Duong, N. D. (2012). Water Body Extraction from Multispectral Image by Spectral Pattern Analysis. *Paper presented at the22nd ISPRS Congress.*

McFeeters, S. K. (1996). The use of the Normalized Difference Water Index (NDWI) in the delineation of open water features. *International Journal of Remote Sensing*, 17(7), 1425-1432. doi: 10.1080/01431169608948714

Qian, J., Zhoua, Q., & Houa, Q. (2007). Comparison of pixel-based and object-oriented classification methods for extracting built-up areas in arid zone. *Paper presented at the ISPRS Workshop on Updating Geo-spatial Databases with Imagery & the 5th ISPRS Workshop on DMGISs.*

Rundquist, D., Lawson, M., Queen, L., & Cerveny, R. (1987). The Relationship between Summer Season Rainfall and Lake-Surface Area. *Water Resources Bulletin*, *23*(3), 493–508. doi:10.1111/j.1752-1688.1987.tb00828.x

Sreenivas, B., & Chary, B. N. (2011, January 18-21). Processing of Satellite Image using Digital Image Processing. *Paper presented at the Geospatial World Forum*, Hyderabad, India.

USGS. (2012). *LANDSAT Data Continuity Mission (LDCM) Level 1 (L1) Data Format Control Book (DFCB).* South Dakota: USGS.

Xu, H. (2002). Spatial Expansion of Urban/Town in Fuquing of China and its Driving Force Analysis. *Remote Sensing Technology and Application*, *17*, 86–92.

Yu, J., Huang, Y., & Feng, X. (2001). Study on water bodies extraction and classification from SPOT image[in Chinese]. *Journal of Remote Sensing*, *5*, 214–219.

Chapter 2
Land Surface Temperature Estimation and Urban Heat Island Detection:
A Remote Sensing Perspective

Abhisek Santra
Haldia Institute of Technology, India

ABSTRACT

Earth's land surface temperature is considered to be very important for modeling the environment. Following the trend of increasing global population, urban areas are expanding in spatio-temporal domain. In this way it is affecting the urban climate and subsequently the global climate. Thus, scientific understanding is required to conceive the knowledge about interaction between urban land use/land cover and the atmospheric conditions prevailing in that area. In this chapter the land surface temperature estimation and urban heat island detection are perceived from remote sensing perspective. The chapter in this context highlights three major aspects, viz. the theoretical background, description about some of the common thermal sensors and widely used algorithms to retrieve surface temperature from these satellite sensors.

DOI: 10.4018/978-1-5225-1814-3.ch002

INTRODUCTION

Land surface temperature (LST) is measured at the Earth's surface. It is considered as the skin temperature of the earth. It is an important quantity for many environmental models like energy and water exchange between atmosphere and surface of the earth; numerical prediction model of weather; global ocean circulation model; climatic variability model, etc. (Dash, Gottsche, & Olesen, 2002; Valor & Caselles, 1996). Urban growth and urban sprawl are the prime factors in regional landscape evolution across the globe. Understanding the interactions among different urban land use and land cover types, atmospheric conditions and land surface temperatures is significant to conceptualize urban climate. Rapid urban expansion due to large scale commercial, manufacturing and transportation development leads to the emergence of Urban Heat Island (UHI) effect (Landsberg, 1981). The urban areas are characterized by higher temperature in comparison with the surrounding rural areas, as shown in Figure 1. The process of urbanization can raise the local temperatures, however, the temperatures of built-up fringed areas will generally remain constant (Resenzweig et al., 2005).

UHI results from differential characteristics of radiation and heat budget in urban landscape. Oke (1982) listed some common causes of the UHI including:

1. High absorption of short wave radiation because multiple reflection.
2. Human induces heat sources.

Figure 1. Urban heat island

3. Attenuated outgoing long wave radiation due to the presence of atmospheric constituents.
4. Low radiative cooling.
5. High sensible heat storage due to increased thermal admittance of building materials.
6. Low evapotranspiration due to construction material.
7. Low total turbulent heat fluxes from wind speed reduction in the city.

However, the UHI effects may be decreased with increasing wind speed and cloud cover. Conversely UHI may raise with increasing population and size of urban areas. The stronger effect of UHI may be visible during summer to late fall season (Oke, 1982). UHIs are responsible for the changes occur in the precipitation characteristics of the urban areas. These changes may be similar or greater than that predicted by global change models (Changon, 1992). Thus, the increasing temperatures associated with UHI s may induce climatic and biophysical hazards in urban environments including heat stress, anxiety and acute response to air pollutants (Resenzweig et al., 2005; Watson, 2012). Similarly, UHIs also have significant impact on human health, energy demand and environmental conditions (Strathopoulou, Cartalis, & Kermatisoglou, 2004). Oke (1991) commented that the intensity of an UHI can be estimated by determining the maximum difference between urban minimum temperature and rural minimum temperature in consideration with the size, population, layout, topography and atmospheric conditions of the city. The UHI map depict the urban locations that produce hot spots within the city This may be considered as useful information to urban planners and policy makers regarding new strategies, which may reduce the adverse impact on urban thermal environment of a city (Strathopoulou & Cartalis, 2007; Strathopoulou et al., 2004; Watson, 2012).

Over the years several methods have been applied to study UHI in relation to land surface Temperature (LST) in small scale to large scale dimension. Some of these studies rely upon ground based observations taken from fixed weather stations or ground based instruments. However, weather station data collected in cities often misinterpret the urban weather conditions as most studies are calibrated to forecast synoptic climatic parameters at places, where the weather parameters are not conformed by anthropogenic activities and urban morphology (Lee, 1993).

The advent of thermal remote sensing systems onboard satellite and aerial platforms provided new avenue to study UHIs and its effects in relation with urban micrometeorology (Voogt & Oke, 2003). Remote Sensing of LST sensors range between 8 and 15 μm. However, passive microwave remote sensing is also used for LST estimation in China (Chen, Chen, Chen, Su, & Li, 2011), USA (McFarland, Miller, & Neale, 1990), Canada (Fily, 2003) and even at global scale (Peterson, Basist, & Williams, 2000; Williams, Basist, Peterson, & Grody, 2000). Satellite based remote sensing provides physical data at high resolution (120m, 60m, 30m, etc.) and low temporal (even twice daily image collection) resolutions (Watson, 2012). However, the use of passive microwave remote sensing is limited as this type of remote sensing typically offers a very coarse resolution (tens of kilometers) (Tomlinson, Chapman, Thornes, & Baker, 2011). In general, satellite remote sensing is capable of monitoring large, inaccessible, remote areas over an extended period of time. Liu et al. (2010), listed three factors, which have important effects on the retrieval of LST, i.e., the atmospheric effect in the infrared radiation transfer, synergetic use of LST and Land Surface Emissivity (LSE) makes a complex situation to LST retrieval and types of sensor to estimate LST. Several researchers developed different algorithms to calculate LST from different satellite sensors so far in the course of time. In the foregoing sections, the theoretical background, common sensors and widely used algorithms of LST estimation are provided.

THEORETICAL BACKGROUND

Earth surface energy budget is composed of radiative and non-radiative components. The radiative components include shortwave incoming and long wave outgoing radiation, whereas sensible and latent heat flux and energy storage change in water comprise non-radiative components (Kim, 2009; Oke, 1982; Terjung & O'Rourke, 1980). Mathematically this can be expressed as:

$$Q = \left(S_{in} + S_{out}\right) + \left(L_{in} + L_{out}\right)$$

Where, Q = net energy flux = sum of down-welling and up-welling radiation

S_{in} = direct and diffuse incoming shortwave solar radiation from surface of the earth

S_{out}= shortwave reflected radiation from surface of the earth

L_{in} = Longwave incoming radiation from clouds and atmosphere

L_{out} = Longwave reflected and emitted radiation from surface of the earth]

Net energy flux is the sum total of downward component directly heats up the surface, while upward radiation is divided into two sub-components – shortwave reflected radiation and long wave re-radiated radiation. These two sub-components are responsible for reducing temperature in urban environment.

Considering the radiative components, albedo is one of the most important factors affecting urban climate (Oke & Cleugh, 1987). Albedo is the fraction of the incident radiation reflected by the earth's surface back into space. It is expressed as a ratio and subject to considerable change in spatio-temporal dimension. The albedo and emissivity characteristics of common urban features are described in Table 1.

The net effect of the non-radiative components (Oke, 1982) can be mathematically expressed as:

$$Q = Q_E + Q_H + \Delta Q_s)$$

Where, Q = net surface radiant flux:

Q_E = latent heat flux which is a function of surface wetness and relative humidity

Q_H= Sensible heat flux which is a function of surface air temperatures

ΔQ_s = Thermal storage, greater in city than the adjacent areas due to higher thermal conductivity and heat capacity of building materials]

From the basic physics, there are three most important ways to transfer energy from source to target—convection, conduction and radiation. Advection is considered as important when energy budget of UHI is dealt with. Convection transfers energy in fluid motion. Earth's surface transfers incoming solar energy to the atmosphere and increases its temperature. As a result, the hot air moves upward by buoyancy. After reaching higher altitudes, the warm air mass cools down and begins to sink. When the cooler air reaches the earth's surface, it is heated again and repeats the transfer of energy between objects that are in physical contact. The process distributes energy from atom to atom within a substance. This transfer occurs mostly in

Table 1. Albedo and emissivity characteristics of common urban features

Materials	Albedo (∞)	Emissivity (ε)
Asphalt	0.05 – 0.20	0.95
Concrete	0.10 – 0.35	0.71 – 0.91
Urban Areas	0.10 – 0.27	0.85 – 0.96
Wet to dry soil	0.05 – 0.40	0.98 – 0
Long to short grass	0.16 – 0.26	0.90 – 0.95

solids, e.g. from asphalt road surface to adjacent concrete buildings. The thermal conductivity is evaluated in germs of Fourier's Law for heat conduction. Radiation transfers energy by emission of electromagnetic radiation and is thus very important from remote sensing perspective. Advection is the horizontal movement of heat by wind and is dependent on the motion and momentum of the wind.

In the atmosphere, energy comes from a primary solid angle per unit time and wavelength and may be expresses as (Kerr, Lagouarde, Nerry, & Ottle, 2004):

$$dI_\lambda / d_s = \left(I_\lambda + B_\lambda\right)\tau_\lambda$$

Where, I_λ = Radiation intensity at wavelength λ crossing from a layer which has absorption and emission

s = length of direction

B_λ = blackbody emissivity of the layer given by the Planck function

T_λ = Optical depth]

Remote estimation of LST is based on Planck's function, which is related to radiative energy emitted by a blackbody (ε = 1) to its temperature. However, most of the natural objects are non-black bodies ($0<\varepsilon_{(\lambda)}<1$). $\varepsilon_{(\lambda)}$ is the spectral emissivity, the ratio between the emitted radiance of an object at wavelength λ and the blackbody emission at the same temperature. Planck's function may be mathematically expressed as:

$$R\left(\lambda,T\right) = \varepsilon_{(\lambda)} .B\left(\lambda,T\right) = \varepsilon_{(\lambda)} \frac{C_1\lambda^{-5}}{\pi[\exp(C_2 / \lambda T) - 1}$$

Where, $R\left(\lambda,T\right)$ =non-blackbody spectral radiance ($Wm^{-2}\mu m^{-1}sr^{-1}$)

$B\left(\lambda,T\right)$=blackbody spectral radiance ($Wm^{-2}\mu m^{-1}sr^{-1}$) (Planck's function)

λ = Wavelength (μm)

T = Temperature (K)

$\varepsilon_{(\lambda)}$ =spectral emissivity of a body at λ

C_1 and C_2 = Universal constants]

Kirchoff's law of radiation determines that a body is a good absorber and emitter using the following relation:

$$\infty(\lambda) \approx \varepsilon(\lambda)$$

Where, $\infty(\lambda)$ = absorpivity]

Kirchhoff's law functions well at local thermodynamic equilibrium condition characterized by a single thermodynamic equilibrium for the earth's surface and the atmosphere prevails up to 50-70 Km (Clarmann et al., 1998). If atmospheric effects (upwelling radiance and reflected down welling irradiance) are separated and emissivity is recognized, the temperature of a Lambertian reflector can be estimated by reversing the original Planck's equation:

$$T = \frac{C_2}{\lambda \ln\left[\frac{\varepsilon(\lambda) C_1}{\pi \lambda^s R} + 1\right]}$$

Remote sensing techniques are helpful to analyze spectral response of different objects to different wavelength of electromagnetic spectrum. The dominant wavelength of blackbody radiation is depended upon its temperature by Wein's displacement law.

$$\lambda_m = A/T$$

Where, λ_m = Wavelength of maximum spectral radiant existence.

A = 2898 µm K

T = temperature in K

To estimate land surface temperature, the thermal infrared (TIR) band (8 - 15µm) is most used. This spectral region forms the atmospheric window, where energy absorption by the atmosphere is minimum and maximum transmission occurs with less attenuation. The first derivative of Planck's law explains the sensitivity to changes in LST.

$$\frac{dT}{dB} = \frac{C_1 C_2 \lambda^{-6} e^{C_2/\lambda T}}{\pi T^2 \left(e^{C_2/\lambda T - 1}\right)^2}$$

Thus to understand the relation between radiance and temperature radiative transfer models are used. These models works under the basic assumptions (Sattari & Hashim, 2014):

1. Local atmospheric thermodynamic equilibrium range is almost acceptable up to about 50-70 Km above from the ground.
2. Clear and non-smoggy sky conditions are considered due to less scattering effect.
3. Earth's surface is considered in between specular reflector and Lambertian surface.

THERMAL SATELLITE SENSORS

There are different thermal satellite sensors providing thermal information of the earth's surface and the atmosphere. These sensor generated thermal datasets are available for a vast spatio-temporal domain covering the whole glove at varying spatial resolution with varying accuracies. The first thermal data was collected by the U.S. Television IR Operational Satellite (TIROS) in 1960 and was used to monitor regional color patterns and frontal movement. NASA took the initiative in 1978 and launched the Heat Capacity Mapping Mission (HCMM), which retrieved 600m spatial resolution thermal IR data in the range of 10.5-12.6μm both at day and night time. In the same year NASA also launched Nimbus 7, which had a Coastal Zone Color Scanner (CZCS) that included a thermal IR sensor to monitor sea-surface temperature. In the successive year NASA and Jet Propulsion Laboratory developed the thermal IR Multispectral Scanner (TIMS) that acquired terrestrial thermal energy in 6 bands at the wavelength interval of <1.0 μm. In the mid-80s of the 20th century, Landsat 4 and 5 were launched and collected 120m thermal infra-red data at the wavelength range of 10.4-12.5 μm (Jensen, 2000). Today Landsat 7 ETM+, 8TIRS NOAA Geostationary Operational Environmental Satellite (GOES), Advanced Very High Resolution Radiometer (AVHRR), Moderate Resolution Imaging Spectro-Radiometer (MODIS), ASTER collect thermal infrared data covering whole or part of the globe at varying spatial resolution ranging from 30m to 8Km. Thus, this section will focus on widely used thermal satellite sensors, which offer good quality datasets ranging from local to global scale.

AVHRR

The National Oceanic and Atmospheric Administration (NOAA) has launched several generations of satellites. The present NOAA-18 and NOAA-19 contains Advanced Very High Resolution Radiometer (AVHRR), which measures earth reflectance in five spectral bands. The first AVHRR was carried on NOAA-15 platform, launched in May, 1998. The latest one is ABHRR/3 onboard NOAA-19. Two thermal spectral channels cover the range 10.3-11.3 μm and 11.5-12.5 μm. The spatial resolution is 1.1K, at Nadir. The sensor scabs earth twice in a day (0230 and 1430 local solar time). The global dataset is acquired through the sun-synchronous orbit. Data are available from NOAA comprehensive Large Array Stewardship System (http://www.nsof.class.noaa.gov/saa/) and High Resolution Picture Transmission Software (http://www.satsignal.eu/software/hrpt.htm) can be useful for analysis. MetOP, EU-METSAT satellite platform also has an AVHRR sensor with an orbital repeat time of 29 days (Tomlinson et al., 2011). Potential application areas of AVHRR data are surface water temperature mapping, snow cover mapping, flood monitoring, vegetation and soil moisture analysis, forest fire detection and fire fuel mapping, dust and sandstorm monitoring, mapping of volcanic eruptions, drainage and terrain feature analysis, etc. Figure 2 describes sea surface temperature information derived from AVHRR. Ottle and Vidal-Madjar (1992), Vasquez, Reyes and Arboledas (1997) offered a comprehensive account of different AVHRR algorithms. Considering its long historical record, AVHRR has been successfully applied to create an 18 year (1981-1998) diurnal LST dataset, which gives monthly, diurnally averaged, minimum and maximum surface temperatures (Jin, 2004; Tomlinson et al., 2011). Another advantage of AVHRR is its long historical record in comparison with other available sensors. However, unavailability of night time images displays a serious weakness of AVHRR (Tomlinson et al., 2011).

GOES

The Geostationary Operational Environmental Satellite (GOES) system is a part of meteorological satellite network located in geostationary orbit around the globe and data is available from http://www.oso.noaa.gov/geostatus/). From its vantage point GOES captures the image of the entire hemisphere. The multispectral sensor

Figure 2. NOAA-AVHRR SST data
(Source: https://en.wikipedia.org/wiki/Advanced_very-high-resolution_radiometer)

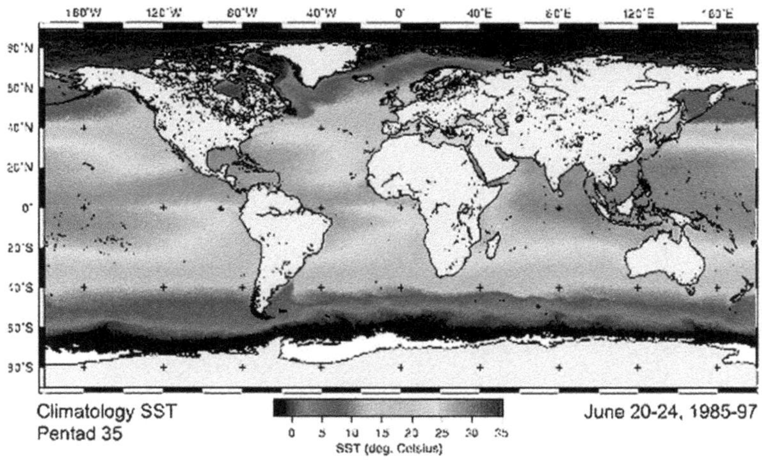

**For a more accurate representation of this figure, please see the electronic version.*

offers five-band system having one visible band (0.55 to 0.75μm) and four TIR bands (3.80 - 4.00 μm, 6.50 - 7.00 μm, 10.20 – 11.20 μm and 11.50 – 12.50 μm). The spatial resolution at nadir of the visible band is 1 Km, and three thermal band is 4 Km (Lillesand, Kiefer, & Chipman, 2004). Sun (2003), Sun, Pinker & Basara (2004) listed the details of the algorithms for dual thermal channel sensors (GOES 8 and 10) and single thermal channel sensors (GOES M-Q). Inamdar, French, Hook, Vaughan & Luckett (2008) linked MODIS data as a calibration source for GOES data, resulting in a 1 KM LST dataset at 30 minutes temporal resolution with accuracy > 2°C. GOES images are used in mostly for meteorological purposes, starting from weather forecasting to assess the regional snow cover, where moderate to low spatial resolution of the data is desired, as shown in Figure 3.

MODIS

The Moderate Resolution Imaging Spectroradiometer (MODIS) sensor is carried on both NASA's Aqua and Terra Satellites to provide comprehensive information about land, ocean and atmospheric processes. The near polar orbits of both the satellites result in two images per satellite per day (1330 and 0130h LST for Aqua and 1030 and 2230h LST for Terra). The spatial resolution is 1Km and the data is available from the United States Geological Survey Land Processes Distributes Archive Centre (https://lpdaac.usgs.gov/). The available Aqua and Terra data products are MYD1LA1 and MOD11A1 respectively (Tomlinson et al., 2011). TIR bands 31

Figure 3. GOES-1 Image
(Source: https://en.wikipedia.org/wiki/Geostationary_Operational_Environmental_Satellite)

(10.78 – 11.28 μm) and 32 (11.77 – 12.27 μm) are used for the LST products. Split window algorithm is used to combine these (Wan & Dozier, 1996). The accuracy is >1°K over homogenous surface (Coll et al., 2005; Tomlinson et al., 2011; Wan, 2002, 2008; Wan, Zhang, Zhang, & Li, 2004). These MODIS LST products were successfully applied to the UHI in different European cities specially in summer months (Cheval & Dumitrescu, 2009; Cheval, Dumitrescu, & Bell, 2009; Pongracz, Bartholy, & Dezso, 2006, 2010), in some Asian megacities (Jin, Dickinson, & Zhang, 2005) and in USA (Imhoff, Zhang, Wolfe, & Bounoua, 2010). In comparison to the sensors having higher spatial resolution but low temporal resolution (Landsat) and sensors having lower spatial resolution but low temporal resolution (Landsat) and sensors having lower spatial resolution with higher repitivity (SEVIRI), MODIS provides daily images with considerable spatial resolution, as shown in Figure 4. The characteristics of MODIS thermal bands are summarized in Table 2.

Figure 4. MODIS daily LST data (MOD11)
(Source: http://disc.sci.gsfc.nasa.gov/uui/datasets?keywords=neespi)

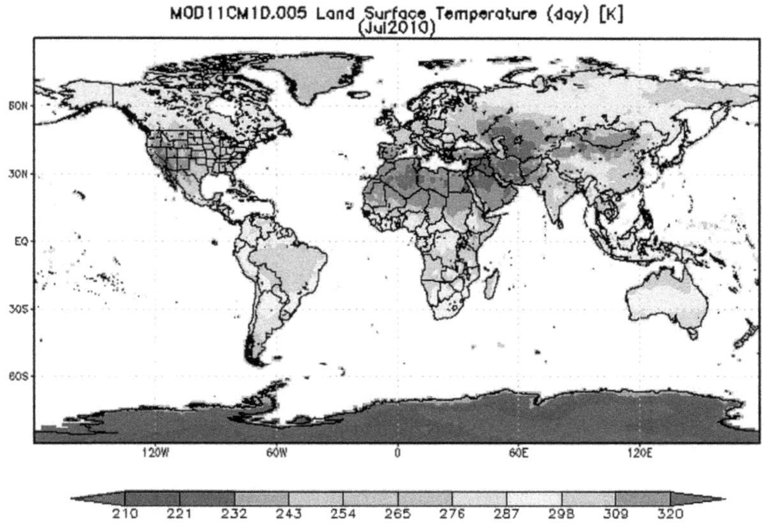

Table 2. MODIS Thermal Bands (After LP DAAC, NASA, USGS)

Thermal Bands	B20-B23	B24-B25	B31-B32
Orbit	705 Km, Terra (Descending Node), Aqua (Ascending Node)		
Swath	2330 Km x 10 Km		
Radiometric Resolution	12 Bit		
Temporal Resolution	1-2 Days		
Spatial Resolution	1 Km		
Spectral Resolution	3.66-3.84 μm (B20) 3.93-3.99 μm (B21) 3.93-3.99 μm (B22) 4.02-4.08 μm (B23)	4.43-4.50 μm (B24) 4.82-4.55 μm (B25)	10.78-11.28 μm (B31) 11.77-12.27 μm (B32)
Major Purpose	Surface / Cloud Temperature Retrieval	Atmospheric Temperature Retrieval	Surface / Cloud Temperature Retrieval
Data Products	MCD 45A1 (500m/Monthly), MCD 11A1 (1Km/Daily), MCD 11A2 (1Km/ Composite), MCD 11B1 (5.6Km/Daily), MCD 11C1 (Daily/(5600m), MCD 11C2 (5600m/Composite), MCD 11C3 (5600m/Monthly), MCD 14 (1km/5 Minutes), MCD 14A1(1Km/Daily), MCD 14A2 (1Km/Composite).		

LANDSAT

The most well-known and longest running space based earth observation system is the Landsat series of satellites. Landsat 4 and 5 Thematic Mapper has one thermal IR band (10.4 – 12.5 μm) at 120m resolution and Landsat 7 Enhanced Thematic Mapper Plus (ETM+) also has one TIR in the same wavelength region at 60m spatial resolution. The temporal resolution of these near polar sun-synchronous satellite is 16 days and they acquire images at 1000hr LST. The data products are available at USGS (http://earthexplorer.usgs.gov/ and http://glovis.usgs.gov) (Tomlinson et al., 2011). However, data from 2003 onwards are having scan line errors. Number of research works using Landsat data was increased after 2008 when USGS started free dissemination of the archival data. The 16 bit Landsat 8 Thermal Infrared Sensor (TIRS) data joined the archive from 2013. The data is having two thermal bands 10 and 11 (10.6 – 11.2 μm and 11.5 – 12.5 μm respectively) with 100m spatial resolution (resampled to 30m). However, after January 2014 due to the higher calibration uncertainty associated with TIRS band 11, reliability of the data was drastically reduced for any quantitative analysis, such as the use of split window algorithm for atmospheric correction or estimation of LST (NASA, 2016). Band 10 data is suggested to e used in conjunction with an atmospheric correction model to estimate surface brightness temperature (Barsi, Barker, & Schott, 2003; Barsi, Schott, Palluconi, & Hook, 2005). The accuracy of the current brightness temperature is ± 1°K over 15 - 35°C targets (NASA, 2016). All the Landsat TM, ETM+, TIRS data are used successfully to identify micro UHI (hotspot areas within the city) (Aniello, Morgan, Busbey, & Newland, 1995; Strathopoulou & Cartalis, 2007; Q. Weng, 2003; Q. Weng, Lu, & Schubring, 2004; Xian & Crane, 2006). Thermal properties of different Landsat sensors are described in Table 3.

Table 3. Thermal band characteristics of different Landsat sensors

Materials	Landsat TM	Landsat ETM+	TIRS
Bands	B6	B6	B10 & B11
Spectral Resolution	10.4 – 12.5 μm	10.4 – 12.5 μm	B10: 10.6 – 11.19 μm B11: 11.5 – 12.51 μm
Spatial Resolution	120 m (Resampled to 30m)	60 m (Resampled to 30m)	100 m (Resampled to 30m)

ASTER

The Advances Spaceborne Thermal emission and Reflection Radiometer (ASTER) is a cooperative venture between NASA and Japan's Ministry of International Trade and industry (Lillesand et al., 2004). The imaging instrument operates at a very high spatial resolution (90m) and calculates LST using Temperature Emissivity Separation (TES) algorithm (Gillespie & Rokugawa, 1998. Detailed information about LST data (ASTO8) product is available at http://asterweb.jpl.nasa.gov/content/03-data/01-Data-Products/STemperature.pdf (Tomlinson et al., 2011). ASTER is having 14 bands operable under three separate instrument subsystems. Each of subsystems works under three broad spectral region (VNIR, SWIR, TIR). Five TIR bands (band 10 - 14) operate at 12 bit quantization level. The details are listed in Table 4.

The TIR subsystem, unlike other instrument subsystems, incorporates a whisk-broom scanning mirror. Each band uses 10 detectors and any point on the glove is accessible at least every 16 days (Lillesand et al., 2004). The technical details are available in Yamaguchi, Kahle, Tsu, Kawakami & Pniel (1998). ASTER data product in earlier is payable and can be adequate on request. This reduces the accessibility and increased the cost of the time series historical data. However, from April 1, 2016, LPDAAC (Land Processes Distributed Active Archive Centre) started distributing Terra ASTER data products over the entire globe at free of cost (LP DAAC, 2016).

Table 4. ASTER sensor characteristics (After JPL, NASA)

Bands	B10	B11	B12	B13	B14
Spectral Resolution/ Band channels in μm	8.125 – 8.475	8.475 – 8.825	8.925 – 9.275	10.25 – 10.95	10.95 – 11.65
Spatial Resolution	90				
Radiometric Resolution	12 bits				
Orbit	705 km altitude, sun synchronous				
Equator Crossing Time	10:30 AM (LST)				
Swath	60 km				
Cross Track Pointing	±8.55 (degree) ± 116 (km)				

ASTER and MODIS images are frequently for validation purposes as they are based on the same satellite platform, so image acquisition occurs at the same time, height and location. The scale of these two images is also complementary in nature (~1 km for MODIS and 90m for ASTER) (Tomlinson et al., 2011). Wang & Liang (2009) compared the long term ground based long wave radiation with the ATER and MODIS LST and emissivity data for the period of 2000 to 2007. Liu & Zhang (2011) compared the LST information derived from ASTER and Landsat TM data for UH studies in Hong Kong. Weng, Rajasekar & Hu (2011) studied the relationship between UHIs impervious surfaces and vegetation abundance using ASTER images in Indianapolis city.

AATSR

The Advanced Along Track Scanning Radiometer (AATSR) carried onboard the European Space Agency (ESA) ENVISAT was launched in 2002. The primary objective of these Along Track Scanning Radiometer (ATSR) series of instruments is to measure sea surface temperature (SST). TIR bands (11.0 – 12.0 µm) of AATSR operational from March, 2004 are used to provide LST at ~1 km spatial resolution (Tomlinson et al., 2011). The temporal resolution is 35 days. AATSR literature includes theoretical science for algorithm development (A. Prata, 2002), evolution of algorithms (Soria & Sobrino, 2007). The developed algorithms are also applied to the earlier ATSR-1 and ATSR-2 data series to retrieve LST information from 1991 onwards (Tomlinson et al., 2011). AATSR data has been used to map monthly LST over Europe (Joan & Cesar, 2009), drought prediction and estimating evapotranspiration (Djepa, 2011; Liu et al., 2010) and to detect snow covered areas (Istomina, von Hoyningen-Huene, Kokhanovsky, & Burrows, 2010).

SEVIRI

The Spinning Enhanced Visible and Infrared Imager (SEVIRI) onboard Meteosat-8 uses a generalized split-window algorithm to calculate LST from two TIR bands (10.8 and 12 µm). The repitivity time is 15 minutes (maximum 96 images per day). However, the coverage area is constant not global. Only pixels within 60° viewing angle are processed for LSY measurement to avoid high atmospheric attenuation. The spatial resolution is 3km at nadir and increases up to 6km away from nadir point. The high temporal resolution of the sensor helps to get cloud free images of a certain area and enables to study diurnal LST UHIs patterns (Tomlinson et al., 2011). After comparing with the MODIS LST products, it is observed that the

METEOSAT LST data are warmer, particularly in day time (Trigo, Monteiro, Olesen, & Kabsch, 2008). Retalis, Paronis, Lagouvardos & Kotroni (2010) captured a significant correlation between the two data products in Athens, Greece. Nieto, Sandholt, Aguado, Chuvieco & Stisen (2011) observed satisfactory correlation between air temperature and Meteosat LST products.

LAND SURFACE TEMPERATURE RETRIEVAL ALGORITHMS

Land surface temperature can be retrieved using various algorithms such as split-window (SW) method (Wan & Dozier, 1996), Multi-channel and multi-angle (MCMA) (Sobrino, Li, Stoll, & Becker, 1996), Temperature/Emissivity Separation (TES) method (Gillespie & Rokugawa, 1998), Mono-Window (MW) method (Qin, Karnieli, & Berliner, 2001), Single Channel (SC) method (Jimenez-Munoz & Sobrino, 2003). Wan & Dozier (1996) proposed a generalized SW method to retrieve LST from LST from AVHRR and MODIS data. From the regression analysis of radiative transfer simulations, it is found that the coefficients of this LST algorithm depend on viewing angle. The sensitivity and error are analyzed over wide ranges of surface temperature, emissivity and atmospheric water vapor abundance. According to the simulations when atmospheric column water vapor increases and viewing angle is >45°, optimization of SW method is necessary by separating the ranges of the atmospheric column water vapor, temperature of lower boundary and skin temperature into well-mannered sub ranges. The lower boundary atmospheric temperature and vertical column water vapor values obtained from MODIS atmospheric sounding channels can be used to determine the range for the optimum coefficients of the SW method. In this way, the accuracy of this viewing angle dependent algorithm may be increased. This is also less sensitive to the uncertainty in the band emissivities of the land surface in the split-window and to the noise of the instrument. However, a major hurdle of using this algorithm is to assign appropriate band emissivities for each pixel in real data processing, as this demands sound knowledge on emissivity of natural terrestrial materials. It also requires strong understanding in atmospheric temperature materials. SW method applies atmospheric correction using a linear combination of radiances from two adjacent IR bands. Initially this method was applied to obtain sea surface temperature (SST). After its success, the SW method was widely used to estimate LST. In 10.5-12.5 µm atmospheric window area, there are broad bands of many operational satellite sensors. The algorithm uses absorption difference between two channels to minimize the atmospheric influence and calculate LST as a linear combination of two brightness temperatures. It is assumed that the contribution functions of the two channels are similar and the difference in transmittance results from different absorption coefficients in two channels and is mainly due to water vapor, not emissivity differences (Sattari & Hashim, 2014). Becker & Li (1990) successfully used a local SW technique extracted from SST SW method for LST estimation. Sobrino, Cool & Caselles (1991) related the coefficients to the atmospheric and surface emissivity condition. Kerr, Lagouarde & Imbernon (1992) considered the effects of mixed pixels and proposed a semi-empirical technique which classifies surfaces either as bare soil or as vegetated area and combines their contributions to obtain effective skin surface temperature of the earth. Becker & Li (1995) modified their earlier algorithm by incorporating atmospheric water vapor content. Sun & Pinker (2003) developed an Advanced SW (ASW) technique where a separate equation is established for each surface type by using 11.0 and 12.0 μm SW. They assumed that when the satellite viewing angle increases, the optical path and atmospheric attenuation increase. Thus a zenith angle correction is added to LST retrieval equation. Qin et al. (2004) tested 17 published SW algorithms for estimating LST from NOAA AVHRR data. However, the essence of these algorithms is similar, that is to develop a linear relationship among the LST and brightness temperatures of two adjacent thermal channels and the algorithm coefficients depend on different land surface emissivities. Jimenez-Munoz & Sobrino (2008) listed a complete set of coefficients to retrieve LST from TIR sensors onboard the most common satellites. Jimenez-Munoz, Sobrino, Skokovich, Matter & Cristobal (2014) applied the SW technique to compute brightness temperature of earth surface from the Landsat 8 Thermal Infrared Sensor (TIRS), which includes two bands in the atmospheric window of 10-12μm. The algorithms were tested with simulated data retrieved from forwarded simulations using atmospheric profile data base and emissivity information obtained from spectral libraries.

SC methods apply radiance data measured in one TIR window and correct the atmospheric effects to determine LST (Price, 1983; Susskind, Rosenfield, Reuter, & Chahine, 1984). The technique needs accurate information about vertical and horizontal distribution of temperature and water vapor (Ottle & Vidal-Madjar, 1992). Radiative Transfer Model (RTM) e.g. LOWTRAN, MODTRAN, 6S, etc. is used in this method to simulate satellite measurements over a range if surface parameters like temperature, elevation and emissivity for the given atmosphere. Atmospheric correction and subsequently estimation of LST can be done by interpolating these forward calculations (Dash et al., 2002; Schroedter, Olesen, & Fischer, 2003) RIMs do not linearize atmospheric effects, variations of surface parameters and viewing angles; This forward calculations using an RIM are laborious and time consuming. In this regard, neural networks are used as faster method to retrieve vertical temperature, humidity profiles and to compute the long wave atmospheric radiation budget and surface radiation fluxes (Cheruy, Chevallier, Morcrette, Scott, & Chedin, 1996; Escobar-Munoz, Chedin, Cheruy, & Scott, 1993; Key, Amano, Collins, & Schweiger, 1999; Schweiger & Key, 1997).

Qin, Karnieli & Berliner (2001) developed the MW method to estimate brightness temperature from Landsat TM band 6 TIR data. Three parameters are necessary to run this algorithm, such as emissivity, transmittance and effective mean atmospheric temperature. Atmospheric transmittance is retrieved from the LOWTRAN 7 Simulation program. Sensitivity analysis of the method was performed under the assumption of clear sky condition. The analysis indicated that the possible ground emissivity error is difficult to estimate and has relatively insignificant impact on the probable LST estimation error. The LST estimation error is sensible to the possible transmittance error and mean atmospheric temperature. Sobrino, Jimenez-Munoz & Paolini (2004) compared the MW and SC methods to estimate LST from Landsat 5 TM band 6 data. However, they found that only one thermal channel processing makes impossible to apply well accepted methods like TES algorithm (Gillespie & Rokugawa, 1998). As an alternative they applied NDVI to obtain LSE. Wang et al., (2015) introduced an Improved Mono-Window (IMW) technique to retrieve LST from Landsat 8 TIRS band 10 data. The essential parameters are similar to that of MW method. However, they proposed a new method to estimate effective men atmospheric temperature from local meteorological data. Whereas, ground emissivity was estimated through the land cover pattern obtained from other bands of Landsat 8. The sensitivity analysis showed that the possible error in estimating the required atmospheric water vapor content has the most important impact on the probable error of LST estimation. The accuracy of the algorithm is ~1.4 K under moderate errors in both water vapor content and ground emissivity. The IMW algorithm was successfully tested over Nanjing, China.

The Multi-Angle (MA) technique assumes a spatially uniform atmospheric column. The measurements are made from different sensors e.g Meteosat, TIROS-N, etc. The emissivity or temperature of elements are apparently different, the temperature element are inversed by the multi-angle radiance data (Klemen & Schroedter-Homscheidt, 2009). Sensors like ATSR observe earth objects under different viewing angles for a given channel since there is differential absorption due to different atmospheric slant-path lengths (Dash et al., 2002). Prata (1993) presented an extensive discussion of dual-angle technique assuming that angular variation of emissivity is negligible ($\Theta<60°$). On its basis Sobrino, Li, Stoll & Becker (1996) proposed an improved dual angle process. They concluded that this MA algorithm has the potential to estimate SST with a standard deviation <0.23 k for error free satellite data. They claimed that the MA method is superior to SW technique if the spectral and angular variations of emissivity of a homogenous surface are of the same magnitude. The technique may become unstable if it does not have significantly longer atmospheric path measurements (A. J. Prata, 1993). The method also demands knowledge of the angular variation of the surface emissivity (Dash et al., 2002). Sobrino & Cuenca (1999) demonstrated that with increasing viewing angle (highest angular dependence over water) emissivity decreases. They set up a goniometric system at 5° interval and reported that this dependency increases for viewing angles >30°. However, considerable amount of research is done to estimate LST and LSE from microwaves (Calvet & Julien, 1996; Dash et al., 2002; Givri, 1997; Pringent, Rossow, & Matthews, 1997; F. Weng & Grody, 1998; Xiang & Smith, 1997). Tropical regions are covered by clouds most of the time in a year. This reduces the scope of LST estimation in tropical region from TIR data at regular interval. Microwave data is in contrast serves as a potential data source in such situations as microwaves penetrate clouds. Microwave radiometry has the ability to estimate LST under all weather conditions except rain (Wentz, Chelle, Smith, & Chelton, 2000). Large portions of the microwave spectrum are nearly transparent with oxygen and water vapor as major absorbers (Dash et al., 2002). SSM/I sensor data onboard DMSP block 5D-2F8 spacecraft, was used to derive brightness temperatures over land surfaces (Kumar, Sahoo, & Singh, 1999). The SSM/I is a conical scanning radiometer with a viewing angle of 53.1°. The sensor measures the terrestrial emission at 19, 22, 37 and 85 GH_z for horizontal and vertical polarization. Only 22 GH_z channel is vertically polarized, while the remaining channels are horizontal (Hollinger, 1991). Calvet & Julien (1996) compared LST data retrieved from bi-polarized SSM/I data at 19 and 37 GH_z channels from AVHRR and MVIRJ thermal IR data for central Spain within the European Field Experiment in a Desertification-threatened Area (EFEDA) campaign. Weng & Grody (1998) developed a physical LST retrieval algorithm based on two closely spaced SSM/I channels at 19 and 22 GH_z to estimate variations in LSE. Non-linear equation was developed and solved by Newtonian iteration in this algorithm. Inclusion of low frequency channels in this technique estimates the requirement of any surface classification. For the same reason it is less affected by ice surface scattering, precipitation and snow covered areas. The average root mean square error (RMSE) is 3.8 K, while higher RMSE values occur in dry atmospheric conditions with total precipitable water of 5.0 mm or less. Givri (1997) compared the LST derived from TIR bands in AVHRR and microwave bands in SSM/I. They found critical influence of the atmosphere and LSE in both the spectral regions. They investigated the application potential of SW method in case of microwave data. The SW method requires an adoption of radiative transfer equations (RTE) to microwave propagation. The microwave radiation is controlled largely by emissivity [$\varepsilon(v)$], surface temperature (T_s) and atmospheric transmittance (τ). Both LSE and sea surface emissivity (SSE) vary from 0.6-1.0 in microwave region. This variation is dependent upon surface conditions, polarization and frequency. After comparing results estimated from TIR and microwave data, Givri (1997) concluded that LST cannot be easily determined from microwave data alone.

CONCLUSION

The chapter has given an overview of the role of remote sensing in land surface temperature (LST) estimation with special emphasis on urban heat island studies. It encircled the theoretical background, description of common thermal remote sensors and different algorithms to retrieve LST from satellite data. Inferring the focus on UHI phenomenon, it is worthy to mention that a serious research gap still exists in understanding the relationship between the remotely sensed LST data and measured air temperature data. This relationship is important from the policy decision making point of view as well. The advantage of satellite remote sensing data and techniques is their ability to capture global information. However, there also exists lack of sufficient ground information, which is required both for LST retrieval and validation. Another serious lacuna is the data resolution. Still most of the TIR sensor data are of coarse to medium resolution in comparison to very high spatial resolution optical data. Strong requirement is felt when we analyze UHI phenomenon. However, the progression of various scientific communities with countless efforts to increase data resolution and to evolve or modify new or existing algorithms shows rays of hope to overcome such problems in future.

REFERENCES

Aniello, C., Morgan, K., Busbey, A., & Newland, L. (1995). Mapping micro urban heat islands using Landsat TM and a GIS. *Computers & Geosciences*, *21*(8), 965–967. doi:10.1016/0098-3004(95)00033-5

Barsi, J. A., Barker, J. L., & Schott, J. R. (2003). An atmospheric correction parameter calculator for a single thetmal band Earth-Sensing instrument. *Paper presented at the IGARSS 03, Centre de Congres Pierre Baudis*, Toulouse, France.

Barsi, J. A., Schott, J. R., Palluconi, F. D., & Hook, S. J. (2005, August). Validation of a web-based atmospheric correction tool for single thermal band instruments. Earth Observing Systems X. *Paper presented at theProc. SPIE*, San Diego, CA, USA (V*ol. 5882)*.

Bartolucci, L. A. (1985). Calibration of Landsat-5 TM Thermal IR data. *Paper presented at theProceedings of the First National Symposium on Remote Sensing*, Baghdad, Iraq.

Bartolucci, L. A., & Chang, M. (1988). Look-up tables to convert Landsat TM thermal IR data to water surface temperatures. *Geocarto International*, *3*(3), 61–67. doi:10.1080/10106048809354167

Becker, F., & Li, Z.-L. (1990). Towards alocal split window method over land surface. *International Journal of Remote Sensing*, *11*(3), 369–393. doi:10.1080/01431169008955028

Becker, F., & Li, Z.-L. (1995). Surface temperature and emissivity at different scales: Definition, measurement and related problems. *Remote Sensing Reviews*, *12*(3-4), 225–253. doi:10.1080/02757259509532286

Calvet, J.-C., & Julien, J.-P. (1996). Land surface temperature retrieval in dry conditions from infrared and microwave satellite radiometry. *Remote Sensing Reviews*, *13*(3-4), 235–255. doi:10.1080/02757259609532306

Changon, S. A. (1992). Inadvertent weather modification in urban areas: Lessons for global climate change. *Bulletin of the American Meteorological Society*, *73*, 621–627.

Chen, S., Chen, X., Chen, W., Su, Y., & Li, D. (2011). A simple retrieval method of land surface temperature from AMSR-E passive microwave data - A case study over Sothern China during the strong snow disaster of 2008. *International Journal of Applied Earth Observation and Geoinformation*, *13*(1), 140–151. doi:10.1016/j.jag.2010.09.007

Cheruy, F., Chevallier, F., Morcrette, J.-J., Scott, N. A., & Chedin, A. (1996). Une methode utilisant les techniques neuronales pour le calcul rapide de la distribution verticale du bilan radiatif thermique terreste. *C. R. Acad. Sci., Paris, t. 322. Serie II b*, 665-672.

Cheval, S., & Dumitrescu, A. (2009). The July urban heat island of Bucharest as derived from MODIS images. *Theoretical and Applied Climatology*, *96*(1-2), 145–153. doi:10.1007/s00704-008-0019-3

Cheval, S., Dumitrescu, A., & Bell, A. (2009). The urban heat island of Bucharest during the extreme high temperatures of July 2007. *Theoretical and Applied Climatology*, *97*(3-4), 391–401. doi:10.1007/s00704-008-0088-3

Clarmann, T. V., Dudhia, A., Echle, G., Flaud, J.-M., Harrold, C., & Kerridge, B. (1998). Study on the simulation of atmospheric infrared spectra: Final Report.

Coll, C., Caselles, V., Galve, J. M., Valor, E., Niclos, R., Sanchez, J., & Rivas, R. (2005). Ground measurements for the validation of land surface temperatures derived from AATSR and MODIS data. *Remote Sensing of Environment*, *97*(3), 288–300. doi:10.1016/j.rse.2005.05.007

Coll, C., Hook, S. J., & Galve, J. M. (2009). Land surface temperatures from the advanced along-track scanning radio meter: Validation over inland waters and vegetated surfaces. *IEEE Transactions on Geoscience and Remote Sensing*, *47*(1), 350–360. doi:10.1109/TGRS.2008.2002912

Dash, P., Gottsche, F. M., Olesen, F. S., & Fischer, H. (2002). Land surface temperature and emissivity estimation from passive sensor data: Theory and practice-current trends. *International Journal of Remote Sensing*, *23*(13), 2563–2594. doi:10.1080/01431160110115041

Djepa, V. (2011). Drought prediction using the Along Track Scanning Radiometer (ATSR2) on board ERS2 satellite. *Advances in Space Research*, *48*(1), 56–60. doi:10.1016/j.asr.2011.01.026

Escobar-Munoz, J., Chedin, A., Cheruy, F., & Scott, N. A. (1993). Multi-layer neural networks for the retrieval of atmospheric variables from satellite-borne vertical sounding. *Comptes Rendus de l'Academic des Sciences, Serie II (Macanique, Physique. Chimie Sciences de la Terre et de l'Univers*, *317*, 911–918.

Fily, M. (2003). A simple retrieval method for land surface temperature and fraction of water surface determination from satellite microwave brightness temperatures in sub-arctic areas. *Remote Sensing of Environment*, *85*(3), 328–338. doi:10.1016/S0034-4257(03)00011-7

Gillespie, A., Rokugawa, S., Matsunaga, T., Cothern, J. S., Hook, S., & Kahle, A. B. (1998). A temperature and emissivity separation algorithm for Advanced Spaceborne Thermal Emission and Reflection Radiometer (ASTER) images. *IEEE Transactions on Geoscience and Remote Sensing*, *36*(4), 1113–1126. doi:10.1109/36.700995

Givri, J. R. (1997). The extension of the split window technique to passive microwave surface temperature assessment. *International Journal of Remote Sensing*, *18*(2), 335–353. doi:10.1080/014311697219105

Hollinger, J. (1991). *DMSP SSM/I Calibration / Validation. Final Report. Parts 1 and 2*. Washington, D.C.: Naval Research Laboratory.

Imhoff, M., Zhang, P., Wolfe, R., & Bounoua, L. (2010). Remote sensing of the urban heat island effect across biomes in the continental USA. *Remote Sensing of Environment*, *114*(3), 504–513. doi:10.1016/j.rse.2009.10.008

Inamdar, A. K., French, A., Hook, S. J., Vaughan, G., & Luckett, W. (2008). Land surface temperature retrieval at high spatial and temporal resolutions over the southwestern United States. *Journal of Geophysical Research*, *113*(D7), 1–18. doi:10.1029/2007JD009048

Istomina, L. G., von Hoyningen-Huene, W., Kokhanovsky, A. A., & Burrows, J. P. (2010). The detection of cloud-free snow-covered areas using AATSR measurements. *Atmospheric Measurement Techniques*, *3*(4), 1005–1017. doi:10.5194/amt-3-1005-2010

Jensen, J. R. (2000). *Remote Sensing of the Environment. An Earth Resource Perspective*. Singapore: Pearson Education, Inc.

Jimenez-Munoz, J. C., & Sobrino, J. A. (2003). A generalized single channel method for retrieving land surface temperature from remote sensing data. *Journal of Geophysical Research*, *108*(D22), 4688. doi:10.1029/2003JD003480

Jimenez-Munoz, J. C., & Sobrino, J. A. (2008). Split-window coefficients for land surface temperature retrieval from low-resolution thermal infrared sensors. *IEEE Geoscience and Remote Sensing Letters*, *5*(4), 806–809. doi:10.1109/LGRS.2008.2001636

Jimenez-Munoz, J. C., Sobrino, J. A., Skokovich, D., Matter, C., & Cristobal, J. (2014). Land surface temperature retrieval methods from Landsat-8 thermal infrared sensor data. *IEEE Transactions on Geoscience and Remote Sensing*, *11*(10), 1840–1843. doi:10.1109/LGRS.2014.2312032

Jin, M. (2004). Analysis of land skin temperature using AVHRR observations. *Bulletin of the American Meteorological Society*, *85*(4), 587–600. doi:10.1175/BAMS-85-4-587

Jin, M., Dickinson, R., & Zhang, D. (2005). The footprint of urban areas on global climate as characterized by MODIS. *Journal of Climate*, *18*(10), 1551–1565. doi:10.1175/JCLI3334.1

Joan, M., & Cesar, C. (2009). Monthly land surface temperature maps over European zone using advanced along track scanning radiometer data for 2007.*Geoscience and Remote Sensing Symposium, 2009, IEEE International, IGARSS 2009, 4*, 292-295. doi:10.1109/IGARSS.2009.5417316

Kerr, Y. H., Lagouarde, J. P., & Imbernon, J. (1992). Accurate land surface temperature retrieval from AVHRR data with use of an improved split window algorithm. *Remote Sensing of Environment*, *41*(2-3), 197–209. doi:10.1016/0034-4257(92)90078-X

Kerr, Y. H., Lagouarde, J. P., Nerry, F., & Ottle, C. (2004). Land surface temperature retrieval techniques and applications: Case of AVHRR. In D. A. Quattrochi & J. C. Luvall (Eds.), *Thermal Remote Sensing in Land Surface Processes* (pp. 33–109). Boca Raton, Florida: CRC Press. doi:10.1201/9780203502174-c3

Key, J. R., Amano, E., Collins, J., & Schweiger, A. J. (1999). FluxNet User's Guide. Technical Report 96-03: Department of Geography, Boston University.

Kim, J. P. (2009). *Land-use planning and the urban heat island effect* (Ph.D. thesis). The Ohio State University.

Klemen, Z., & Schroedter-Homscheidt, M. (2009). Parameterization of air temperature in high temporal and spatial resolution from a combination of the SEVIRI and MODIS instruments. *ISPRS Journal of Photogrammetry and Remote Sensing*, *64*(4), 414–421. doi:10.1016/j.isprsjprs.2009.02.006

Kumar, S., Sahoo, P. K., & Singh, R. P. (1999). Monitoring of brightness temperature over Indian and adjoining regions using SSM/I data. *International Journal of Remote Sensing*, *20*(12), 2305–2307. doi:10.1080/014311699212038

Landsberg, H. E. (1981). *The Urban Climate*. New York: Academic Press.

Lee, H. (1993). An application of NOAA thermal data to the study of urban heat islands. *Atmospheric Environment*, *27B*, 1–13.

Lillesand, T. M., Kiefer, R. W., & Chipman, J. W. (2004). Remote Sensing and Image Interpretation. Singapore: John Wiley & Sons (Asia), Pte. Ltd.

Liu, H., Lisheng, X., Ding, J., Zhuoma, B., Deng, X., & Liu, Z. (2010, August 28-31). Atmospheric correction and land surface temperature retrieval method for FY-3 IR observations. *Paper presented at theSecond IITA International Conference on Geoscience and Remote Sensing (IITA-GRS)*, Qingdao.

Liu, L., & Zhang, Y. (2011). Urban heat island analysis using the Landsat TM data and ASTER data: A case study in Hong Kong. *Remote Sensing*, *3*(12), 1535–1552. doi:10.3390/rs3071535

Liu, R., Wen, J., Wang, X. L. W., Tian, H., & Zhang, T. T. (2010). Actual daily evapotranspiration estimated from MERIS and AATSR data over the Chinese Loess Plateau. *Hydrology and Earth System Sciences*, *14*(1), 47–58. doi:10.5194/hess-14-47-2010

LP DAAC. (2016). ASTER data: Unlimited public access. Retrieved from http://lpdaac.usgs.gov/user_resources/outreach_materials/aster_no_charge_promo_

McFarland, M. J., Miller, R. L., & Neale, C. M. U. (1990). Land surface temperature derived from SSM/I passive microwave brightness temperatures. *IEEE Transactions on Geoscience and Remote Sensing*, *28*(5), 839–845. doi:10.1109/36.58971

NASA. (2016). Landsat Science. Retrieved from http://www.landsat.gsfc.nasa.gov/?page_id=5377

Nieto, H., Sandholt, I., Aguado, I., Chuvieco, E., & Stisen, S. (2011). Air temperature estimation with MSG-SEVIRI data: Calibration and validation of the TVX algorithm for the Iberian Peninsula. *Remote Sensing of Environment*, *115*(1), 107–116. doi:10.1016/j.rse.2010.08.010

Noyes, E., Soria, G., Sobrino, J. A., Remedios, J., Llewellyn-Jones, D. T., & Corlett, G. K. (2007). AATSR land surface temperature product algorithm verification over a WATERMED site. *Advances in Space Research*, *39*(1), 171–178. doi:10.1016/j.asr.2006.02.024

Oke, T. R. (1982). The energetic basis of the urban heat island. *Quarterly Journal of the Royal Meteorological Society*, *108*, 1–24.

Oke, T. R., & Cleugh, H. A. (1987). Urban heat storage derived as energy balance residuals. *Boundary-Layer Meteorology*, *39*(3), 233–245. doi:10.1007/BF00116120

Oke, T. R., Johnson, G. T., Steyn, D. G., & Watson, I. D. (1991). Simulation of surface urban heat islands under heat islands under ideal conditions at night part 2: Diagnosis of causation. *Boundary-Layer Meteorology*, *56*(4), 339–358. doi:10.1007/BF00119211

Ottle, C., & Vidal-Madjar, D. (1992). Estimation of land surface temperature with NOAA 9 data. *Remote Sensing of Environment*, *40*(1), 27–41. doi:10.1016/0034-4257(92)90124-3

Peterson, T., Basist, A., Williams, C., & Grody, N. C. (2000). A blended satellite in-situ near-global surface temperaure dataset. *Bulletin of the American Meteorological Society*, *81*(9), 2157–2164. doi:10.1175/1520-0477(2000)081<2157:ABSSNS>2.3.CO;2

Pongracz, R., Bartholy, J., & Dezso, Z. (2006). Remotely sensed thermal information applied to urban climate analysis. *Advances in Space Research*, *37*(12), 2191–2196. doi:10.1016/j.asr.2005.06.069

Pongracz, R., Bartholy, J., & Dezso, Z. (2010). Application of remotely sensed thermal information to urban climatology of Central European cities. *Physics and Chemistry of the Earth*, *35*(1-2), 95–99. doi:10.1016/j.pce.2010.03.004

Prata, A. (2002). Land Surface Temperature Measurement from Space: AATSR Algorithm Theoretical Basis Document (pp. 1-34): Contract Report to ESA. CSIRO Atmospheric Research: Aspendale.

Prata, A. J. (1993). Land surface temperatures derived from the Advanced Very High Resolution Radiometer and the Along-Track Scanning Radiometer 1. Theory. *Journal of Geophysical Research*, *98*(D9), 16689–16702. doi:10.1029/93JD01206

Price, J. C. (1983). Estimation of surface temperatures from satellite thermal infrared data - a simple formulation for the atmospheric effect. *Remote Sensing of Environment*, *13*(4), 353–361. doi:10.1016/0034-4257(83)90036-6

Pringent, C., Rossow, W. B., & Matthews, E. (1997). Microwave land surface emissivities estimated from SSM/I observations. *Journal of Geophysical Research*, *102*(D18), 21867–21890. doi:10.1029/97JD01360

Qin, Z., Karnieli, A., & Berliner, P. (2001). A mono-window algorithm for retrieving land surface temperature from Landsat TM data and its application to the Israel-Egypt border region. *International Journal of Remote Sensing*, *22*(18), 3719–3746. doi:10.1080/01431160010006971

Qin, Z., Xu, B., Zhang, W., Li, W., Chen, Z., & Zhang, H. (2004). Comparison of split window algorithms for land surface temperature retrieval from NOAA-AVHRR data. *Paper presented at theGeoscience and Remote Sensing Symposium IGARSS'04.*

Resenzweig, C., Solecki, W. D., Paschall, L., Chopping, M., Pope, G., & Goldberg, R. (2005). Characterizing the urban heat island in current and future climates in New Jersey. *Global Environmental Change Part B: Environmental Hazards*, *6*(1), 51–62. doi:10.1016/j.hazards.2004.12.001

Retalis, A., Paronis, D., Lagouvardos, K., & Kotroni, V. (2010). The heat wave of June 2007 in Athens, Greece - part 1: Study of satellite derived land surface temperature. *Atmospheric Research*, *98*(2-4), 458–467. doi:10.1016/j.atmosres.2010.08.005

Sattari, F., & Hashim, M. (2014). A brief review of land surface temperature retrieval methods from thermal satellite sensors. *Middle-East Journal of Scientific Research*, *22*(5), 757–768.

Schroedter, M., Olesen, F. S., & Fischer, H. (2003). Determination of land surface temperature distributions from single channel IR measurements: An effective spatial interpolation method for the use of TOVS, ECMWF and radiosonde profiles in the atmospheric correction scheme. *International Journal of Remote Sensing*, *24*(6), 1189–1196. doi:10.1080/01431160110115087

Schweiger, A. J., & Key, J. R. (1997). Estimating surface radiation fluxes in the Arctic from TOVS brightness temperatures. *International Journal of Remote Sensing*, *18*(4), 955–970. doi:10.1080/014311697218854

Sobrino, J. A., Cool, C., & Caselles, V. (1991). Atmospheric correction for land surface temperature using NOAA-11 AVHRR channels 4 and 5. *Remote Sensing of Environment*, *38*(1), 19–34. doi:10.1016/0034-4257(91)90069-I

Sobrino, J. A., & Cuenca, J. (1999). Angular variation of thermal infrared emissivity for some natural surfaces from experimental measurements. *Applied Optics*, *38*(18), 3931–3936. doi:10.1364/AO.38.003931 PMID:18320001

Sobrino, J. A., Jimenez-Munoz, J. C., & Paolini, L. (2004). Land surface temperature retrieval from Landsat TM 5. *Remote Sensing of Environment*, *90*(4), 434–440. doi:10.1016/j.rse.2004.02.003

Sobrino, J. A., Li, Z. L., Stoll, M. P., & Becker, F. (1996). Multi-channel multi-angle algorithms for estimating sea and land surface temperature with ATSR data. *International Journal of Remote Sensing*, *17*(11), 2089–2114. doi:10.1080/01431169608948760

Soria, G., & Sobrino, J. A. (2007). ENVISAT/AATSR derived land surface temperature over a heterogeneous region. *Remote Sensing of Environment*, *111*(4), 409–422. doi:10.1016/j.rse.2007.03.017

Strathopoulou, M., & Cartalis, C. (2007). Daytime urban heat islands from Landsat ETM+ and CORINE land cover data: An application to major cities in Greece. *Solar Energy*, *81*(3), 358–368. doi:10.1016/j.solener.2006.06.014

Strathopoulou, M., Cartalis, C., & Kermatisoglou, I. (2004). Mapping micro urban heat islands using NOAA/AVHRR images and CORINE land cover: An application to coastal cities of Greece. *International Journal of Remote Sensing*, *25*(12), 2301–2316. doi:10.1080/01431160310001618725

Sun, D. (2003). Estimation of land surface temperature from a Geostationary Operational Environmental Satellite (GOES-8). *Journal of Geophysical Research*, *108*(D11), 1–15. doi:10.1029/2002JD002422

Sun, D., Pinker, R., & Basara, J. (2004). Land surface temperature estimation from the next generation of geostationary operational environmental satellites: GOES M-Q. *Journal of Applied Meteorology*, *43*(2), 363–372. doi:10.1175/1520-0450(2004)043<0363:LSTEFT>2.0.CO;2

Sun, D., & Pinker, R. J. (2003). Estimation of land surface temperature from Geostationary Operational Environmental Satellite (GOES-8). *Journal of Geophysical Research*, *108*(D11), 4326. doi:10.1029/2002JD002422

Susskind, J., Rosenfield, J., Reuter, D., & Chahine, M. T. (1984). Remote sensing of weather and climate parameters from HIRS2 / Msu on TIROS-N. *Journal of Geophysical Research*, *89*(D3), 4677–4697. doi:10.1029/JD089iD03p04677

Terjung, W. H., & ORourke, P. A. (1980). Influences of physical structures on urban energy budgets. *Boundary-Layer Meteorology*, *19*(4), 421–439. doi:10.1007/BF00122343

Tomlinson, C. J., Chapman, L., Thornes, J. E., & Baker, C. (2011). Remote sensing land surface temperature for meteorology and climatology: A review. *Meteorological Applications*, *18*(3), 296–306. doi:10.1002/met.287

Trigo, I. F., Monteiro, I. T., Olesen, F. S., & Kabsch, E. (2008). An assessment of remotely sensed land surface temperature. *Journal of Geophysical Research*, *113*(D17), 1–12. doi:10.1029/2008JD010035

Valor, E., & Caselles, V. (1996). Mapping land surface emissivity from NDVI: Application to European, African and South American areas. *Remote Sensing of Environment*, *57*(3), 167–184. doi:10.1016/0034-4257(96)00039-9

Vasquez, D. P., Reyes, F. J. O., & Arboledas, L. A. (1997). A comparative study of algorithms for estimating land surface temperature from AVHRR data. *Remote Sensing of Environment*, *62*(3), 215–222. doi:10.1016/S0034-4257(97)00091-6

Voogt, J. A., & Oke, T. R. (2003). Thermal remote sensing of urban climates. *Remote Sensing of Environment*, *86*(3), 370–384. doi:10.1016/S0034-4257(03)00079-8

Wan, Z. (2008). New refinements and validation of the MODIS land-surface temperature / emissivity products. *Remote Sensing of Environment*, *112*(1), 59–74. doi:10.1016/j.rse.2006.06.026

Wan, Z., & Dozier, J. (1996). A generalized split-window algorithm for retrieving land-surface temperature from space. *IEEE Transactions on Geoscience and Remote Sensing*, *34*(4), 892–905. doi:10.1109/36.508406

Wan, Z., Zhang, Y., Zhang, Q., & Li, Z. (2002). Validation of the land-surface temperature products retrieved from terra moderate resolution imaging spectroradiometer data. *Remote Sensing of Environment*, *83*(1-2), 163–180. doi:10.1016/S0034-4257(02)00093-7

Wan, Z., Zhang, Y., Zhang, Q., & Li, Z. (2004). Quality assessment and validation of the MODIS global land surface temperature. *International Journal of Remote Sensing*, *25*(1), 261–274. doi:10.1080/0143116031000116417

Wang, F., Qin, Z., Song, C., Tu, L., Karnieli, A., & Zhao, S. (2015). An improved mono-window algorithm for land surface temperature retrieval from Landsat 8 thermal infrared sensor data. *Remote Sensing*, *7*(4), 4268–4289. doi:10.3390/rs70404268

Wang, K., & Liang, S. (2009). Evaluation of ASTER and MODIS land surface temperature and emissivity products using long-term surface long wave radiation observations at SURFRAD sites. *Remote Sensing of Environment*, *113*(7), 1556–1565. doi:10.1016/j.rse.2009.03.009

Watson, C. (2012). *Analysis of urban heat island climates along the I-85 / I-40 corridor in central North Carolina* [Ph.D. thesis]. University of North Carolina at Greensboro.

Weng, F., & Grody, N. C. (1998). Physical retrieval of land surface temperature using special microwave imager. *Journal of Geophysical Research*, *103*(D8), 8839–8848. doi:10.1029/98JD00275

Weng, Q. (2003). Fractal analysis of satellite-detected urban heat island effect. *Photogrammetric Engineering and Remote Sensing*, *69*(5), 555–566. doi:10.14358/PERS.69.5.555

Weng, Q., Lu, D., & Schubring, J. (2004). Estimation of land surface temperature-vegetation abundance relationship for urban heat island studies. *Remote Sensing of Environment*, *89*(4), 467–483. doi:10.1016/j.rse.2003.11.005

Weng, Q., Rajasekar, U., & Hu, X. (2011). Modeling urban heat islands and their relationship with impervious surface and vegetation abundance by using ASTER images. *IEEE Transactions on Geoscience and Remote Sensing*, *49*(10), 4080–4089. doi:10.1109/TGRS.2011.2128874

Wentz, F. J., Chelle, C., Smith, D., & Chelton, D. (2000). Satellite measurements of sea surface temperature through clouds. *Science*, *288*(5467), 847–850. doi:10.1126/science.288.5467.847 PMID:10797004

Williams, C. N., Basist, A., Peterson, T. C., & Grody, N. (2000). Calibration and verification of land surface temperature anomalies derived from the SSM/I. *Bulletin of the American Meteorological Society*, *81*(9), 2141–2156. doi:10.1175/1520-0477(2000)081<2141:CAVOLS>2.3.CO;2

Xian, G., & Crane, M. (2006). An analysis of urban thermal characteristics and associated land cover in Tampa Bay and Las Vegas using Landsat satellite data. *Remote Sensing of Environment*, *104*(2), 147–156. doi:10.1016/j.rse.2005.09.023

Xiang, X., & Smith, E. A. (1997). Feasibility of simultaneous surface temperature-emissivity retrieval using SSM/I measurements from HAPEX-Sahel. *Journal of Hydrology (Amsterdam)*, *188-189*, 330–360. doi:10.1016/S0022-1694(96)03165-4

Yamaguchi, Y., Kahle, A. B., Tsu, H., Kawakami, T., & Pniel, M. (1998). Overview of advanced spaceborne thermal emission and reflection radiometer (ASTER). *IEEE Transactions on Geoscience and Remote Sensing*, *36*(4), 1062–1071. doi:10.1109/36.700991

Zhang, J. H., Li, X., Yao, F. M., & Li, X. H. (2009). The progress in retrieving land surface temperature based on thermal infrared and microwave remote sensing technologies. *Spectroscopy and Spectral Analysis*, *29*(8), 2103–2107. PMID:19839318

Chapter 3
A Review of Outline on Spectroscopic vis-a-vis Geochemical Study of Earth Analogues for Mars

Shovan L. Chattoraj
Indian Institute of Remote Sensing, India

ABSTRACT

It is widely recognized that interpretations about the planet Mars must begin by using Earth as a reference, because Earth analogues can provide ground truth to constrain interpretations on the geological history of Mars. Fortunately, Deccan Basalt and its derivatives, exposed in Kutch area of Gujarat, can be considered as best candidate from Indian subcontinent to approximate this analogue in terms of petrology, stratigraphy and geochemical similarity. It is believed that Deccan Volcanic province with its extensive volcanic plains, cones and craters as well as abundant hydrous sulphates of secondary origin, approximates the geological, geomorphological and environmental conditions on the Mars.

DOI: 10.4018/978-1-5225-1814-3.ch003

BACKGROUND OF THE STUDY AND IMPORTANCE AT THE NATIONAL LEVEL

Siddaiha and Kumar (2009) reported occurrence of Minamiite (a Ca-variety of Jarosite) from Matanomadh area in Kutch for the first time. Jain et al. 2011 confirmed the occurrence of Jarosite from the same area. It was concluded that Jarosite from this area can be considered as a potentialanalogue for Martian hydrated sulphates as it has been formed through hotspot activities and having a similar geological set up as that of Tharsis and other regions on Mars.

Biswas (1992) placed Matanomadh Formation immediately overlying Deccan Basalt. So high chances are there that Jarosite formed from basalt or its derivative only. Though they are reported from Paleocene sandstones of Kutch but their occurrence in shales is also ubiquitous (Chattoraj 2011). See Figure 1. Chitale and Guven (1987) related the formation of natro-alunite (end member of Jarosite family) to the post-diagenetic formation of sulfuric acid-rich solutions by the oxidation of pyrite in the black shales of the Matanomadh basin. Understanding the processes involved in the formation of hydrous sulfates in the Deccan volcanic province of Kutch would therefore, help in inferring about the similar processes that might have been operated and/or operational in the Mars.

PROBLEM STATEMENT

Spectral vis-à-vis geochemical analyses of the hydrous sulfates from the Deccan Volcanic Province (DVP) of Kutch is proposed to be carried out in order to get insights into the fluid-rock interaction under extreme condition of high Eh and low pH in a basaltic terrain similar to that observed in the Meridiani Planum and other regions on Mars.

The niche research area is focused on stature of Jarosites derived Deccan Trap in Kutch as terrestrial analogues of Mars. Its geochemical signature is yet to be established to compare with that of Mars. More particularly, the source mineral and the process involved are to be proved compatible with Jarosites reported from Mars.

Chattoraj et al (2009) reported that some of the some framboidal pyrites in shales in Matanomadh Formation which indicates a marshy marine/estuarine or coastal environment. This kind of environment is reported to be hostile for Jarosite formation as the increased humidity will lead to its transformation of ferric oxy-hydroxides. Thus the optimal condition for Jarosite formation is to be defined and then can be extrapolated to Martian environment. Nonetheless, the Deccan basalts in Kutch is a promisinganalogue site for improving our understandingof the Martian surface aswell as for comparing the geologicprocesses on the two planets.

Figure 1. (a & b) Graphic log of Matanomadh Formation in Matanomadh and other areas (Green stars indicate occurrence of Jarosite in the horst rock); (c) Field photograph showing trough cross-stratified sandstone overlying carbonaceous shale; (d) Vertical section showing sandstibefacies overlying grey shale (pen length is 12 cm); Notice the presence of Yellow clayey balls of Jarosite in both sandstone and shale (marked by yellow arrow)

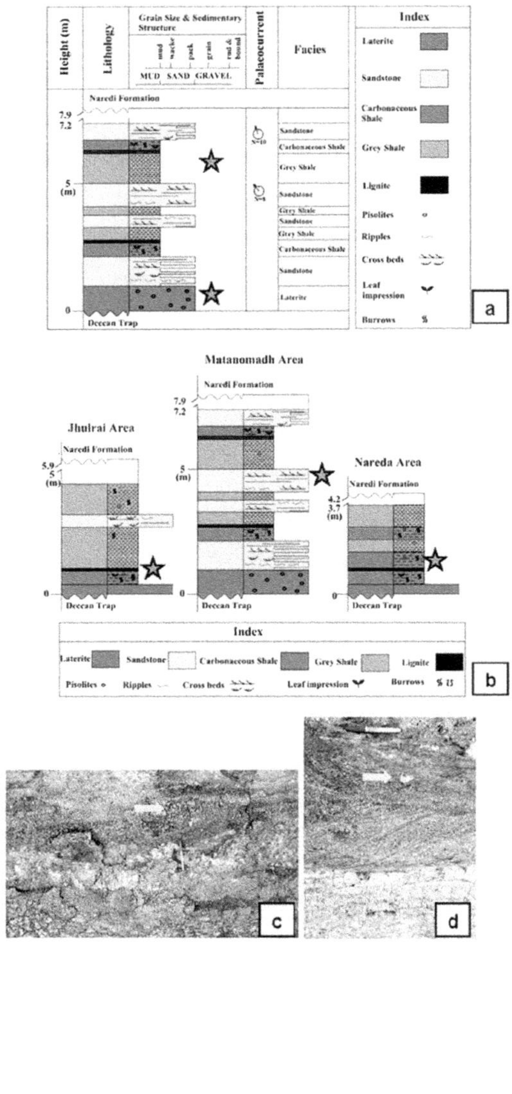

SCOPE AND OBJECTIVE OF THIS STUDY

Scope

1. Characterization of source mineral of Jarosite occurring in Matanomadh formation in Kutch.
2. Establishment of geochemical process/reaction involved in formation of Jarosites.
3. Characterization of extent of geochemical and temporal maturation of Jarosites.
4. Establishment of geochemical compatibility of Jarosites of Matanomadh Formation and Jarosites reported from Mars.

Objectives

1. Locate and describe the source of elements of Jarosite.
2. Unearthing the limiting Eh-pH condition for Jarosite formation.
3. To determine geochemical maturity of Jarosites.
4. To establish the approximation of Terrestrial and Martian Jarosites.

NEED ASPECT AND PROGRAM LINKAGE

It is internationally recognized that Earth should be considered as the standard reference in order to start interpreting Mars, as Earth analogues can give access to ground truth to constrain interpretations on geological history of Mars. However, there are few documentations of this fact is available. In particular, establishment of suitability of terrestrial mineralogical analogues for other planets is still in budding stage in India. The research till now is concentrated mainly on identification of Jarosites and its hydration states in visible and near-infrared (VNIR) reflectance spectra. However, reflectance in infra-red region gets us closer to mineralogical/elemental maturation in Jarosites. IR reflectance identification of Jarosites can fruitfully be integrated with geochemistry of the mineral focusing on zeroing in on the precursor mineral and chemical reaction leading to optimal elemental concentrations in Jarosite. The outputs of the proposed work can thus be utilized to strengthen our ideas on the then environmental condition prevailing on Mars and its similarity with Earth as far as surface geochemical reactions are involved. In context of ISRO`s successful Trans-Mars Injection of Mars Orbiter Mission- 'Manglayaan', understanding of the geological processes of the planet is going to be crucial for future explorations of Mars.

The data required for this project are:

- Image and field reflectance data: Aster and Hyperion.
- Bulk Rock geochemistry/REE data: ICP-MS.
- Major elemental data: XRD, EPMA.

METHODOLOGY

The hydrous sulfates at Matanomadh and its surrounding areas primarily occur as spheroidal and ovoid bodies in trap derivative laterites and in Paleocene shales/ sandstones. Any surface manifestation of this mineral and its assemblage will be mapped using hyperion and aster data to delineate the study area. Both in-situ and lab-based hyperspectral data sets in the visible to shortwave infrared (SWIR) domain (0.35-2.5 μm) is to proposed be analyzed in conjunction with the X-Ray Diffraction (XRD) and Electron probe micro-analyzer data of the samples collected from the field to detect the hydrous sulfate phases present and to get insights into the fluid-rock interactions and aqueous alteration processes associated with the formation of such ahydrous sulfate assemblage. IR reflectance is also to be utilized to give insights about absorption peaks of the extended spectral region. ICP-MS data with bulk rock geochemistry and REE analysis of the host rock will be carried out to compliment elemental analysis. Through petrographic studies will unearth any relict trace of precursor mineral so as to get us closer to the beginning of the geochemical reaction leading to formation of Jarosites. The methodology is described in Figure 2.

KEY OUTCOME

The project is envisaged to characterize the Jarosite present on Mars. It will be intriguing to note the reaction behind formation of this mineral on Martian surface, thereby throwing light on the then prevailing Eh-pH condition. This will give crucial information on geochemical maturity of Jarosite which in turn will make it comparable with terrestrial Jarosites of Deccan volcanic province. The methodology described here is dedicated to address the approximation of terrestrial and Martian Jarosites. The ground verification of Martian Jarosites are not plausible, albeit.

Figure 2. A flow chart of methodology for spetroscopical vis-à-vis geochemical study of Jarosite and Goethite from the Paleocene of Kutch in India illustrates the extent of suitability as terrestrial mineralogical/geochemical analogue to Mars

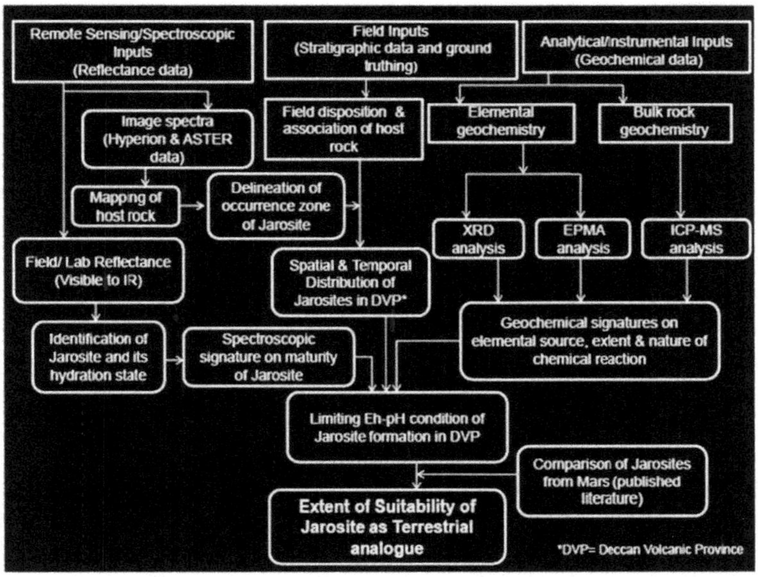

REFERENCES

Biswas, S. K. (1992). Tertiary stratigraphy of Kutch. *Journal of the Palaeontological Society of India*, *37*, 1–29.

Chattoraj, S. L. (2011). *Glauconite formation in stratigraphic framework of Palaeogene succession in Kutch* [Unpublished Ph.D. thesis]. Indian Institute of Technology Bombay.

Chattoraj, S. L., Banerjee, S., & Saraswati, P. K. (2009). Glauconites from the Late Palaeocene EarlyEocene Naredi Formation, western Kutch and their genetic implications. *Journal of the Geological Society of India*, *73*(4), 567–574. doi:10.1007/s12594-009-0040-x

Chitale, D. V. (1987). Natroalunite in a laterite profile over Deccan Trap basalt at Matanumadh Kutch India. *Clays and Clay Minerals*, *35*(3), 196–202. doi:10.1346/CCMN.1987.0350305

Farrand, W., Timothy, D. G., Rice, J. W., Hurowitz, J. A., & Swayze, E. (2009). Discovery of jarosite within the Mawrth Vallis region of Mars: Implications for the geologic history of the region. *Icarus*, *204*(2), 478–488. doi:10.1016/j.icarus.2009.07.014

Jain, N., Bhattacharya, S., Chauhan, P., Ajai. (2011). Hyperspectral study of hydrous sulphate minerals from Deccan Volcanic Province of Kutch, India: Implications for aqueous processes on Mars. *Proceedings of EPSC-DPS '11*.

Madden, M. E., Bodnar, R. J., & Rimstidt, J. D. (2004). Jarosite as an indicator of water-limited chemical weathering on Mars. *Nature*, *431*(7010), 821–823. doi:10.1038/nature02971 PMID:15483605

Siddaiah, N. S., & Kumar, K. (2009). Discovery of minamiite from the Deccan Volcanic Province, India: Implications for Martian surface exploration. *Current Science*, *97*(11), 1664–1669.

Swayze, G, A., Desborough, G, A., Smith, K.S., Lowers, H, a., Hammarstrom, J, M, Diehl, S.F, Leinz, R. W., & Driscoll, R.L. (2008). Understanding Jarosite-From Mine Waste to Mars. *USGS Circular*, 1328, 8-13.

Chapter 4
Improving the Efficiency of Image Interpretation Using Ground Truth Terrestrial Photographs

Gennady Gienko
University of Alaska Anchorage, USA

Michael Govorov
Vancouver Island University, Canada

ABSTRACT

Researchers worldwide use remotely sensed imagery in their projects, in both the social and natural sciences. However, users often encounter difficulties working with satellite images and aerial photographs, as image interpretation requires specific experience and skills. The best way to acquire these skills is to go into the field, identify your location in an overhead image, observe the landscape, and find corresponding features in the overhead image. In many cases, personal observations could be substituted by using terrestrial photographs taken from the ground with conventional cameras. This chapter discusses the value of terrestrial photographs as a substitute for field observations, elaborates on issues of data collection, and presents results of experimental estimation of the effectiveness of the use of terrestrial ground truth photographs for interpretation of remotely sensed imagery. The chapter introduces the concept of GeoTruth – a web-based collaborative framework for collection, storing and distribution of ground truth terrestrial photographs and corresponding metadata.

DOI: 10.4018/978-1-5225-1814-3.ch004

INTRODUCTION

Emerging innovations in optics, electronics and computers have led to the launch of a number of earth-observation systems which transmit terabytes of visual data every day. Some of the data are free, and pricing for the rest is declining due to competition among image providers. This provides great opportunities for institutions, research groups and organizations worldwide to use remotely sensed imagery in their projects in both the social and natural sciences.

Yet users often encounter difficulties working with aerial or satellite images, because these images differ from conventional photos in at least three important ways: objects are portrayed from an overhead (and unfamiliar) position, images are taken at scales most people are unaccustomed to seeing, and images are recorded over a range of wave lengths on multiple channels which are often beyond the visible spectral zone.

Along with computer-assisted automated classification of low-resolution multispectral imagery, visual image interpretation of large-scale aerial photographs and high-resolution satellite images is still widely used not only by non-technically driven geographers, biologists and other scientists, but by experienced professionals, as well.

While geographers at many universities study the basics of visual image interpretation, other professionals (*e.g.* biologists and environmental specialists) are generally not aware of image interpretation techniques until they become involved in a project and are forced to acquire such skills. The most practical way to acquire such experience is for observers to go into the field, locate themselves in an overhead image, observe the landscape and find corresponding features in the overhead image. Visual links between the natural view of a particular object and its representation in an overhead image create a particular association in the user's brain. The wider the variety of such mental images, the more experienced an image interpreter becomes. These mental pictures, along with analytical abilities for establishing such associations between objects and sites are the essence of visual image interpretation.

In many cases, natural observations could be substituted for terrestrial photographs taken from the ground with conventional cameras. If the coordinates of a terrestrial photograph are known, the position of the photographer can be located in an overhead image, making it possible for users to use conventional terrestrial photographs as a ground truth to establish a visual link between an object and its representation in an overhead image without traveling to the field.

There have been a few attempts to create repositories of sample objects, but they either represent images acquired by a particular sensor without ground truth (Remote Sensing Guides, 2008) or they provide a ground truth photograph and represent very specific objects (i.e. types of vegetation) in a limited area (Ochi & Takagi, 1996; Zion National Park, 2015). Any such repository will always have certain limits, because the compilation of feature identification keys with ground truth for diverse geographic objects worldwide is too immense a task for any single project or organization. Users could, however, compile such repositories by themselves. Many such users have had, and will have, field trips in which they take terrestrial photographs of different landscapes, geographic objects and phenomena; many use GPS and are willing to share their data over the Internet. A good example of such activity is The Degree Confluence Project (www.confluence.org). This example suggests that enthusiasts can form a virtual community driven by a common interest. The same can happen among professionals who use aerial photographs and satellite imagery. Such a utilitarian Virtual Professional Community would obviously be driven by common professional interests and could eventually be seeded around a Web-based data portal which allows any user to share geographically-referenced terrestrial photographs. Such a mechanism could provide the ability to store and explore geo-referenced data and metadata to allow any user to search and download terrestrial photographs pertaining to the user's area of interest and use them as a ground truth for interpretation of remotely sensed images.

This chapter introduces the concept of an open-source GeoTruth system – a web-based, open-community, participatory mechanism for the collection, storing and distribution of geo-referenced ground truth terrestrial photographs and corresponding metadata. The chapter discusses the value of terrestrial photographs as a substitute for field observations, elaborates on issues of data and metadata collection, and presents results of experimental estimation of effectiveness of the use of terrestrial ground truth photographs in interpretation of remotely sensed imagery.

GROUND TRUTH AND THE INTERPRETATION OF REMOTELY SENSED IMAGERY

The interpretation of remotely sensed imagery can be divided into two major subcategories: visual image interpretation and automated (computer-assisted) image analysis, with a hybrid approach as an intermediary. These wide and well-studied areas could be more refined in depth and width, but a principal point is the level of involvement of a human operator at different stages of the image analysis process. It is normal practice both in visual and automated classification to assess the correctness of the derived information by implementing an accuracy assessment.

Accuracy assessments determine the quality of information derived from remotely sensed data. Assessment methods are a complex subject and can be either qualitative or quantitative. While qualitative assessment is usually a first-cut assessment, quantitative assessments attempt to identify and measure errors of image classification. In most cases, the accuracy assessment is based on statistical evaluation of classification results with reference to the ground truth data (where ground truth data is assumed to be 100% correct), or sample areas on other remotely sensed imagery (reference information) (Buckland & Elston, 1994; Nishii & Tanaka, 1999; Stehman & Czaplewski, 1998; Liu et. al., 2007; Congalton & Green, 2008; Foody, 2010; Mather & Koch, 2011). In order to assess classification accuracy, there must be perfect or near-perfect registration between the reference information (or ground truth data) and classified images (Steven, 1987; Campbell & Browder, 1995; Kyriakidis & Dungan, 2001; McCoy, 2005; Strahler et. al, 2006; Wulder et. al., 2006; Rozenstein & Karnieli, 2011; S Bhagat, 2012).

Apart from being used for assessment purposes, ground-based data has been widely and successfully used to develop different feature interpretation keys (Paine & Kiser, 2003; Dahdouh-Guebas et. al., 2006; Lillesand & Kiefer, 2015;) and reference information – ground objects identified in overhead images by other experts and used as samples for image interpretation (Remote Sensing Guides, 2015; Zion National Park, 2015). Identified in a particular image, acquired by a particular sensor in particular conditions, these sample image fragments and feature identification keys can have almost no value for images acquired by another sensor or at another time. However, when supported by the ground truths, these sample references and identification keys become the most accurate and valued information for image interpretation and can be applied to any remotely sensed image (Thu & Popolus, 2007; Backes & Jacobi, 2006). Another valuable application of ground truth data is inter-calibration of different remotely sensed imagery and sensors (Johnson et. al., 1996; Teillet et. al., 2006; Steven et. al., 2007; Pe'eria et. al., 2013).

TERRESTRIAL PHOTOGRAPHS AS A SUBSTITUTE FOR GROUND TRUTH OBSERVATIONS

While broad aspects of visual perception of natural landscapes in terms of landscape architecture are widely discussed in the literature (Stamps, 1990; Hunt, 1984; Shang & Bishop, 2000; Rodrigues et.al, 2010), the use of terrestrial photographs to simulate observation of natural landscapes for environmental studies have been less studied (Cherem & Driver, 1983; Herzog, 1989; Herzog, 1995; Kushwaha & Singh, 2007).

Stewart et. al. (1984) pointed out the methodological weakness of previous empirical approaches and offered a new method for the use of photographs as surrogates for field observations for visual environments, but restricted their studies to the investigation of visual air quality, mostly for meteorological purposes. Henderson (1992) published a detailed theoretical study of visual-cognitive processes while viewing natural visual scenes. He elaborates on issues of determining the scene's semantic category, constituent objects, and spatial relations; he also offers an alternative framework for understanding scene processing, which he calls the local-processing hypothesis. This topic has been developed further in works of Palmer & Hoffman (2001), Palmer (2004), Uuemaa et. al. (2009), Frank et. al. (2013), and many others.

Scientific proofs of the applicability of photographic surrogates of natural landscapes for interpretation of remotely sensed images should be based on comprehensive studies of visual perception of geospatial imagery, but little has been done in this area (Lowe, 1985; Hoffmann & Markman, 2001; Gienko & Chekalin, 2004; Gardin et. al, 2011). Nevertheless, analysis of the available literature suggests that, despite theoretically important differences between perception of a still photograph and a real three-dimensional landscape scene, terrestrial photographs could be successfully used as a substitute for ground truth observations for the interpretation of remotely sensed imagery.

CONCEPT OF THE GEOTRUTH SYSTEM

In general, there are two major stages of image interpretation: formation of a set of models (mental images) for predefined classes of objects, and identification (recognition) of similar objects by matching mental models against perceived visual patterns. Once this matching is done, a final conclusion is made as to whether or not the target objects belong to the same class.

While this general technology is applicable both for visual and automated (computer-aided) image interpretation, this study illustrates the application of terrestrial photographs for assisting visual image interpretation. This does not mean that terrestrial landscape photographs may not be used to support automated image interpretation. To the contrary, our preliminary studies suggest that they have even more value for computer-aided image analysis, especially for supervised classification of low- and medium- resolution multispectral images and spectral signature-based analyses of images acquired by hyperspectral airborne and satellite sensors.

The concept of the GeoTruth system is based on the assumption that natural observations of the environmental landscape can be substituted with terrestrial photographs. Geographically referenced, time-stamped, and annotated landscape photographs could be uploaded to a web-based media data engine and then be made accessible to the wider community for identification of ground features during the interpretation of remotely sensed imagery. A distinctive element of this repository is that the participants can upload not only terrestrial photographs, but also fragments of aerial or satellite imagery for the study area. This would actually double the value of these data, as other users could use these pairs as a training set. Of course, Google Earth by Google and Virtual Earth by Microsoft provide global coverage of high-resolution satellite imagery and, in rare cases, aerial photographs in natural colors. These images can be successfully used for general study of the territory, but cannot yet surpass the value of multi- or hyper-spectral data samples provided by users for the study area.

In the GeoTruth system, the web-based media engine processes the uploaded source data and generates the output image along with the corresponding metadata file. Along with location and time stamp, the metadata file also contains annotation by the author of the source image, as well as comments added later by any other user, either in text, graphic or audible form.

Once such a data set is downloaded from the server, coordinates of a terrestrial photograph can be used to calculate the position of the photographer in a satellite image. If, in addition to the focal length and image resolution, the bearing of the photo is also known, these values can be used to determine a visibility cone which covers the area shown in the photograph. Thus, the user is simultaneously provided with two views: one from on the ground and one from above it. Looking at the terrestrial photograph, a user could imagine himself standing on the spot and observing the landscape. Looking at the "overhead" image at the same time, the user can associate the "natural" view of the landscape in a terrestrial photo with features in an aerial or satellite "footprint" of the same area. See Figure 1. In other words, the user can use conventional terrestrial photographs to establish a visual link between an object and its representation in an overhead image, which is an essential component in the process of acquiring image interpretation skills. While this general approach has been described in terms of applicability to visual image interpretation, it can have even more value for computer-aided image analysis, es-

Figure 1. Geo-linked terrestrial photo and satellite "footprint" of the same area, showing positions of all available terrestrial images (blue circles) and visibility cone for the current photo (green circle)

pecially for supervised classification of low- and medium-resolution multispectral images and spectral signature-based analyses of images acquired by hyperspectral airborne and satellite sensors.

Proposed as a web-based participatory mechanism for professional collaboration in remote sensing and image interpretation, the GeoTruth system is aimed at wide and active participation of users. The more users share their geo-referenced and annotated terrestrial photographs of different landscapes, terrain features and geographical phenomena acquired around the globe, the more knowledge can be acquired by a particular user and the more users can benefit from the system. Once users start sharing their data and knowledge, they actually seed a purpose-driven Professional Virtual Community which lives and behaves according to a certain code of conduct and ethics (Katzy & Ma, 2002; Goodfellow, 2003; Forster, 2004; Caby-Guillet et. al, 2006, Gienko et. al., 2007; Schiavone & Agrifoglio, 2012; Agrifoglio & Metallo, 2015).

THE VALUE OF LANDSCAPE PHOTOGRAPHS FOR IMAGE INTERPRETATION

The value of one or another landscape photograph as a clue for image interpretation can vary significantly depending on many methodological and technical factors, including the following:

- Geometric accuracy of geographic positioning.
- Scale and spatial resolution of the image to be interpreted.
- Pictorial representativeness and descriptiveness.
- Comprehensiveness and correctness of an annotation.
- Method by which the ground truth photographs were taken.
- Method of image interpretation.

The *geometric accuracy* of geographic positioning directly relates to the scale and, if in digital form, the spatial resolution of the image to be interpreted. Since the majority of recreational grade GPS receivers are accurate within 10 to 15 meters, it is very important for a photographer to choose the right location with a relatively uniform land cover within a reasonably large area. The size of the "relatively uniform land cover area" is a very subjective matter, but based on a 30 m spatial resolution of Landsat imagery, an area of 100x100 m can be used as general guidance for the size of a Minimum Mapping Unit (MMU).

The following experiment illustrates importance of geo-referencing accuracy relative to the spatial resolution of a sensor. Two adjunct pixels were picked in a Landsat ETM+ image for one of the ground truth test areas. Spectral responses for both pixels in tabular and graphic form are almost identical in all bands apart from bands 5 and 7 (825.00 nm and 918.75 nm respectively). See Figure 2.

These two adjunct pixels have been used as samples for supervised classification. As seen in Figure 3, classified pixels have formed well-defined clusters, providing clear evidence that these sample pixels are not variations within the same spectral class, but representatives of two distinct and self-consistent classes. Thus, if a terrestrial photograph used to assign a class label to the group of pixels is not accurately referenced, the classification results can be interpreted incorrectly.

Figure 2. Spectral responses for two adjunct pixels in a Landsat ETM+ image

Band	Wavelength, nm	Spectral response, pixel 1	Spectral response, pixel 2
1	450.00	53	49
2	543.75	36	34
3	637.50	31	24
4	731.25	69	65
5	825.00	54	20
7	918.75	34	11
6a	1012.50	118	117
6b	1106.25	125	125
8	1200.00	27	19

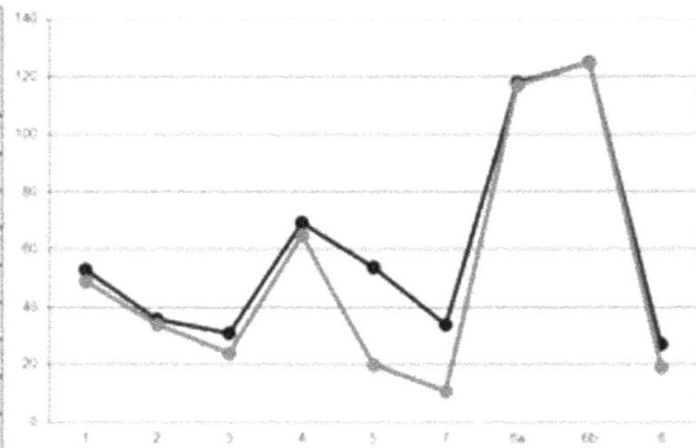

Figure 3. Two adjacent pixels (band 5, magnified 16x) and corresponding results of supervised classification (over band 5, magnified 2x)

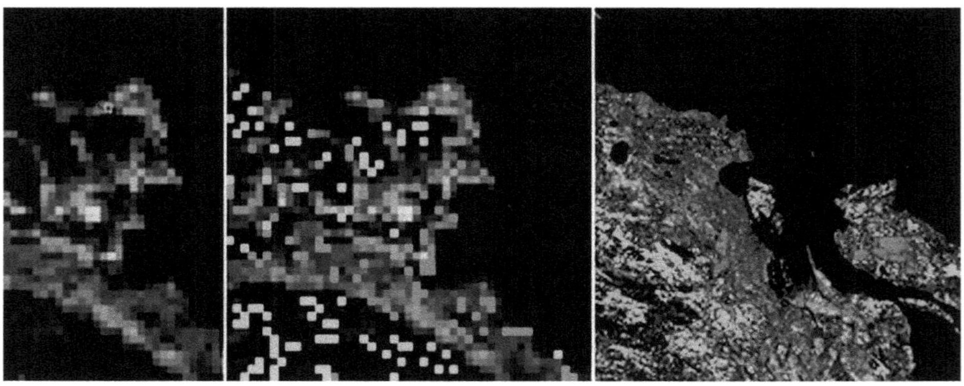

*For a more accurate representation of this figure, please see the electronic version.

Pictorial representativeness, and descriptiveness, along with the method used to take the ground truth photographs, should ensure adequacy of visual perception of the terrestrial photograph to real field observations. The resolution of a digital camera should be sufficient to depict the scenery in sufficient detail to allow the user to make an affirmative decision regarding the depicted objects (camera should have at least one megapixel image resolution). Although most modern digital cameras can acquire pictures in automatic mode balancing aperture and shutter speed to get the optimal exposure, the photometric quality of pictures still very much depends on the photographer's perception of the space and lighting (especially in case of a bright sky or against the light source).

Comprehensiveness and correctness of image annotation is the key issue defining the usefulness of the whole GeoTruth concept. The value of terrestrial photographs may be significantly deteriorated by incomplete or, even worse, by incorrect description of the ground truth area. In many cases, collaborative efforts of the community members may be required to evaluate the correctness of descriptions of photographs and make corresponding alterations.

The chosen *method of image interpretation* defines a particular approach to the use of georeferenced landscape photographs. In visual (manual) image interpretation of aerial photographs and high-resolution satellite imagery, terrestrial photographs taken in a ground truth sample area can be directly used as a surrogate for real field observations. The *number of photographs, their orientation and coverage of the sample area* should create a visual environment which is adequate or very close to real visual perception of the scene. In case of supervised or unsupervised automated classification of multi- or hyperspectral imagery, coordinates of the ground truth sample area can be used to locate a particular pixel in a digital image which can be then used as a sample point to launch a classification algorithm. The decision of assigning a particular class to that pixel (or corresponding cluster) can be made by analyzing the whole scene depicted in corresponding terrestrial photographs in a way similar to that that of visual image interpretation.

GEOTRUTH IMPLEMENTATION ISSUES: FIELD DATA COLLECTION

Ground truth terrestrial photographs can be taken either occasionally by passing by "an interesting object," or by intentionally photographing during dedicated field trips. This leads us to distinguish Occasion-Driven (OD) georeferenced landscape photographs from Purpose-Driven (PD) photographs that were captured as part of a deliberate effort to populate the system. Of course, these OD photographs may not contain the whole set of information compiled during Purpose-Driven field work, but still many OD photographs may have incidental value as proxies for field studies.

While developing methodology for Purpose-Driven field data collection, the following properties of the landscape and conditions for picture acquisition should be taken into account:

1. Types of ground truth sample area by visibility (obscurity):
   a. Open-space area (without high vegetation cover, e.g. grassland, low bush).
   b. Closed-space area (e.g. forested, dense tall bush).
2. Types of ground truth sample area by relief:
   a. Hilly to mountainous areas.
   b. Flat to moderately hilly areas.
3. Types of terrestrial photograph by direction:
   a. Single spot photograph.
   b. Omni-directional photographs.
   c. Panoramic photograph.
4. Types of terrestrial photograph by orientation and distance to the sample site:
   a. In-site photograph (inside the ground truth sample area).
   b. Off-site photograph (perspective, aside from the ground truth sample area).

Purpose-Driven photographs, both in open-space and closed-space areas, can be collected using either a Target-Facing (TF) or Target-Reverse (TR) orientation. Both methods have their merits and supplement each other. While Target-Facing photographs can be mostly used for off-site photography for perspective images, Target-Reverse photographs are most useful for taking multi-directional in-site pictures. Figure 4 provides an example of an open-space area test site and terrestrial photographs acquired by the omni-directional Target-Reverse method with azimuthal orientation (north, east, south, and west).

Figure 5 illustrates the value of using both in-site and off-site terrestrial photographs at the same time. The off-site perspective photograph shows the general pattern of the landscape and significantly contributes to general understanding of the local ecosystem.

In-site panoramic images probably have the highest value, as they provide a 360 degree view of the scene. Figure 6 gives an example of a panoramic image taken in a forested area of the Greene Valley Forest Preserve, Naperville, IL (courtesy of David E. Bodenstab, http://www.bodenstab.org/panorama/index.html), and a corresponding high resolution overhead image (source: live.maps.com).

Figure 4. Open-space area test site and position of the photographer pointed with green arrow on high-resolution overhead imagery (maps.google.com), and corresponding omni-directional Target-Reverse ground truth terrestrial photographs, taken in the east, north, south and west directions in Zimbabwe. (Terrestrial photographs by Terence Garde and Stuart Garde, www.confluence.org)

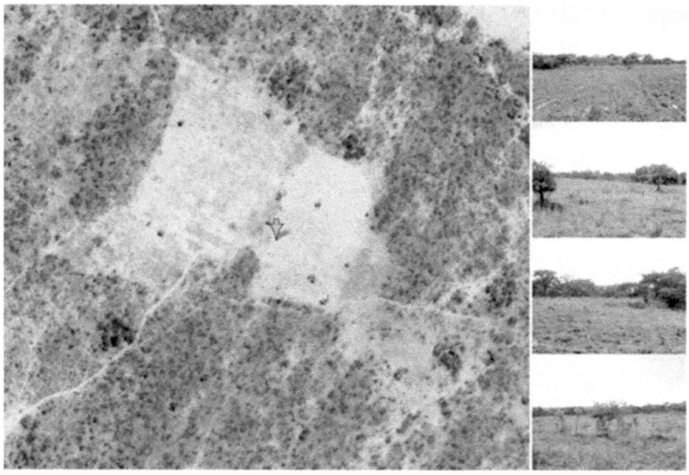

Figure 5. In-site (red circle) and off-site (yellow circle) terrestrial photographs in Nadromai, Fiji Islands of a ground truth test area and corresponding high resolution satellite image (source: maps.google.com). (Terrestrial photographs by Gennady Gienko)

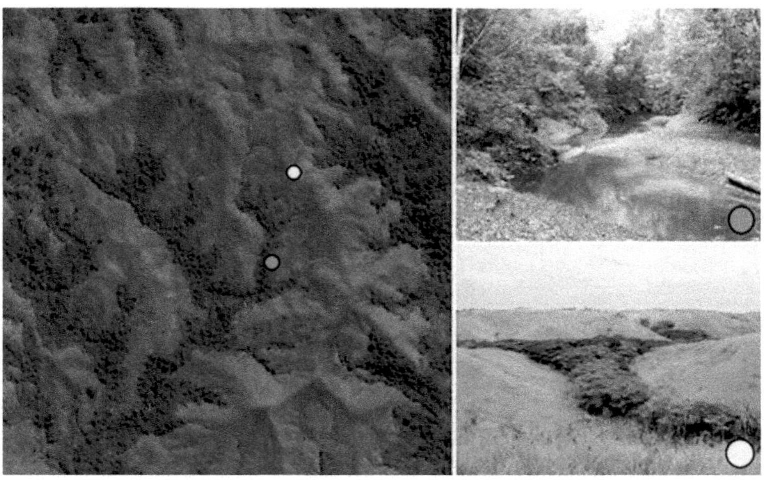

There are a number of web-based solutions that can be implemented for online scrolling of panoramas (www.bodenstab.org/panorama/index.html), e.g., SCCplay viewer by Stu Nicholls (www.cssplay.co.uk).

Off-site photographs can be distinguished by range. Close-range photographs provide a close look at a target from a relatively short distance, and they are usually Target-Facing with precise targeting. Long-range or perspective terrestrial photographs are mostly aimed at depicting the general landscape in the sample target area, and contain objects in the sample ground truth area, but may not be precisely targeted. Figure 7 illustrates close-range off-site Target-Facing photographs and their locations and orientation to the target on the high resolution overhead imagery.

TEST DATA SETS AND PRACTICAL EXPERIMENTS

Although, intuitively, it is clear that terrestrial photographs can be useful for image interpretation, it is quite difficult to quantitatively estimate the level of such usefulness. Due to strong correlation with the expert's expertise and knowledge of a particular region, direct comparison of image classification results with and without

Figure 6. High-resolution overhead image (live.maps.com) and 360-degree panoramic photo taken in forested area in Naperville, Illinois (courtesy of David E. Bodenstab)

Figure 7. High-resolution overhead image in Nanaimo, Canada showing positions and orientation of off-site close-range Target-Facing omni-directional photographs (blue symbols) relative to the target (red symbol). (Terrestrial photographs by Brad Maguire)

the use of terrestrial photographs may not be fully conclusive. Nevertheless, such analysis can illustrate the general level of an interpreter's confidence in identification of objects in overhead images.

A set of satellite imagery has been compiled to carry out a feasibility study of the usefulness of terrestrial ground truth photographs. Fragments of satellite imagery (source: maps.google.com) at different spatial resolutions have been chosen for 10 test areas in different regions of the world (Table 1). Test areas represent different land covers: natural vegetation (forested and woody areas), agriculture (crop fields), and desert areas. For comparative study of scale and generalization effect, satellite photographs for two sites have been simultaneously provided in two spatial resolutions – 1 m and 30 m.

Geo-tagged terrestrial ground truth photographs for corresponding test sites have been downloaded from the Degree Confluence Project (www.confluence.org, individual photographs retain copyrights of their respective authors). Terrestrial

Table 1. Details for test sites satellite imagery, and ground truth terrestrial photographs

Site	Region	Lat	Long	Spatial Resolution
A	Benin	11N	01E	30 m
B	Philippines	11N	125E	1 m
C	India	12N	76E	1 m
D	Egypt	24N	28E	30 m
E	India	29N	76E	1 m
F	Argentina	34S	65W	30 m
G	Tennessee, USA	35N	85W	1 m
H	New Zealand	42S	172E	1 m, 30 m
I	Mongolia	49N	91E	1 m, 30 m
J	Sweden	59N	18E	1 m

photographs on each site have been taken in four directions (north, east, south, and west); some have additional off-site or panoramic photographs.

Three groups of users have been involved in experimental studies: 1) professional image interpreters (5 experts, with 10 to 30 years of experience), 2) post-baccalaureate students enrolled in a GIS program and taking Remote Sensing course (15 students, general background, but no real image interpretation experience), and 3) undergraduate students taking an introductory GIS course (40 students, no background and experience). The experiment was designed to run in three stages, where the users were given the following data and information:

1. Satellite image only
2. Satellite image plus region name and coordinates where the image was acquired
3. Terrestrial photographs in addition to the above

The users have been asked to classify the marked objects consecutively in all 10 satellite images and estimate the level of their confidence, by percentage, for each stage. A statistical summary of the experiment is provided in Table 2 and further illustrated in Figure 8.

The results of the experiment show that overall confidence increases as more information becomes available to the interpreters. Confidence steeply increased at Stage 3 when the interpreters used terrestrial photographs to support image interpretation. At the same time, the standard deviations decrease from step to step, and decisions become more uniform after the use of terrestrial photographs.

Table 2. Dynamics of confidence in image interpretation

Level of Expertise	No. of Participants	Confidence and Standard Deviation, per Cent		
		1st Stage	2nd Stage	3rd Stage
Experts	6	68.9	76.1	93.7
		8.5	6.9	6.1
Some background	18	52.4	62.4	89.9
		23.0	17.1	11.0
Novice	40	42.2	51.3	80.8
		23.6	23.3	16.7

Figure 8. Confidence and standard deviation in image interpretation

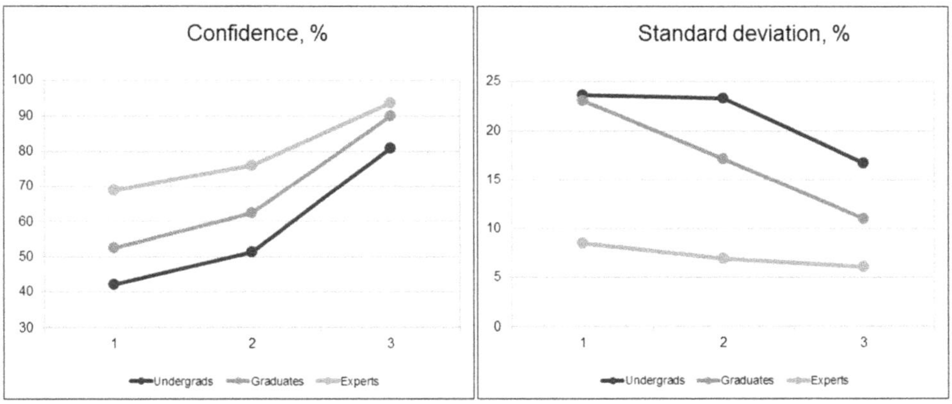

While the absolute values of confidence have to be interpreted in a very careful manner (as the novices tended to overestimate their confidence and the experts, quite the contrary, to understate it), the experiment has confirmed the hypothesis on the general tendency of changing of confidence in image interpretation with support of additional information and, especially, with terrestrial ground truth photographs.

GEOTRUTH IMPLEMENTATION ISSUES: DATA AND METADATA

The usefulness of the GeoTruth system (and the existence of its corresponding Virtual Professional Community) is directly related to the completeness, comprehensiveness and quality of the data acquired and stored in the warehouse. Several data-related components define the core content of the GeoTruth repository:

1. Source data submitted by the authors and data contributed by the community members:
    a. Terrestrial photographs.
    b. Video footage.
    c. Geo-positional data.
    d. Field-collected spectral signatures.
    e. Fragments of aerial or satellite photographs for sample area.
    f. Auxiliary geospatial data (GIS layers, DEM, LiDAR data, point clouds generated from aerial and terrestrial photographs).

In addition to terrestrial photographs as the main type of visual data, the system can be populated with video footage taken in ways similar to those described above for acquisition of still landscape photographs. Geo-positional data should be submitted as geographic coordinates (longitude and latitude) in WGS84 to ensure consistency and compatibility of outputs, and should contain some estimation of positional accuracy. Geo-positional data can be acquired by using GPS, Unmanned Aerial Systems (UAS) or, in some cases, from Web-based map services which provide high-resolution satellite imagery, such as Google Maps, Google Earth and Microsoft Virtual Earth, etc., where users can locate the visited ground truth sample area by relying on their spatial and visual memory.

If field spectral measurements are available, they can be also placed in the system. Details of spectrometer and acquisition technique should also be entered as metadata. Users can contribute to the system by supplying fragments of aerial and other remotely sensed data along with results of their interpretation. These can be used as case studies for many novices in image interpretation; the value of such composite data for this purpose cannot be overestimated.

The inclusion of a fine- or moderate-resolution DEM can extend the functionality of the system considerably, allowing the user more transitions between overhead satellite views and terrestrial photographs. Users can view an overhead view, one or more views of the draped satellite image from different angles, and, finally, a terrestrial view. In addition, view-shed calculations can be used to determine the farthest extent of the camera field of view in many cases, i.e., where it intersects the terrain. With images obtained during Purpose-Driven collection for which there are accurate location, bearing and inclination for the photograph, it may be possible to map individual pixels on the satellite image to particular areas on the terrestrial photograph. This would provide users of the satellite image a new level of understanding, because they would immediately be able to understand why particular pixels on the satellite image appear the way they do. Using relevant GIS layers can be very supportive and would significantly increase user confidence in making the final decisions in image interpretation.

2. Annotations to the data, consisting of a source description by the author and comments added by community members:
   a. General characteristics of the landscape and vicinity.
   b. Description of the ground truth area.
   c. Description of the components of the ground truth area (type of vegetation, plants, species composition, height and density of canopy and understory, type of material for man-made objects, condition, other object characteristics).

Annotation of the data is an essential inherent component of the GeoTruth system. The consistency and adequacy of the description of the ground sample area defines the value of the entire GeoTruth concept. The description of components and elements of the ground truth area and the general characteristics of the surrounding landscape can be the weakest information provided in annotations if no special efforts are made to design a structured questionnaire.

As a starting point, it is proposed to use one of the land use and land cover classification systems - for example, the international FAO Land Cover Classification System (Di Gregorio & Jansen, 2005) used by the TC211 Technical Committee of the International Organization for Standardization as the base for development of the ISO 19144-2 standard "Geographic information – Classification Systems – Part 2: Land Cover Classification System LCCS", or, as an alternative, the European classification system CORINE (Bossard et. al., 2000), or any national or regional system, such as U.S. National Land Cover Data (Anderson et. al., 1976) or Australian ESCAVI (ESCAVI, 2003). The advantage of using a standardized system is that the classes and their hierarchy are already defined, allowing a high level of interoperability in exchange of data and metadata. The appropriateness of such a formal approach should be judged by the community, and, if necessary, an alteration of an existing system or a completely new alternative approach can be developed by collective efforts of members.

3. The GeoTruth system may support extraction of the following datasets:
   a. GIS layers in an interchange or major proprietary formats for export of georeferenced photographs into GIS and image processing packages (ArcGIS, MapInfo, ERDAS, ENVI, etc.).
   b. Geotags in KLM or other interchange format to export georeferenced photographs into Web-GIS applications and systems (Google Earth, Virtual Earth, World Wind, etc.).
   c. Spectral signatures extracted from available sample imagery.

Output data generated by the GeoTruth system should provide an easy way for users to use downloaded photographs in their GIS or Remote Sensing and image analysis packages, whether a stand-alone commercial package or an open-access Web-based application. Depending on the type of output data, there might be a need to develop and use additional tools or plug-ins to browse the data within a particular software package. Those plug-ins can be developed and submitted by community members as the need arises.

4. The GeoTruth system should support metadata of stored datasets, such as:
   a. Details of source data acquisition.
   b. Details of annotation.
   c. Details of generated derivatives.
   d. History of updates.

The metadata files will be the first and foremost source of information about the ground truth sample area for storing and searching in the system. The content of this file should provide non-ambiguous information for users to make decisions about whether they need to download the whole data set or not. The structure and format of the metadata files need to be carefully designed in a way to ensure cross-platform interoperability and compliance with the most current standards for geospatial data and metadata.

GEOTRUTH VIRTUAL PROFESSIONAL COMMUNITY

During the last decade, activity in cyberspace has grown in an explosive manner. The World Wide Web and other telecommunication technologies have opened up new possibilities for collaboration. People form Virtual Communities (VC) over the networks (internet, mobile, etc.), each of which is driven by a particular interest or professional need to access and mediate knowledge and information remotely.

One more efficient way to facilitate the accumulation of data for a terrestrial photograph repository and to disseminate RS knowledge and applied skills, is to establish a Virtual (or online) Professional Community. Such a Remote Sensing VPC can be public or member-initiated by design, and can include a professional by relationship-orientation (Porter, 2004).

A Virtual Professional Community (VPC) encompasses a combination of the concepts of virtual community (Rheingold, 1998) and professional community. Professional communities ("communities of practice") provide environments for professionals to share the body of knowledge of their professions, such as similar working cultures, perceptions of problems, problem-solving techniques, professional values, and professional behaviour (Katzy & Ma, 2002; Caby-Guillet, et. al., 2006; Schiavone & Agrifoglio, 2012; Agrifoglio & Metallo, 2015). A professional virtual community is a manifestation of Collaborative Networks. Public professional communities bring together audiences on specific topics, such as product application or product development (e.g., GeoTruth VPC).

Preliminary analysis of the various characteristics of virtual communities (Porter, 2004; Porter, 2008; Messinger, et. al., 2009; Zhou, 2011) leads us to propose the following attributes for a proposed Remote Sensing VPC:

- **Purpose:** Accumulate and share knowledge, skills and new approaches for effective use of remotely-sensed images.
- **Primary Audience:** Researchers, organizations, students, and project managers, both in social and natural sciences (e.g., environmental specialists, foresters, biologists etc.), who are using remotely-sensed imagery in their multidisciplinary projects.
- **Places:** The community's "space" encompasses the whole globe; the community's "place" is to be useful for researchers across many disciplines which use space imagery; the socio-cultural "space" provides the opportunity to meet and communicate without prejudice.
- **Platform Hybrid:** Where the community has elements of both asynchronous (web GUI for downloading and uploading images and metadata, forum, email etc.) and synchronous communication (direct data transfer etc.).
- **Population Interaction Structure:** A Network-based Community, where members are geographically and socially dispersed and focus on the functional benefits of the community, such as imagery acquisition and related problem solving. Relationships in networked-based communities are often of short duration and driven by utilitarian needs, and social ties are generally weak among members, yet they can demonstrate supportive and reciprocal behaviour, despite the fact that they are socially and/or physically distal (Porter, 2004; Zhou, 2011). Within this networked-based community, members can form clusters with similar interests in particular study areas and thus create stronger ties among members.
- **Driving Motivations:** The search for information and new techniques, but also to collaborate with many colleagues in different parts of the world, as well as the symbolic value of making a contribution to the member-generated image content.
- **Degree of Organization:** High interaction is not the norm in such VPC, and would not occur without intentional efforts on the part of a moderator(s). The community can exercise "technological control" over access to preserve quality of the imagery information, but requires some system of control over

unwanted behaviour. Stressful interactions in such community are likely, but have low frequency. It is important to keep a balance among privacy, functionality needs, security, content, etc.

- **Profit Model:** The GeoTruth VPC is projected as a non-profit community.
- **Main Technical Requirements:** Strong, distributed repository; an information warehouse; a user-friendly Web 2.0-based interface.
- **Psychological Sense of Community (PSOC):** The RS community will likely have a low level of PSOC (Forster, 2004), but members of the community still need feelings of belonging, trust, expected learning, recognition and obligation, which can be created by introducing roles, protocols, policies and/or norms for the VPC.
- **Some Challenges:** VPC must be intellectually accessible; users need to share only what he/she knows to be true.

The projected attributes of the GeoTruth VPC are based on a review of literature, and observations of standard design principles and activities in actual virtual communities. These attributes were used for strategic planning and assessment of the GeoTruth community and user needs.

PARTICIPATORY MECHANISMS FOR VIRTUAL PROFESSIONAL COMMUNITIES

Internet utilization is a typical and most effective foundation for Virtual Professional Communities nowadays. In several publications and VC projects (Porter, 2004; Katzy & Ma, 2002; Forster, 2004; Zhou, 2011 etc.), the Internet was used as the main Participatory Mechanism tool. The GeoTruth project seeks a better way to utilize the Internet, based on the following criteria:

- Flexibility and independence from proprietary vendors (will be discussed later).
- Reliability (maturity of implemented technology).
- Costs of implementation, technological and organizational support and sustainability.
- Support by a Virtual Professional Community (self-support).
- Openness for community.
- Availability of services for Virtual Professional Communities (technologically).
- Equity of all participants.
- Readily distributed and collaborative environment.
- Usability (user-friendly interface).

VPC participatory mechanisms typically utilize a client-server-database architecture and respective components to support a Web 2.0-based interface for all services and resources (Beaujardière et. al., 2000; Aggett & McColl, 2006; Jankowski, 2011; Slotterback, 2011, etc.). The GeoTruth system deals with geospatial datasets and needs to utilize Geographic Information Systems (GIS) technologies within a client-server-database architecture. Nowadays, almost all commercial and public GIS systems support web-based services and interfaces to get access for the GIS information. To design the GeoTruth system, different technologies of the multi-tiered architecture can be utilized to implement its client, server and back-end components (Govorov et. al, 2005).

In the sections that follow, the attributes, operational principles, and architecture for the GeoTruth system are elaborated based on the above-mentioned criteria.

THE GEOTRUTH VPC: LIFE CYCLE DEVELOPMENT AND OPERATIONAL PRINCIPLES

As a Virtual Professional Community, the GeoTruth system cannot emerge spontaneously, and needs to be nurtured during its creation and development until it is able to sustain itself. Most researchers adopt five stages of group dynamics for the life cycle of professional community development (Preece, 2000; Katzy & Ma, 2002; Palloff & Pratt, 2007; Jankowski, 2011). The authors have adapted this group dynamics concept for creation of the GeoTruth Virtual Professional Community, founded on an operational principal for each phase.

Phase 1: Initiation

The idea of a GeoTruth VPC originated from remote sensing professionals residing in three countries, and the idea was shared amongst initial members of the professional community. Members of a multidisciplinary group (including representation from Remote Sensing, GIS, IT, Computer Sciences and Physiology fields) set up a team, which identified a GeoTruth VPC purpose and assessed the potential of RS community and user needs. During this phase, the team exchanged opinions with domain experts and potential users, analyzed public web sites, investigated RS academic associations and networks professional lists (which may have an impact on the GeoTruth VPC) and relevant literature, and carried out preliminary data interpretation and methodological case studies.

Phase 2: Strategic Planning - Assessment of Community and User Needs

The core team is responsible for defining guidelines by which the community will operate, including community values, standards and an organizing structure. Following are GeoTruth guidelines:

- **Values:** The strategic objectives of the GeoTruth community are to develop an open access participatory mechanism as a set of tools, which, from a qualitative view, will support users worldwide in accumulating and sharing knowledge, skills and new approaches for effective use of remotely-sensed images, and, from a quantitative view, will save user time and money by reducing numbers of field trips.
- **Standards:** Community policies which describe the requirements and access procedures, are articulated in a public forum, open for anyone's use, but with technological control over information contributions and content. This online community requires that newcomers demonstrate some level of "knowledge" only for *inputting* new information. A system and "community" controls over unwanted behaviour is also established to regulate interactions among facilitators and users, and users themselves.
- **Organizational Structure:** A few moderators, together with a community manager, will encourage communication, moderate collaboration among members, and facilitate new members to join the community. The moderators' roles include: maintenance of online forum tools and their contents, technological help and control over input of images and their metadata, methodology, data (raster and vector general reference base), software, hardware and polices upgrade and support, knowledge sharing with VPC users, and organization of meetings which help to cultivate trust among participants, etc.
- **Sustainability Model:** Development of the GeoTruth system at the implementation and nurturing stages could be mostly funded through research and industrial foundations, as well as supported by individual members. At the implementation stage, GeoTruth engine contents and source code will be open for public use. Functioning of the GeoTruth at the operational stage has to be self-sustainable.

Phase 3: Technical Implementation of Online Environment

Technical implementation of the GeoTruth community represents the conclusion of the previous two phases. Principles of usability of the virtual environment and sociability of the interaction within were considerations for design and implementation of GeoTruth mechanism. The following design dimensions of "usability" and "sociability" are considered (Preece, 2000) and reflected in Table 3:

Several of Kim's "social scaffolding" principles (Kim, 2006) have also been adopted for the system design and implementation as follows: building flexible, extensible gathering places; creating meaningful and evolving member profiles; encouraging appropriate etiquette; promoting cyclic events and facilitating member-run subgroups.

Technical implementation implies the design, implementation and testing of a prototype of the GeoTruth online platform, and subsequently refining its sociability and usability (Goodfellow, 2003; Schiavone & Agrifoglio, 2012).

This study uses a rapid prototyping methodology for development of the GeoTruth system. *Prototyping* is a way of quickly developing an incomplete working system from a design model in order to test various aspects of a design, illustrate ideas or features and gather early user feedback (Crinnion, 1991; Grimm, 1998; Stephen et. al., 2006). Prototyping is often treated as an integral part of the SW Engineering process, which reduces project risk and cost, testing and implementation time, and improves the software quality. This usually requires several iterations to create a complete real working system.

Phase 4: Welcoming and Nurturing the Community

In this phase, the community begins to operate within the defined structure and processes. A community committee will be established to promote the GeoTruth system, and will interact with the available community. Critical success factors in this phase are clear and visible community purposes, membership benefit, and structures and policies (Goodfellow, 2003).

Table 3. Preece's design issues for online community

Usability	Sociability
interaction dialogue	membership policy
navigation	codes of conduct
registration	security
representation of users	privacy
message format	copyright
archives	free speech
support tools	moderators

In this phase, the efficiency of management activities, technological tools and methodology are verified by using a feedback system and through analysis of possible deviations from the strategic plan.

Phase 5: Operation, Evolution Management, Sustainability and Adaptation

In this stage, the GeoTruth repository will be filled with information and subscribe a number of members to reach stable levels. The community needs to encourage members to participate in the on-line activities. The community also needs to balance privacy, functionality needs and content. The monitoring of community activities, the evaluation of its results and achievements, and its adaptation to new requirements and needs has to be carried out by the community committee and the community itself.

GEOTRUTH SERVICES, ARCHITECTURE, AND PORTAL PROTOTYPE

There is a well-established literature in the field of web-based map systems related to the work proposed here. The following publications provide information on different aspects of web map systems: Dykes et. al., 1999; Peng & Tsou, 2003; Tsou, 2004; Zhang & Tsou, 2009; Wright & Wang, 2011; Wang, et. al., 2013; The Open Geospatial Consortium, 2015; System Design Strategies, 2015, etc. The participatory mechanism of the GeoTruth VPC can be implemented in the form of a Public Participation Web GIS (PPGIS) which possesses principles such as accessibility, understandability, and accountability (Craig et. al., 1999; Jankowski & Nyerges, 2001; Jankowski, 2011), and provides the following group services and functions:

- Raster and image data can be shared and manipulated (add, delete, or update raster) as Image Service.
- Web map visualization of vector and raster data (terrestrial and remote-sensed images) and backdrops can be implemented as Web Map Tile Services (WMTS).
- Vector data (e.g., georeferenced points with terrestrial photographs attributes) can be accessed and manipulated as Web Feature Service (WFS).
- Geoprocessing services (signature development, band collection statistics, transformations, etc. and data-format conversion tools) can be implemented via Web Processing Service (WPS).

- Metadata catalogue (metadata display, data preview, data download, metadata search and indexing of photographs) can be implemented as Metadata services.
- There are a number of innovative approaches to indexing photographs for effective user search: by keywords, based on common items in adjoining photos; digital signatures of similar photographs; spatial attributes, etc. (Snavely et. al., 2006; Kennedy & Naaman, 2008; Snavely et. al., 2008; Luo et. al., 2011; Flickr, 2015; Photosynth, 2015).
- User asynchronous interaction (user discussion forum, email list, chat) can be implemented by using standard API modules/libraries.
- Administrative (security and user control).

An outline of the GeoTruth software architecture is displayed in Figure 9.

Client Side of GeoTruth

One important aspect which will affect the engagement of communities and citizens' involvement with PPGIS applications is usability of a graphical user interface (GUI) and respective integrated tools. When applied to an online community, GUI usability relates to how easily and effectively your online community can use your web site and other online spaces. There are four key elements which can be used to rate PPGIS GUI instrument utility: (i) user-friendliness, (ii) flexibility, (iii) transparency, and (iv) ability to incorporate and communicate stakeholder interests (Aggett & McColl, 2006; Chen et. al., 2008, Jankowski, 2011).

Visualization tools are clearly a key component because they provide a common language to which all community members can relate (Jankowski & Nyerges, 2001;

Figure 9. Architecture of the GeoTruth System

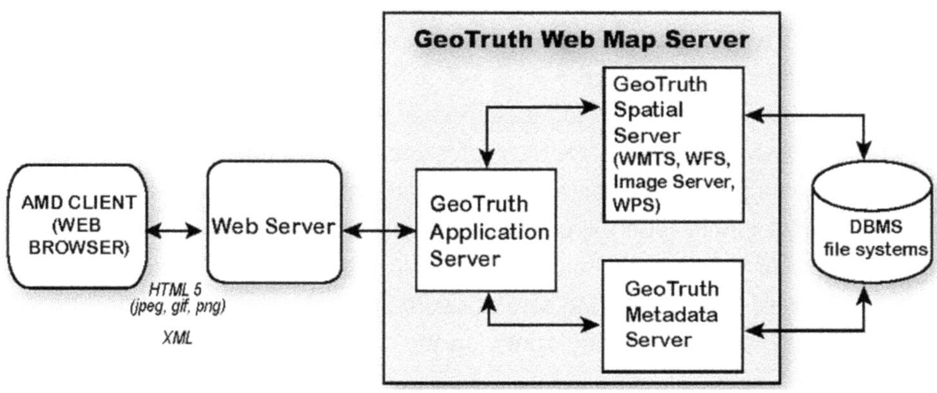

Jankowski, 2011). Similarly, design of GUI and implementation of an initial set of tools for the GeoTruth system are very important tasks for attracting members and to support an integrated knowledge environment.

Kazman & Komeniuk (1993) discuss five principles of good UI design. These include:

- Centering the application around the user.
- Making the interface visible.
- Providing a unifying theme.
- Making object interaction direct.
- Providing many open doors.

Because the system should have two levels of users (amateur and professional), it makes sense to offer two levels of User Interface (UI). This addresses Kazman and Komeniuk's (1993) concept of centering the application around the user. In order to avoid connotations of "amateur" and "expert" users, these User Interfaces will be simply called the "Photo" UI and the "Remote Sensing" UI.

The Photo UI is concerned with the contribution, georeferencing, organization, and retrieval of photographs. This UI should be as simple to use as possible.

A lot of work has gone into the design of the World-Wide Media eXchange (WWMX) client software user interface (Toyama et. al., 2003; Naaman et. al., 2004; Peca et. al., 2011; Zheng et. al., 2011; Fang et. al., 2013). The primary UI is map-based, but the system also allows the retrieval of photographs to be constrained by time. There is also a tight coupling between the thumbnails, the full-sized images, and the maps, which Toyama refers to as a "Reflective UI." This User Interface makes the interface visible and makes the object interactions direct, since every click on an item highlights all related items shown on the screen.

The WWMX client clearly has an "expert level" UI, in which a great deal of functionality is revealed to the user at one time. Although very easy to use, the number of panes in the WWMX client can be quite intimidating. The authors propose a tab-based UI for this project, with separate tabs to search by geographic location, keyword, or date of acquisition.

The Remote Sensing UI would incorporate all of the features of the Photo UI, but would include further tools for the professional user of this system. The Remote Sensing UI would allow for the overlay of one or more remotely sensed images onto the map, together with the calculated fields of view for the terrestrial photographs. Users of the Remote Sensing UI would be able to locate satellite images by location or keyword, and find sets of overlapping terrestrial photographs. The terrestrial photographs would be drawn on top of the satellite image, showing the camera field of view.

The architecture of the GeoTruth client can be based on an AJAX generic application model. It is a Web application model that employs partial screen update and asynchronous communication (Holzner, 2006). Specifically, the GeoTruth client can be implemented based on the Asynchronous Module Definition (AMD) specifications of AJAX.

AMD defines an Application Programming Interface (API) specifying a mechanism for defining modules such that the module and its dependencies can be asynchronously loaded. This is particularly well suited for the GeoTruth concept, where synchronous loading of modules such as terrestrial photographs, their referenced point locations, and geoprocessing result on top of a background image impacts performance and usability.

Server Side of GeoTruth

GeoTruth server components can be built from a few services such as WMTS, WFS, Image Service, WGS, and catalog capabilities of the respective services.

The following services can be supported for a user's request from the client side to the server side of GeoThruth:

- Users can send a request from an Internet web client to the GeoTruth server to search, download or upload terrestrial photographs and their metadata.
- The Web Server provides the access to GeoTruth information from the Internet by using HTTP or over HTTP protocols.
- The Application Server is a middle tier server, providing an interface between the Web-server and the GeoTruth Servers. The Application Server enables the Web-server and Servlet Engine to intercommunicate with GeoTruth Servers. It also handles the load distribution of incoming requests.
- The core of the GeoTruth Server is the Spatial Server, which hosts WMTS, WFS and WPS, and supports functionality such as visibility modeling, custom object and metadata uploading, editing and deleting, querying attribute data for spatial and tabular queries, and data-format conversion. It provides the functional capabilities for filtering, element generation and rendering maps and data into the appropriate format before sending the map or data back to a client.
- The Web Map Tile Service is responsible for cartographic visualization of imagery (e.g., multispectral Landsat imagery, terrestrial photographs and post-processing results) and vector data (locations of terrestrial photographs). Maps and their legends are generated on the server and sent to clients as images.

- The Web Image Service provides the capability to display, download, edit and work with metadata of raster datasets. This service can be used to work with satellite imageries and their pixel sets.
- The Web Feature Service provides requested services to access a feature geometry collection in XML/GML, and is capable of updating positions of terrestrial photographs in the vector file stored in the back end of server.
- The Web Processing Service defines an interface which facilitates the publishing of geospatial processes and clients' discovery of, and binding to, those processes. A user can use these web services for preprocessing the background imagery, e.g., to find seed pixels, view shed area of a terrestrial photograph, etc. These types of analyses can assist with imagery interpretation and help to better understand a study area.
- The Metadata Service provides an engine for publishing, sharing and searching metadata on terrestrial photographs available within the GeoTruth repository. The Metadata Service is a geographic search engine which can be used when a user wants to find terrestrial photographs and other associated resources.
- Geographical data and terrestrial photographs with metadata are stored in a relational database on the data side of the GeoTruth server.

LEGAL AND AUTHORSHIP ISSUES IN THE GEOTRUTH

Two key issues need to be addressed for the system: determination of ownership and copyright issues surrounding photographs that are submitted to the system, and the management and protection of the system from vandalism.

As concerns ownership and copyright issues, there are really only two options: either the ownership and copyright are retained by the owner, or they are assumed by the GeoTruth team when photographs are submitted to the system. The first option is used by the Flickr photo sharing website (Flickr, 2015). Flickr allows individuals to assign a Creative Commons (http://creativecommons.org/) license to each of their photographs, in which copyright is retained, but permission is granted for limited, non-commercial personal use (https://policies.yahoo.com/us/en/yahoo/terms/utos/index.htm). This is a better option than having the GeoTruth team assume copyright, since it allows contributors to retain control over their work, and it does not discourage contributions.

Although the system is intended to facilitate open participation by members of the public and the remote sensing community, the system will not be self-managing. There will be ongoing requirements to maintain the website, perform backups, check new entries, and fix or prevent vandalism.

Recent experience by the users of the Wikipedia community-open encyclopedia project has shown that vandalism does occur (Reuters, 2005; Morris, 2013). The interesting side of this is that there is a significant degree of self-policing among users/contributors. Wikipedia contributors rapidly correct entries made by vandals, often within minutes (Arrington, 2005; Harpalani, 2010).

With the GeoTruth website, vandalism can be reduced by allowing open contribution, but offering limited editing and replacement. Once an image has been submitted and approved by a moderator, access will then be restricted. This will prevent content from being destroyed. The moderator's job will be to ensure that only appropriate images are submitted to the site, that copyrighted material is not submitted, and ensure that changes to the site are legitimate. The moderator will also be able to respond to any claims of copyright infringement by removing the images in question from the site until the copyright of the material is established.

CONCLUSION

The concept of the GeoTruth system goes far beyond an ordinary data repository. Offering an opportunity to annotate images by the author and later by any user, the GeoTruth system provides a truly participatory mechanism, not only to supply the data, but also to incorporate textual and audio comments from a wide community of users. Such a mechanism should inspire users to actively participate in the life of the GeoTruth Virtual Professional Community, and ultimately encourage sharing not only the data, but the knowledge.

The formalization of accumulated knowledge in a standard, interchangeable form defines the usability of the system, and, in the end, the existence of it. Our future research will be focused on the detailed development of the GeoTruth concept with formalization and representation of data and metadata as the key topics. This will include a thorough analysis of the types of data to be used and their specifications, development of a methodology and techniques for field data collection, and analysis of possible platforms and Web tools for implementation of the GeoTruth engine. The idea of the GeoTruth professional community needs to be disseminated by informing members of relevant online professional societies, publishing in scientific journals and professional magazines, and carrying out workshops and presentations. This strategy could expand the GeoTruth community, enrich the image repository, and excite users of remotely sensed imagery to exercise and enhance the GeoTruth participatory mechanism.

REFERENCES

Aggett, G., & McColl, C. (2006). Evaluating decision support systems for PPGIS applications. (Public Participatory Geographic Information Systems). *Cartography and Geographic Information Science*, *33*(1), 77–92. doi:10.1559/152304006777323163

Agrifoglio, R., & Metallo, C. (2015). Preserving Knowledge Through Community of Practice: A Multiple Case Study. In From Information to Smart Society (pp. 103-111). Springer International Publishing. doi:10.1007/978-3-319-22234-9_3

Anderson, J. R., Hardy, E. E., Roach, J. T., & Witmer, R. E. (1976). *A land use and land cover classification system for use with remote sensor data*. USGS.

Arrington, M. (2005). *The Ugly (but necessary?) side of Wikipedia*. Retrieved from http://www.techcrunch.com/2005/09/07/the-ugly-but-necessary-side-of-wikipedia/

Backes, M., & Jacobi, J. (2006). Classification of weed patches in QUICKBIRD images: verification by ground truth data. In EARSeL ePROCEEDINGS (pp.173-178).

Beaujardière, J. L., Mitchell, H., Raskin, R. G., & Rao, A. (2000). *The NASA Digital Earth Testbed* (pp. 47–53). ACM-GIS.

Bossard, M., Feranec, J., & Otahel, J. (2000). *CORINE land cover technical guide - Addendum 2000*. European Environment Agency.

Buckland, S. T., & Elston, D. A. (1994). Use of groundtruth data to correct land cover area estimates from remotely sensed data. *International Journal of Remote Sensing*, *15*(6), 1273–1282. doi:10.1080/01431169408954160

Caby-Guillet, L., Crave, S., & Ladame, S. (2006). How ICT can facilitate trust inside networks of SME: the role of Professional Virtual Communities. *Proceedings of theInternational conference on Concurrent Engineering,Milano*. doi:10.1109/ICE.2006.7477102

Campbell, J. B., & Browder, J. O. (1995). Field data collection for remote sensing analysis: SPOT data, Rondonia, Brazil. *International Journal of Remote Sensing*, *16*(2), 333–350. doi:10.1080/01431169508954399

Chen, J., He, C., Jiang, J., & Han, G. (2008). Reconciliation of inconsistent perspectives in collaborative GIS. *Cartography and Geographic Information Science*, *35*(2), 77–89. doi:10.1559/152304008784090586

Cherem, G. J., & Driver, B. L. (1983). Visitor employed photography: A technique to measure common perceptions of natural environments. *Journal of Leisure Research*, *15*(1), 65–83.

Congalton, R., & Green, K. (2008). Assessing the Accuracy of Remotely Sensed Data: Principles and Practices. 2nd ed. CRC/Lewis Press, Boca Raton, FL. doi:10.1201/9781420055139

Craig, W., Harris, T., & Weiner, D. (1999). Empowerment, marginalization and public participation GIS. In *Report of Varenius Workshop* (pp. 22–24). Santa Barbara, NCGIA.

Crinnion, J. (1991). *Evolutionary Systems Development, a practical guide to the use of prototyping within a structured systems methodology*. New York: Plenum Press.

Dahdouh-Guebas, F., Verheyden, A., Kairo, J. G., Jayatissa, L. P., & Koedam, N. (2006). Capacity building in tropical coastal resource monitoring in developing countries: A re-appreciation of the oldest remote sensing method. *International Journal of Sustainable Development and World Ecology*, *13*(1), 62–76. doi:10.1080/13504500609469662

Di Gregorio, A., & Jansen, L. (2005). *Land Cover Classification System. Classification concepts and user manual. Software version 2*. Rome: Food and Agriculture Organization of the United Nations.

Dragicevic, S., & Balram, S. (2004). A Web GIS collaborative framework to structure and manage distributed planning processes. *Journal of Geographical Systems*, *6*(2), 133–153. doi:10.1007/s10109-004-0130-7

Dykes, J., Moore, K., & Wood, J. (1999). Virtual environments for student fieldwork using networked component s. *International Journal of Geographical Information Science*, *13*(4), 397–416. doi:10.1080/136588199241274

ESCAVI. (2003). *Executive Steering Committee for Australian Vegetation Information (ESCAVI). Australian Vegetation Attribute Manual: National Vegetation Information System, Version 6.0*. Canberra: Department of the Environment and Heritage.

Fang, Q., Sang, J., & Xu, C. (2013). Giant: Geo-informative attributes for location recognition and exploration.*Proceedings of the 21st ACM international conference on Multimedia* (pp. 13-22). ACM. doi:10.1145/2502081.2502088

Flickr. Retrieved June 09, 2015, from http://www.flickr.com

Foody, G. M. (2010). Assessing the accuracy of land cover change with imperfect ground reference data. *Remote Sensing of Environment*, *114*(10), 2271–2285. doi:10.1016/j.rse.2010.05.003

Forster, P. M. (2004). Psychological Sense of Community in Groups on the Internet. *Behaviour Change*, *21*(2), 141–146. doi:10.1375/bech.21.2.141.55421

Frank, S., Fürst, C., Koschke, L., Witt, A., & Makeschin, F. (2013). Assessment of landscape aesthetics—validation of a landscape metrics-based assessment by visual estimation of the scenic beauty. *Ecological Indicators*, *32*, 222–231. doi:10.1016/j.ecolind.2013.03.026

Gardin, S., van Laere, S. M., van Coillie, F. M., Anseel, F., Duyck, W., de Wulf, R. R., & Verbeke, L. P. (2011). Remote sensing meets psychology: A concept for operator performance assessment. *Remote Sensing Letters*, *2*(3), 251–257. doi:10.1080/01431161.2010.516280

Gienko, G., Govorov, M., Maguire, B., Khmelevsky Y., & Gienko, A. (2007). A Virtual Professional Community to Support Effective Use of Remotely Sensed Imagery. *The South Pacific Journal of Natural Science*, 25.

Gienko, G., & Chekalin, V. (2004), Neurophysiological features of human visual system in augmented photogrammetric technologies.*Proc. of XXth ISPRS Congress*, Istanbul.

Goodfellow, R. (2003). *Virtual Learning Communities. A Report for the National College for School Leadership*. Open University.

Govorov, M., Khmelevsky, Y., Ustimenko, V., & Khorev, A. (2005). Security for GIS N-Tier Architecture. In P. Fisher (Ed.), *Developments in Spatial Data Handling* (pp. 71–83). Springer Verlag. doi:10.1007/3-540-26772-7_6

Grimm, T. (1998). The Human Condition: A Justification for Rapid Prototyping. *Time Compression Technologies*, 3(3).

Harpalani, M., Phumprao, T., Bassi, M., Hart, M., & Johnson, R. (2010). Wiki Vandalysis - Wikipedia Vandalism Analysis. *Proceedings of the 4th International Workshop on Uncovering Plagiarism, Authorship, and Social Software Misuse (PAN-10)*, Padua, Italy.

Henderson, J. M. (1992). Object identification in context: The visual processing of natural scenes. *Canadian Journal of Psychology*, *46*(3), 319–341. doi:10.1037/h0084325 PMID:1486550

Herzog, T. (1995). A cognitive analysis of preference for urban nature. In *Readings in Environmental Psychology: Landscape Perception* (pp. 65-81).

Herzog, T. R. (1989). A cognitive analysis of preference for field and forest environments. *Landscape Research*, *9*(1), 10–16. doi:10.1080/01426398408706092

Hoffman, R., & Markman, A. (2001). *Interpreting Remote Sensing Imagery: Human Factors*. Lewis Publishers.

Holzner, S. (2006). Ajax: Your visual blueprint for creating rich Internet applications. Wiley Visual Imprint.

Hunt, M. E. (1984). Environmental knowing without being there. *Environment and Behavior*, *16*(3), 307–334. doi:10.1177/0013916584163002

Jankowski, P. (2011). *Designing public participation geographic information system. The SAGE Handbook of GIS and Society* (pp. 347–360). London: Sage Publications Ltd.

Jankowski, P., & Nyerges, T. (2001). *Geographic Information Systems for Group Decision Making: Towards a participatory, geographic information science*. New York: Taylor & Francis.

Johnson, B. C., Bruce, S. S., Early, E. A., Housten, J. M., O'Brian, T. R., Thompson, A. K., Mueller, S. L. (1996). *The Fourth SeaWiFS Intercalibration Round-Robin Experiment (SIRREX-4)*. NASA Tech Memo 104566 37.

Katzy, B. R., & Ma, X. (2002). Virtual Professional Communities - Definitions and Typology. *CeTIM Publications*. Retrieved from http://www.researchgate.net/publication/228808964_Virtual_Professional_Communities-Definitions_and_Typology

Kazman, R., & Kominek, J. (1993). Information Organization in Multimedia Resources. Association for Computing Machinery. *Proceedings of the 11th Annual International Conference on Systems Documentation SIGDOC '93* (pp. 149-162). New York: ACM Press.

Kennedy, L. S., & Naaman, M. (2008). Generating diverse and representative image search results for landmarks.*Proceedings of the 17th international conference on World Wide Web* (pp. 297-306). ACM. doi:10.1145/1367497.1367539

Kim, A.-J. (2006). *Community-Building on the Web*. Peachpit Press.

Kushwaha, G. S., & Singh, V. (2007). Environmental Psychology in Land. *Society, sustainability, and environment*, 184.

Kyriakidis, P. C., & Dungan, J. L. (2001). A geostatistical approach for mapping thematic classification accuracy and evaluating the impact of inaccurate spatial data on ecological model predictions. *Environmental and Ecological Statistics*, *8*(4), 311–330. doi:10.1023/A:1012778302005

Lillesand, T. M., Kiefer, R. W., & Chipman, J. (2015). *Remote sensing and image interpretation* (7th ed.). New York: John Wiley & Sons.

Liu, C., Frazier, P., & Kumar, L. (2007). Comparative assessment of the measures of thematic classification accuracy. *Remote Sensing of Environment, 107*(4), 606–616. doi:10.1016/j.rse.2006.10.010

Lowe, D. G. (1985). *Perceptual Organization and Visual Recognition*. Boston: Kluwer Academic Publishers. doi:10.1007/978-1-4613-2551-2

Luo, J., Joshi, D., Yu, J., & Gallagher, A. (2011). Geotagging in multimedia and computer vision—a survey. *Multimedia Tools and Applications, 51*(1), 187–211. doi:10.1007/s11042-010-0623-y

Mather, P., & Koch, M. (2011). *Computer processing of remotely-sensed images: an introduction*. John Wiley & Sons. doi:10.1002/9780470666517

McCoy, R. M. (2005). *Field Methods in Remote Sensing*. New York: Guilford Press.

Messinger, P. R., Stroulia, E., Lyons, K., Bone, M., Niu, R. H., Smirnov, K., & Perelgut, S. (2009). Virtual worlds—past, present, and future: New directions in social computing. *Decision Support Systems, 47*(3), 204–228. doi:10.1016/j.dss.2009.02.014

Morris, K. (2013). How vandals are destroying Wikipedia from the inside. Retrieved from http://www.dailydot.com/society/wikipedia-hoax-vandalism-legolas2186/

Naaman, M., Song, Y. J., Paepcke, A., & Garcia-Molina, H. (2004). Automatic organization for digital photographs with geographic coordinates. *Proceedings of the 2004 Joint ACM/IEEE Conference onDigital Libraries* (pp. 53-62). IEEE. doi:10.1145/996350.996366

Nishii, R., & Tanaka, S. (1999). Accuracy and Inaccuracy Assessments in Land-Cover Classification. *IEEE Transactions on Geoscience and Remote Sensing, 37*(1), 491–498. doi:10.1109/36.739098

Ochi, S., & Takagi, M. (1996). *Ground Truth Database with Portable GPS for Education*. Retrieved from http://www.a-a-r-s.org/aars/proceeding/ACRS1996/Papers/EDC96-6.htm

Open Source Initiative (OSI). (n. d.). Retrieved from http://www.opensource.org

Paine, D., & Kiser, J. (2012). *Aerial Photography and image interpretation* (3rd ed.). John Wiley & Sons, Inc.

Palloff, R., & Pratt, K. (2007). *Building learning communities in cyberspace: effective strategies for the online classroom* (2nd ed.). San Francisco: Jossey-Bass.

Palmer, J. F. (2004). Using spatial metrics to predict scenic perception in a changing landscape: Dennis, Massachusetts. *Landscape and Urban Planning*, *69*(2), 201–218. doi:10.1016/j.landurbplan.2003.08.010

Palmer, J. F., & Hoffman, R. E. (2001). Rating reliability and representation validity in scenic landscape assessments. *Landscape and Urban Planning*, *54*(1), 149–161. doi:10.1016/S0169-2046(01)00133-5

Peeria, S., McLeoda, A., Lavoiea, P., Ackermanb, S., Gardnera, J., & Parrishc, C. (2013). Field calibration and validation of remote-sensing surveys. *International Journal of Remote Sensing*, *34*(18), 6423–6436. doi:10.1080/01431161.2013.800655

Peca, I., Zhi, H., Vrotsou, K., Andrienko, N., & Andrienko, G. (2011). Kd-photomap: Exploring photographs in space and time. *Proceedings of the2011 IEEE Conference onVisual Analytics Science and Technology (VAST)* (pp. 291-292). IEEE.

Peng, Z., & Tsou, M. (2003). *Internet GIS: Distributed Geographic Information Services for the Internet and Wireless Network*. John Wiley & Sons, Inc. Photosynth. Retrieved from https://photosynth.net/about.aspx

Porter, C. E. (2004). A Typology of Virtual Communities: A Multi-Disciplinary Foundation for Future Research. *Journal of Computer-Mediated Communication*, *10*(1).

Porter, C. E., & Donthu, N. (2008). Cultivating trust and harvesting value in virtual communities. *Management Science*, *54*(1), 113–128. doi:10.1287/mnsc.1070.0765

Preece, J. (2000). *Online Communities: Designing Usability, Supporting Sociability*. John Wiley & Sons Ltd. Retrieved from http://www.amnh.org/our-research/center-for-biodiversity-conservation/biodiversity-informatics/remote-sensing-guides

CNET. (2005). Wikipedia to tighten editorial rules. Retrieved from http://news.cnet.com/Wikipedia-to-tighten-editorial-rules/2100-1025_3-5820655.html

Rheingold, H. (1998). *The Virtual Community: Homesteading on the Electronic Frontier*. New York, Addison-Wesley. Retrieved from http://www.rheingold.com/vc/book

Rodrigues, M., Montañés, C., & Fueyo, N. (2010). A method for the assessment of the visual impact caused by the large-scale deployment of renewable-energy facilities. *Environmental Impact Assessment Review*, *30*(4), 240–246. doi:10.1016/j.eiar.2009.10.004

Rozenstein, O., & Karnieli, A. (2011). Comparison of methods for land-use classification incorporating remote sensing and GIS inputs. *Applied Geography (Sevenoaks, England)*, *31*(2), 533–544. doi:10.1016/j.apgeog.2010.11.006

Bhagat, S, V. (. (2012). Use of remote sensing techniques for robust digital change detection of land: A review. *Recent Patents on Space Technology*, *2*(2), 123–144. doi:10.2174/1877611611202020123

Schiavone, F., & Agrifoglio, R. (2012). Communities of practice and practice preservation: a case study. In Information Systems: Crossroads for Organization, Management, Accounting and Engineering (pp. 331-338). Physica-Verlag HD. doi:10.1007/978-3-7908-2789-7_37

Shang, H., & Bishop, I. D. (2000). Visual thresholds for detection, recognition and visual impact in landscape settings. *Journal of Environmental Psychology*, *20*(2), 125–140. doi:10.1006/jevp.1999.0153

Slotterback, C. S. (2011). Planners perspectives on using technology in participatory processes. *Environment and Planning-Part B*, *38*(3), 468–485. doi:10.1068/b36138

Snavely, N., Seitz, S. M., & Szeliski, R. (2008). Modeling the world from internet photo collections. *International Journal of Computer Vision*, *80*(2), 189–210. doi:10.1007/s11263-007-0107-3

Snavely, N., Seitz, S. M., & Szeliski, R. (2006). Photo Tourism: Exploring Photo Collections in 3D. *Proceedings of ACM SIGGRAPH*. doi:10.1145/1179352.1141964

Stamps, A. E. III. (1990). Use of photographs to simulate environments: A meta-analysis. *Perceptual and Motor Skills*, *71*(3), 907–913. doi:10.2466/pms.1990.71.3.907

Stehman, S. V., & Czaplewski, R. L. (1998). Design and analysis for thematic map accuracy assessment: Fundamental principles. *Remote Sensing of Environment*, *64*(3), 331–344. doi:10.1016/S0034-4257(98)00010-8

Stephen, H., Cummings, M., McCubbrey, D., Pinsonneault, A., & Donovan, R. (2006). *Management Information Systems: For the Information Age* (3rd Canadian ed.). New York: McGraw-Hill Ryerson.

Steven, M. D. (1987). Ground truth: An underview. *International Journal of Remote Sensing*, *8*(7), 1033–1038. doi:10.1080/01431168708954745

Steven, M., Malthusb, T., & Baretc, F. (2007). Intercalibration of Vegetation Indices - an Update. *Proceedings of the10th International Symposium on Physical Measurements and Signatures in Remote Sensing (ISPMSRS07)*, Davos, Switzerland.

Stewart, T. R., Middleton, P., Downton, M., & Ely, D. (1984). Judgments of photographs vs. field observations in studies of perception and judgment of the visual environment. *Journal of Environmental Psychology*, *4*(4), 283–302. doi:10.1016/S0272-4944(84)80001-8

Strahler, A. H., Boschetti, L., Foody, G. M., Friedl, M. A., Hansen, M. C., Herold, M., & Woodcock, C. E. (2006). *Global land cover validation: Recommendations for evaluation and accuracy assessment of global land cover maps* (p. 51). Luxembourg: European Communities.

System Design Strategies. (2005). *ESRI Technical Reference Document*. ESRI Press.

Teillet, P. M., Markham, B. L., & Irish, R. R. (2006). Landsat cross-calibration based on near simultaneous imaging of common ground targets. *Remote Sensing of Environment*, *102*(3-4), 264–270. doi:10.1016/j.rse.2006.02.005

The Open Geospatial Consortium, Inc. (n. d.). Retrieved from http://www.opengeospatial.org/

Thu, P. M., & Populus, J. (2007). Status and changes of mangrove forest in Mekong Delta: Case study in Tra Vinh, Vietnam. *Estuarine, Coastal and Shelf Science*, *71*(1-2), 98–109. doi:10.1016/j.ecss.2006.08.007

Toyama, K., Logan, R., Roseway, A., & Anandan, P. (2003). Geographic Location Tags on Digital Images.*Proceedings of the Eleventh ACM International Conference on Multimedia,*Berkeley, California, USA (pp. 156-166). doi:10.1145/957013.957046

Tsou, M. (2004). Integrating Web-based GIS and image processing tools for environmental monitoring and natural resource management. *Journal of Geographical Systems*, *6*(2), 155–174. doi:10.1007/s10109-004-0131-6

Uuemaa, E., Antrop, M., Roosaare, J., Marja, R., & Mander, Ü. (2009). Landscape metrics and indices: An overview of their use in landscape research. *Living Reviews in Landscape Research*, *3*(1), 1–28.

Wang, S., Anselin, L., Bhaduri, B., Crosby, C., Goodchild, M. F., Liu, Y., & Nyerges, T. L. (2013). CyberGIS software: A synthetic review and integration roadmap. *International Journal of Geographical Information Science*, *27*(11), 2122–2145. doi:10.1080/13658816.2013.776049

Wright, D. J., & Wang, S. (2011). The emergence of spatial cyberinfrastructure. *Proceedings of the National Academy of Sciences of the United States of America*, *108*(14), 5488–5491. doi:10.1073/pnas.1103051108 PMID:21467227

Wulder, M. A., Franklin, S. E., White, J. C., Linke, J., & Magnussen, S. (2006). An accuracy assessment framework for large-area land cover classification products derived from medium-resolution satellite data. *International Journal of Remote Sensing*, *27*(4), 663–683. doi:10.1080/01431160500185284

Zhang, T., & Tsou, M. H. (2009). Developing a grid-enabled spatial Web portal for Internet GIServices and geospatial cyberinfrastructure. *International Journal of Geographical Information Science*, *23*(5), 605–630. doi:10.1080/13658810802698571

Zheng, Y. T., Zha, Z. J., & Chua, T. S. (2011). Research and applications on georeferenced multimedia: A survey. *Multimedia Tools and Applications*, *51*(1), 77–98. doi:10.1007/s11042-010-0630-z

Zhou, T. (2011). Understanding online community user participation: A social influence perspective. *Internet Research*, *21*(1), 67–81. doi:10.1108/10662241111104884

Zion National Park. (n. d.). USGS - NPS Vegetation Mapping Program. Retrieved from http://biology.usgs.gov/npsveg/zion/index.html

KEY TERMS AND DEFINITIONS

Ground Truth: Interpretation of images is not always a straight-forward process. Both visual and computer-assisted methods of interpretation can introduce a great deal of ambiguities and uncertainties in identifying, recognizing, and discriminating between objects in overhead images. An accuracy assessment is always needed to evaluate correctness and completeness of image interpretation, and the most effective method for doing this is ground truthing. During ground truthing, an image interpreter visits the area of study and, by inspecting the actual landscape, identifies questionable objects depicted in the image. Ground truthing can be implemented before interpretation of image, which allows for exploration of the area to identify training sites, or after an image has been interpreted, to resolve uncertainties and maintain quality control.

Image Interpretation: The process of extracting information from aerial and satellite imagery by identifying objects and their characteristics, generally for mapping purposes. Depending on the type of images, interpretation may be visual or computer-assisted. Visual interpretation utilizes the human ability to discriminate between objects on the ground by viewing their pictorial representations in images. Accuracy of visual interpretation is greatly dependent on the expertise of the interpreter. Computer-based interpretation, also called image classification, can either mimic the human visual interpretation by employing descriptors such as shape, color, texture, pattern, etc., or by utilizing properties of objects which reflect differently in different zones of an electro-magnetic spectrum (i.e. comparing spectral responses of objects to identify their level of similarity).

Public Participatory Mechanisms: These are internet tools to provide communication between members of virtual communities. Most of the tools are portals and platforms utilizing the Web 2.0 concept, which allows users to interact and collaborate with each other in a social media dialogue in a virtual community between creators of user-generated content. These mechanisms range from simple forums to exchange textual information, to portals capable of storing and disseminating various datasets; the most advanced solutions can provide options for querying, processing, data mining, and visualization tools.

Remote Sensing: This is a field of geo-spatial sciences dedicated to learning the Earth's processes and mapping environmental phenomena using aerial and satellite images. At an earlier stage, images were limited to black-and-white film aerial photography, with color and infra-red films introduced later; studying Earth's processes and mapping its surface was done by trained experts, visually interpreting printed photographs. Advances in technology have since introduced multispectral (and later, hyperspectral) imaging, allowing images to be acquired both from aerial platforms and satellites. This new type of imagery triggered an entirely new approach to image interpretation: analysis of spectral responses registered in multiple image bands. With the introduction of digital imaging sensors, it became possible to implement computer-assisted automatic image classification.

Terrestrial Photographs: These are pictures of land cover and land objects taken from the ground using conventional cameras for the purpose of substituting real-world observations. When supported by location information (coordinates from GPS) and time stamps, those terrestrial photographs can significantly assist in the interpretation of overhead images. These pictures have more value when taken according to certain guidelines which define optimal conditions for image acquisition (time of day, direction, distance, lighting, number of pictures, field notes, etc.).

Virtual Professional Community: These professional communities ("communities of practice") provide environments for professionals to share the body of knowledge of their professions, including similar working cultures, perceptions of problems, problem-solving techniques, professional values, and professional behaviour. A virtual professional community is a collaborative network of professionals, communicating mostly via the Internet. Due to this means of communication, professional virtual communities have distinctive features, which include the geographical spread of their membership, international audience, language barriers, lag in response times, communication etiquette, etc. Within this networked-based community, members can form clusters with similar interests in particular study areas, and thus create stronger ties among members.

Web Feature Service (WFS): A web service that on request extract geographic features or vector data from the spatial data storage and deliver them to a thick client to render them on a client display. WFS delivers spatial data to the user as streamed features or as XML/GML coded geometry.

Web Map Service (WMS): A web service that on request, dynamically produces a map of spatially referenced data in the digital image file format. WMS delivers map content to a client application (browser) as JPEG, PNG, or GIF images.

Chapter 5
Development of Web GIS Based Framework for Public Health Management System Using ERDAS Apollo 2010

Shipra Verma
Motilal Nehru National Institute of Technology, India

ABSTRACT

This paper proposes a framework for an essential creation of a public health information visualization platform, for Japanese Encephalitis (JE) disease outbreaks in the Gorakhpur district, India. The Web GIS technology is used with ERDAS Apollo 2010 software at customized level, to develop architecture for Web GIS-based public health information systems. A GUI has been created using Java Server Pages (JSP) for its customization. This will help in extending the benefit of GIS and Web technology for public availability in the area for preparation of the health plan in multitier system.

DOI: 10.4018/978-1-5225-1814-3.ch005

INTRODUCTION

Both government and private sector organizations are looking for ways to sustain and improve the health of the public in the world (Najafabadi & Pourhassan. 2009). In this very modern time, health information needs to be updated and shared with people in a quick and efficient way. Web technology is growing a suite for using spatial data on the Internet portals by using web integration tools. This provides new opportunities to advance disease surveillance, control, and public empowerment, to ensure the access of spatial data at local level, for taking health decision for the public (Lu, 2004). Public health information contains geo-referenced data, such as specific location, area code, latitude and longitude, street address, and geo-political boundaries, which can be visualized through GIS distribution maps (Lu. 2005). It is of practical importance to construct a visible public health information system. The use of GIS technologies for the health planning and management has been studied by many investigators in recent years (Kamel et al. 2001, Johnson & Johnson 2001). GIS has been used for public heath area for long time. As early as in 1854, Dr. John Snow had used a map to track the original area where the cholera disease erupted in Lon-don in first time. From that time, map had been used in the infectious disease controlling and preventing, and it becomes an essential tool in protecting the public health environment (Kistemann, T, 2001). Japanese encephalitis (JE) is a vector-borne disease that occurs in Asia and is the cause of major public health problems in India. The application of Geographic Information Technology (GIS) plays a vital role in the surveillance and control of vector-borne diseases, by providing on information on visualization as public health information.

The Web GIS using Java technologies were explored in implementing GIS systems (WHO, 2003). Generally speaking, the research on Web-based GIS for public health planning and management has attracted much attention. The Web GIS system framework is developed by using adopted multi-tier system architectures, which consists of the server tier and the front tire. The server tier adopts the J2EE-based architecture, while the front tier uses GIS maps to show public health information with geo-referenced information. GIS Mapping and management have become a necessity utilizing modern technological tools. With this data accompanies a great fear within the health domain on how to get by and receive the selective source of information on the JE disease, from local to global level. With an increase in problems created by JE, there is an increasing demand by research on health planners to spread the disease information and provide facilities at public health spheres. This study is for Gorakhpur, Uttar Pradesh, India and the approach, as an effective bottom-up method, should increase the capabilities of planners to discover and solve JE disease health problems.

BACKGROUND

Web GIS technologies, with public health, uses a customized framework of ERDAS Apollo 2010, which has an extensive application prospect in the field of protecting public health and security, by exchanging real-time information. Public health information is added and updated through the Internet with participation of public.

Japanese Encephalitis (JE)

It is found that three billion people live in Japanese Encephalitis Virus (JEV) endemic regions, and the disease incidence is about 50,000 cases and 10,000 deaths annually (Fischer et al., 2008). As per the World Health Organization (Khan, 2011) estimate, JE claims 10,000 to 15,000 lives a year. The epidemiology of JE virus is not well understood and only little research has been done (Upadhyayula et al., 2012). An epidemic of viral encephalitis was reported from July through November 2005 in Gorakhpur, Uttar Pradesh, India (Parida et al., 2005). It was the longest and most severe epidemic in three decades; 5,737 people were affected in seven districts of Eastern Uttar Pradesh, while 1,344 people died (WHO, 2006).

JE spreads directly from person to person. It is transmitted by mosquito vectors between wild and domestic birds and pigs. Mosquitoes are vectors that are the crucial intermediate replicative hosts through birds and pigs (Mackenzie et al., 2004). Man is the dead end (Solomon et al., 2000).

Number of population have grown increasing and concern with acute and chronic care of disease control. It is important to reduce public health problems and develop strategies to reduce the disease burden. GIS through the Internet, is a new technological development in the field of Web GIS and are increased in the use of the Internet is creating new standards and challenges. For public health applications, geospatial databases created for the Web will have the added requirement to meet confidentiality safeguards, to ensure the anonymity of individuals from disclosure (Richards et al., 1999). GIS and Web technologies offer opportunities to analyze complex geospatial data, solve problems, and visualize data that can be easily accessed by public health planners and decision makers.

Web GIS Technology

There is a need to update the JE health information on the Web. Web GIS the techniques that setup this information on the Internet. It also provides the GIS function on the website. People can obtain the space data, its integrated information, and maps through the Web. Web technology integrated with GIS can be developed at a very low cost and gives huge amount of spatial information easily, with little time and maintenance work.

Various kinds of geographical space data attribute data, and pictures can be obtained through the Web. With the rapid development of Web GIS, GIS-based applications can be developed with low costs and little maintenance work (Alesheikh, 2002). Web GIS-based system has been gaining popularity on a large scale. The JE disease incidence data of Gorakhpur district at block level is integrated in GIS environment (Verma & Gupta, 2014) and can be updated using Web GIS.

ERDAS Apollo 2010

ERDAS Apollo is customizable, equipped to fully understand, competently manage and rapidly serve an organization's large volumes of geospatial data. It implements an out-of-the-box Service Oriented Architecture (SOA) that provides to publish, find and bind workflow for any data type. With a Spatial Data Infrastructure (SDI), ERDAS Apollo provides an interactively-connected framework to support the management and serving of spatial data, meta-data, and maps.

The ERDAS Apollo 2010 software is used to develop a customized framework for the public health management system. The software has released an OGC Web Mapping Services (WMS), Web Feature Service (WFS), Web Coverage Service (WCS), or Catalog Service Web (CSW). So, here is some information about this service and what ERDAS has done to build the most powerful sharing information (ERDAS, 2010).

OBJECTIVE OF THE PRESENT STUDY

This paper proposed an infrastructure for public health management system for district Gorakhpur, based on Web-GIS ERDAS Apollo 2010, customizable framework. It provides updated information to the district for JE disease health planning. The JE disease data is stored and developed in spatial and dynamic form of sharing which is important for the collaborative preparation, response and recovery stages of disease control and health planners in a public health management system. The system aims to offer a new efficient means for the disease management system, so that different department people could share the information, planning, and monitoring the area which is highly affected. This paper described research work on the design of the system infrastructure, system functions, Web-GIS, on the framework of ERDAS Apollo 2010 customizable framework.

Methodologies Adopted

The Web GIS based software ERDAS Apollo 2010 is used in the present study. It provides an elegant solution to the management of geospatial data, by cataloging the geospatial health data that users have access to, making it accessible over the Internet. The Web GIS framework for public health management deals with the sharing of geospatial data and non-spatial data on the Web GIS platform, as shown in Figure. 1.

It demonstrates the procedure adopted in the customization of the ERDAS Apollo 2010 Web client. This methodology includes the creation of health spatial database and non-spatial database for Gorakhpur. A geo-database has been created in the GIS environment, and it shows the high incidence of disease risk-prone areas and predicts these areas, respectively, in the district. The geo-database is stored on the ERDAS Apollo Server, via, ERDAS Apollo Data Manager. The ERDAS APOLLO 2010 Web Client has been customized and integrated with the developed Public Health Management Interface for imparting geospatial web capabilities at the client end. Moreover, the integrated system is deployed on the same Application Server, i.e.,

Figure 1. The methodology adopted

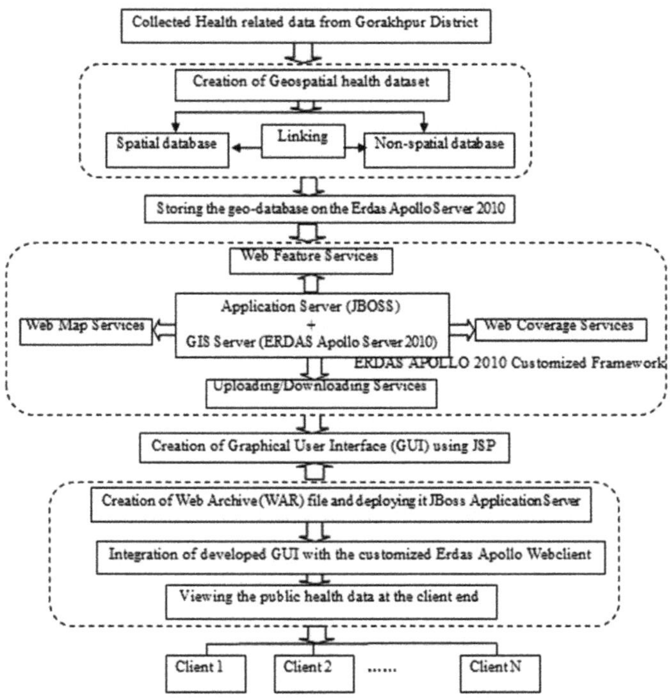

JBoss Application Server, also some authentication is required at the client end to access the customized Web Client.

Waterfall Model

The purpose of public health system development requires a software development model for making Web GIS frameworks, for Public Health Information and management to layout the work in a systematic manner. The software development model provides a sequence of activities, from the requirements of a client to the end of output level of understanding.

There are a number of general models for software processes; in this work, the iterative waterfall model is followed to access the system method. Figure 2 describes the process of development of the Web GIS framework for Public Health mapping and management.

The waterfall model is a widely used software development model and has been used in this work. It is simple to understand, use, and is suitable only for well-understood problems.

Feasibility Study and Requirement Analysis

The main purpose of adopting a software development model for making a Web GIS framework for Public Health information and management is to mould the work in a systematic manner. As per the first phase, a feasibility analysis is done. Next, the economical feasibility is verified. In requirement analysis, requirements in terms of input data, such as topographical maps, software, hardware, etc., are listed out.

Figure 2. Waterfall development life cycle

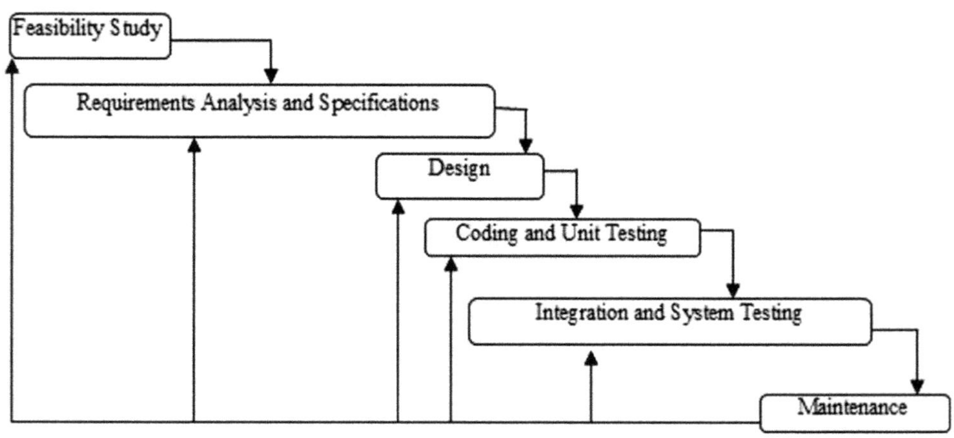

In this phase of implementation, the output of the previous feasibility study and requirement analysis is taken as the input. Figure 3 shows the interface design and working of the system.

Design

The user has to first register and obtains a User ID to access the developed system. Users can then login and access geographic information of different blocks. Data adding, deleting, and editing functionalities are reserved at the administrator end. Clients can just access and view the health-related information of the areas and plan the management strategy accordingly. Some reports regarding the management of various JE health problems are also made available to the client, for assistance in planning.

The user has to first register and obtain a User ID to access the developed system. Users can then login and access geographic information of different blocks. Data adding, deleting, and editing functionalities are reserved at the administrator end. Clients can just access and view the health-related information of the areas and plan the management strategy accordingly. Some reports regarding the management of various JE health problems are also made available to the client, for assistance in planning.

Figure 3. User interface diagram of the public health management system

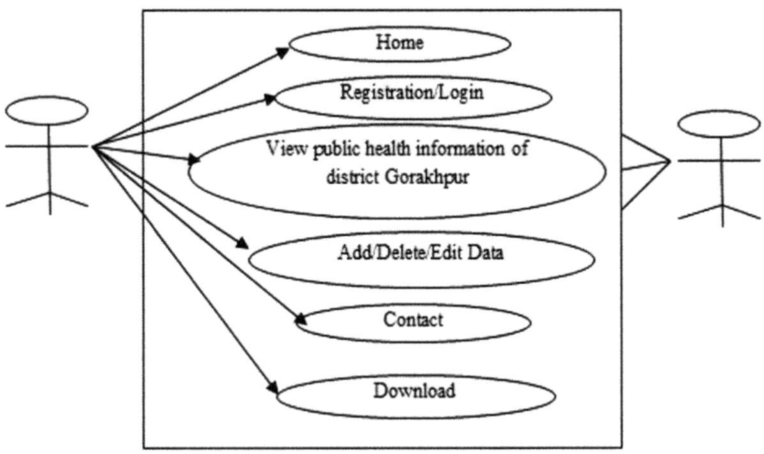

Coding and Unit Testing

Coding is done for the customization of ERDAS Apollo Web Client and creating a web-based Graphical User Interface (GUI), for public health information and management using JSP. The GUI has five sub-modules, as shown, which are discussed subsequently. Each and every module has been coded properly, and unit testing is carried out as per the development module. The design and requirement documents are referred properly, to follow the implementation in a proper manner.

Integration and System Testing

After the successful completion of the above phases of development, the customized ERDAS Apollo Web Client and the GUI are integrated and connected to ERDAS Apollo Server 2010, to ensure that all modules work properly and the end user can access the health information. This is important for the health management and planning of a district.

SYSTEM ARCHITECTURE

The system adopts client server architecture, as shown in Figure 4; the main calculation and the majority of the inquiries are executed at the server side, while the client side is only used for client-computer interaction.

As shown in Figure 4, the customized ERDAS Apollo Web Client is used to make human-computer interaction. The requests are directed to the JBoss Web Application Server, which in turn communicates with the ERDAS Apollo Server for data retrieval. The ERDAS Apollo Server works as the map server in this architecture and fetches the stored spatial data via the ERDAS Apollo Data Manager, which has access to the geo-database consisting of Public Health related spatial information. The required spatial data is retrieved along with the non-spatial attributes and returned to the client.

PUBLISHING THE GEOSPATIAL DATABASE ON APOLLO SERVER

The prepared geospatial database is stored on the ERDAS Apollo Server via the ERDAS Apollo Data Manager. Only the Apollo Administrator is authenticated to add/delete/edit data on the server. Users can just access the data stored on the server.

Figure 4. Architecture of the develop prototype for web based public health information and management system

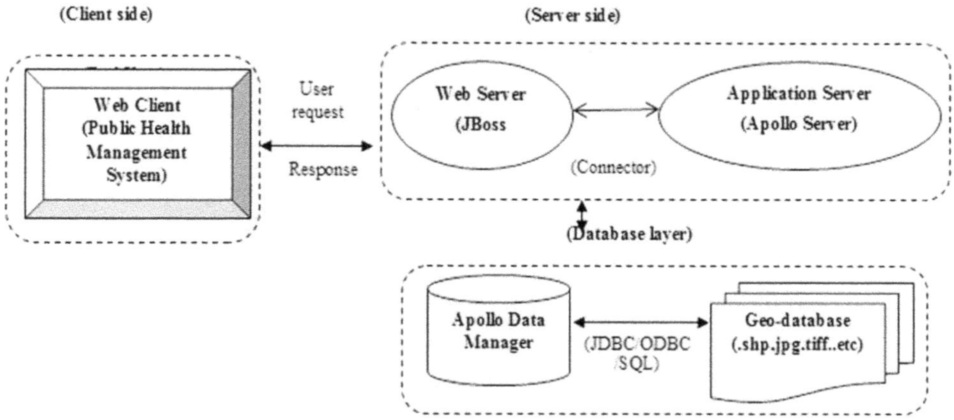

After starting the Apollo log server, vector services are required. Figure 5 shows creation of a vector service. The service is created via catalog of vector and raster services for Gorakhpur district in the form of shapefile. It shows the added service as the map. It also gives the details of the service registered. Similarly, other vector services have been created and stored on the server via Data Manager.

DEVELOPMENT OF THE CUSTOMIZED FRAMEWORK

ERDAS Apollo 2010 is customized by modifying the frameworks and existing tools in the desired field of public health. By modifying the existing tools, made it easy to achieve specific health information. The Graphical User Interface (GUI) prepared and provides easy access to public health information and management system for the district of Gorakhpur. See Figure 6 and Figure 7.

In the preceding screenshot, the first link of GUI explains about public health and domain of public health in research. The second link (in Figure 7) displays Public Health Information, which consists of district health planning units. The collection of health data from Primary Health Centers (PHCs), Community Health Centers (CHCs) to Sub-Centers (SCs), and also with private and government hospitals, can share the health data district to state surveillance unit and further central surveillance unit.

Figure 7 is the link of About Gorakhpur District, which includes in two domain administrative information and history Introduction links and about the district area location map in Study Area links. The new user may fill the registration form

Figure 5. Apollo server log

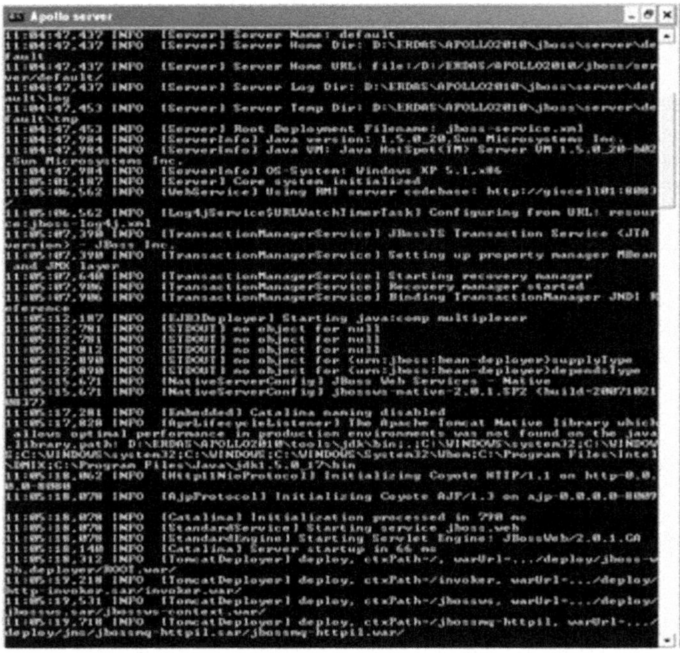

Figure 6. Viewing the homepage on the browser

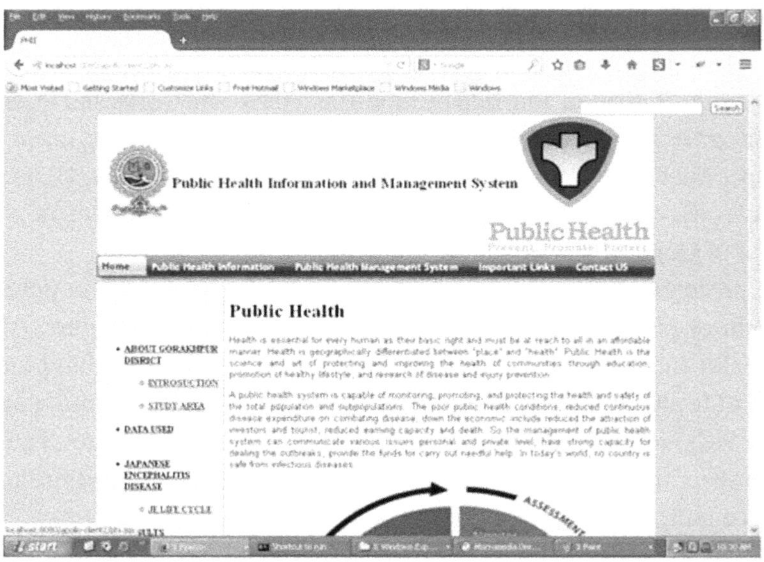

Figure 7. Viewing the public health information on the browser

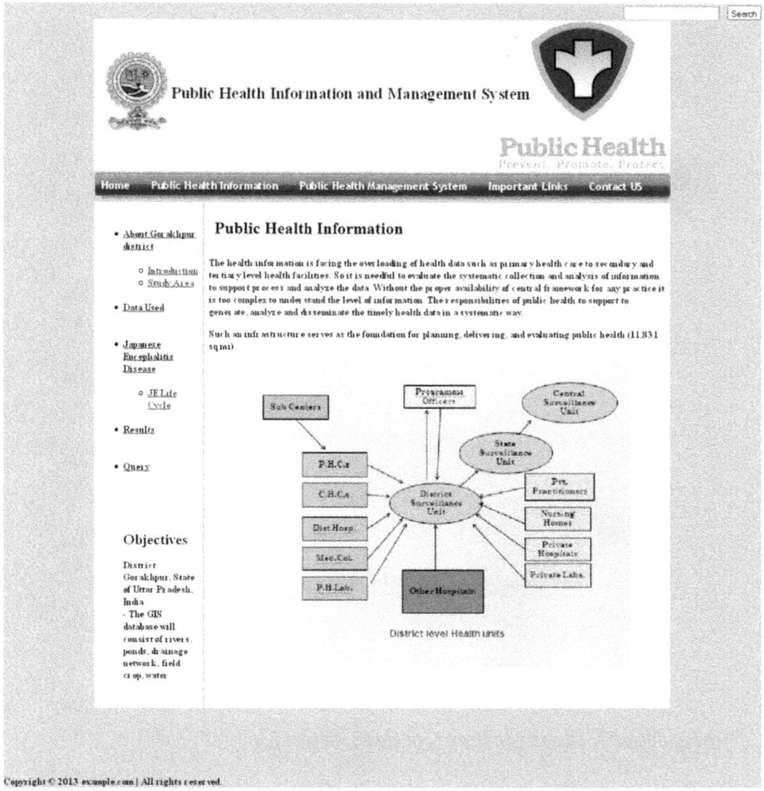

and get the login ID and password to enter the Public Health Management system and various generated links to access the customized framework of APOLLO Web Client, as shown is Figure 8.

The new user may fill the registration form and get the login ID and password to enter the Public Health Management system and various generated links to access the customized framework of APOLLO Web Client, as shown is Figure 9. The Customized Apollo client 2010 is shown in Figures 10, 11 and 12, by importing the collected information of the JE disease.

The number of health centers are present in each block is shown in Figure 10. Similarly, the spatial information of pig number of pig population in each block is shown in Figure 11.

Integration of all data within a web-based GIS framework therefore represents a useful approach for performing more complex analysis, which can support the decision-making processes during the diffusion of the disease. On successful registration, the user can login with his username and password on the Public Health

Figure 8. Login interface to the customized APOLLO web client

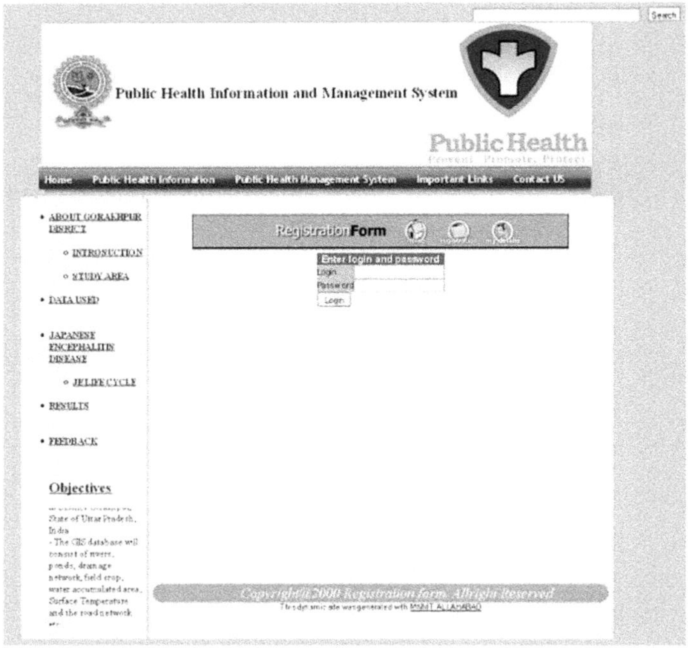

Figure 9. Public health management system browser

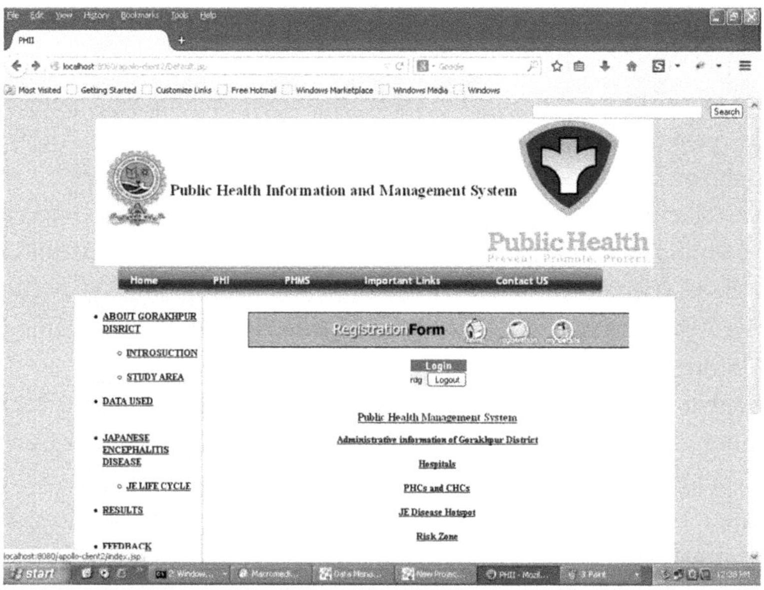

Figure 10. Customized APOLLO web client showing number of hospitals in blocks

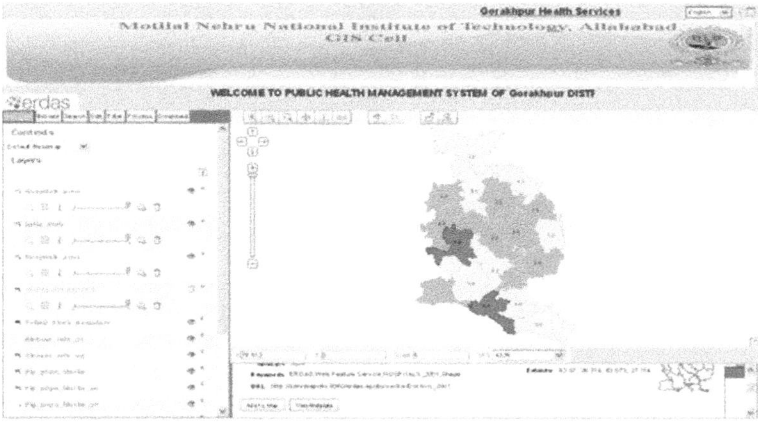

Figure 11. Customized APOLLO web client showing pig population

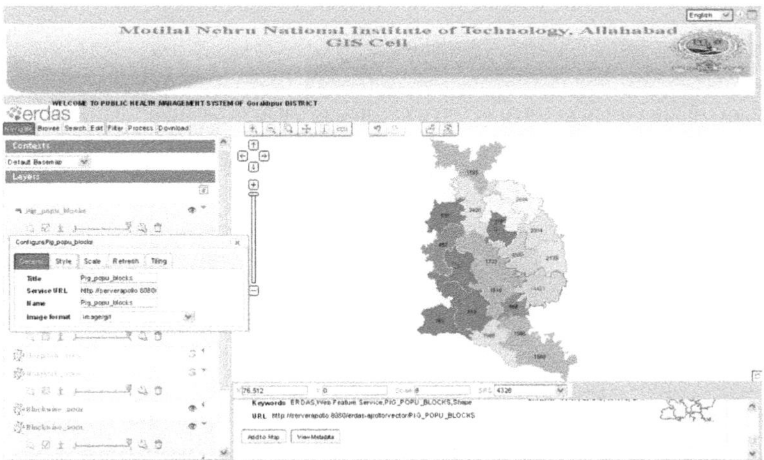

Figure 12. Customized APOLLO web client showing JE disease risk zone area in Gorakhpur district

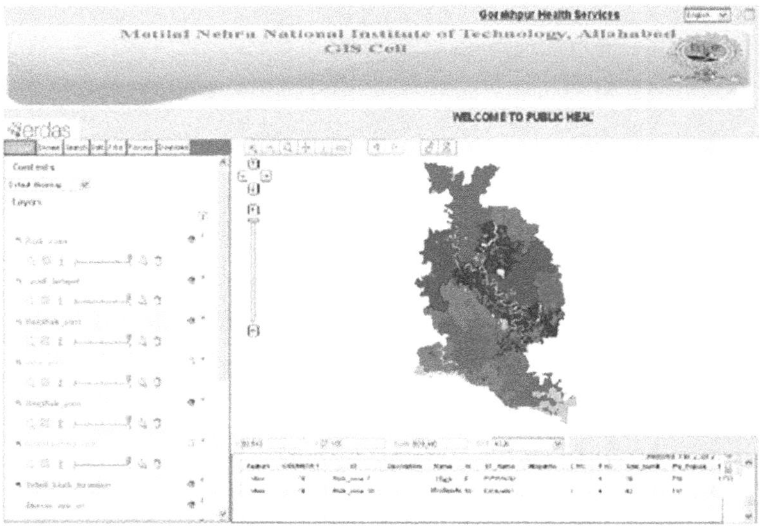

Management System (Figure 9 to 12). The number of pig population distribution in blocks (Figure 11) is categorized in five distinct categories lower to higher number of pig population. The center part of district has a highly number of pig population. Figure 12, represents the location of highly affected disease risk zone in district.

DISCUSSION

Though the prototype of public health system for Gorakhpur district is implemented to access the JE disease health problem for people, there still is a lot of work that needs to be done, to research in the field of urgent health care rescue for JE disease prevention. Following are some key technologies needed and to be investigated in future works:

- **The J2EE Application:** Java GIS is the key part in the system to implement the customized framework using ERDAS Apollo 2010, to display the geometrical information on the Web. The software supports all data formats of GIS, and stores the GIS data with a connection, i.e., JDBC/ODBC and postgresql. Further research needs to utilize and customize more system software functioning at the customized level.

- **GIS and JE Disease Health Data Integration:** GIS with JE disease data integration and generated GIS, maps to the public health system at Gorakhpur district block level. The information would be generated at village level. In the proposed work, GIS maps and risk-zone models were constructed from the topographical maps, with the integration of field data health information, by using the desktop software. In future research works, a more precise map at village level will be used in the system.
- **JE Disease Risk Zone Model:** The risk zone model of JE disease spreading development is the most important method for future works at reach in danger level. It needs to be more practically investigated for public health surveys in further works.

CONCLUSION

The public health system correspondence, planning, prevention, remedies, and relief require precise maps with spatial and non-spatial data. This includes the development of an integrated geo-database consisting of various thematic maps of disease incidence areas under the GIS environment with attribute data. GIS plays important role for management of as user can store, retrieve and model the spatial data about disease growth and factors. Web GIS will be helpful to manage the consumption and feature exploitation of health related factors in the form of spatial data. The study deals with creation of geodatabase of JE disease factors using ERDAS APOLLO 2010 to deals with capability to huge amount of raster as well as vector database. This study emphasizes the capability of ERDAS APOLLO 2010 software for Web GIS based applications in developing public health system.

The developed GUI in JSP form of public health system provides the district health planning managers an effective tool to access and retrieve spatial information needed to handle JE disease outbreaks at block level. The system also provides information about the degree of vulnerability of a particular area. Once the degree of a threat to a particular area, along with the type of Public Health it is prone to, is identified, emergency mangers can respond and take appropriate action to mitigate the effects of Public Health. The emergency managers need not be experts in the field of GIS, as all the information can be accessed and analyzed through a user-friendly GUI.

ACKNOWLEDGMENT

Author gratefully acknowledges the support and encouragement received from the Dr. R.D. Gupta, Professor, Department of Civil Engineering, Motilal Nehru National Institute of Technology, Allahabad, of this research work and expressed my deepest gratitude to Dr. Harnam Singh, Regional sales manager at Intergraph SG & I India for continuous cooperation and suggestions for incorporating the Web GIS framework. I am also very thankful to Dr. K.N. Pandey, D.M.O. Gorkhpur, for help in contribution of disease data collection around the district Gorakhpur.

REFERENCES

Alesheikh, A. A., Helali, H., & Behroz, H. A. (2002). Web GIS: Technologies and Its Applications. *Proceedings of theSymposium on Geospatial Theory, Processing and Applications*, Ottawa.

Fischer, M., Hills, S., Staples, E., Johnson, B., Yaich, M., & Solomon, T. (2008). Japanese encephalitis prevention and control: advances, challenges, and new initiatives. In W. M. Scheld, S. M. Hammer, & J. M. Hughes (Eds.), *Emerging Infections 8* (pp. 93–124). Washington, DC: ASM Press. doi:10.1128/9781555815592.ch6

Parida, M., Dash, P.K., Tripathi, N.K., Ambuj., Sannaragaiah, S., … Sekhar, K. (2006, September). Japanese Encephalitis Outbreak, India, 2005. *Emerging Infectious Diseases*, *12*(9), 1427-1430.

Johnson, C. P., & Johnson, J. (2001, February 7–9). GIS: A Tool for Monitoring and Management of Infectious.*Proceedings of Map India 2001, the 4th Annual International Conference and Exhibition*, New Delhi, India.

Kamel Boulos, M. N., Roudsari, A. V., & Carson, E. R. (2001). Health Geomantic: An Enabling Suite of Technologies in Health and Healthcare (Mythological Review). *Biomed Inform*, *34*(3), 195–219. doi:10.1006/jbin.2001.1015

Khan S. A., Dutta P., Khan A.M., Topno R., Chowdhury P., Borah J., Mahanta J., (2011). Japanese encephalitis epidemiology in Arunachal Pradesh, a hilly state in northeast India. *Asian Pacific Journal of Tropical Disease*, 2011, 119-122.

Kistemann, T. (2001, September 19–20). GIS for communicable disease control: perspectives and pitfalls.*Proceedings of the First European Conference for Geographic Information Sciences in Public Health*, Sheffield, UK.

Upadhyayula, S.M., Mutheneni, S.R., Nayanoori, H.K., Natarajan, A., Goswami, P. (2012). Impact of weather variables on mosquitoes infected with Japanese encephalitis virus in Kurnool district, Andhra Pradesh. *Asian Pacific Journal of Tropical Medicine*, 5(5), 337-341.

Mackenzie S John., Gubler J Duane., Petersen R Lyle. (2004). Emerging flaviviruses: the spread and resurgence of Japanese encephalitis, West Nile and dengue viruses. *Nature Medicine Supplement*, 10(12).

Marc, F.L.N., & Staples, E.J. (2010). Morbidity and Mortality Weekly Report. Recommendations and Reports, 59(RR-1).

Najafabadi, A.T., & Pourhassan, M. (2009). Web GIS and Public Health. *Online Journal of Health and Allied Sciences*, 8(4).

Richards, T. B., Croner, C. M., Rushton, G., Brown, C. K., & Fowler, L. (1999). Geographic Information Systems and Public Health: Mapping the Future. *Pub Health Rep*, *114*(4), 359–373. doi:10.1093/phr/114.4.359 PMID:10501137

Shipra, V., & Gupta, R.D. (2014, December). Spatial and Temporal Variation of Japanese Encephalitis Disease and Detection of Disease Hotspots: A Case Study of Gorakhpur District, Uttar Pradesh, India. Proceedings of Symposium ISPRS Annals, Commission VIII, WG VIII/2, Hyderabad, India.

Tom, S., Nguyen, D. M., Rachel, K., & Mary, G. (2000). Japanese encephalitis. *Journal of Neurology, Neurosurgery, and Psychiatry*, *68*(4), 405–415. doi:10.1136/jnnp.68.4.405 PMID:10727474

WHO, (2003). Communicable Disease Surveillance and Response: Severe acute respiratory syndrome (SARS): Status of the outbreak and lessons for the immediate future.

WHO. (2006). Preventing disease through healthy environments towards an estimate of the environmental burden of disease. WHO Press.

Xiaolin, L. (2005). A Framework of Web GIS Based Unified Public Health Information Visualization Platform. Proceedings of ICCSA 05, *LNCS* (Vol. *3482*, pp. 256–265).

Xiaolin, L. (2009). WebGIS based Information Visualization for Infectious Disease Prevention. *Proceedings of the2009 Third International Symposium on Intelligent Information Technology Application*.

Section 2
Applications of Remote Sensing and GIS for Environmental Monitoring

Chapter 6
Debris Flow Modelling and Risk Assessment of Selected Landslides from Uttarakhand– Case Studies using Earth Observation Data

Shovan L. Chattoraj
Indian Institute of Remote Sensing, India

ABSTRACT

Presently demand for process based modelling of mass movements encompassing snow avalanche, debris flows, landslides, mud flows and rock falls has increased manifold due to their devastating effect and mitigation challenges of disasters caused by such phenomena. Debris flows are multi-phase gravity-driven flows consisting of randomly dispersed interacting phases and therefore, are, extremely challenging to predict the dynamics, run-out distance and area of inundation related to such events which will facilitate mitigation as well as evaluation of simulated scenarios. The numerical simulation model predicts the motion of a geophysical mass movement from head to base in three dimensions. These process-based 3-D models can be utilized for better understanding of vulnerability of a complex mountainous terrain and design appropriate civil engineering structures to withstand the impact of potential flows like Ukhimath, Uttarkashi and Kedarnath.

DOI: 10.4018/978-1-5225-1814-3.ch006

INTRODUCTION

Himalayas, now, have almost become synonymous with landslides. The lower Himalayas have been susceptible to landslides since ages because of tectonically fragile and sensitive mountainous terrain (Ghosh & Suri, 2005). However, the susceptibility to landslides has increased many fold in recent past due to encroachment, infrastructure development, exploitation of natural resources, and so as the vulnerability because of unprecedented human settlements in Himalayan region. The man has put the placid and serene environment of Himalayas in quite strained conditions, thereby disturbing already stressed fragile environment and ecological balance. Consequently, the nature has retaliated through some of the catastrophic disasters in the form of landslides, floods, lake breaches etc. Kedarnath disaster of June 2013 is the most recent one of the many to name. Statistically, approximately 15% of Indian territory is prone to various degrees of landslide hazard where 300 human lives are swallowed and about Rs. 300 Crores property is lost every year.

Landslides in its firm sense are the movement of a mass of rock, debris or earth down slope, due to gravitational pull, and in general are triggered by a variety of external factors such as intense rainfall, earthquake shaking, water level change, storm waves and rapid stream erosion etc. (Dai et al., 2002). These triggering factors have an influence in increasing the shear stress and decreasing shear strength of slope forming materials beyond a threshold limit and which eventually cause failure and hence landslide. Though, landslides encompass a wide range of movements such as Debris flow, Creep, Slump, rock fall etc., however emphasis, here, has been restricted to debris flow only. Debris flows are often described as multi-phase gravity-driven flows consisting of randomly dispersed unsorted material ranging from clay to boulders. These multiple phases interact in a very complex way to lead to extremely hazardous events. According to (Varnes, 1978), which is a part of landslide classification, is commonly used and states that flow are rapid movements of materials as a viscous mass where inter-granular movements predominate over shear surface movement. These phenomena could be debris flow, mudflows or rock avalanches, depending upon the nature of the materials involved in the movement. Debris flow can further be classified into two types: hill-slope (open-slope) debris flow and channelized debris flow. Between these two hill-slope debris flows creates their own path down the valley slope as tracks or sheets, depositing their materials on lower slope gradient (Cruden & Varnes, 1996).

Pertinently, in the recent past, devastating landslides have occurred in the state of Uttarakhand such as Ukhimath in 2012 and Varunabat Parbat, Uttarkashi in 2003, Lambagarh in 2004, La landslide in 2009 and numerous landslides in 2013 including Kedarnath. In order to model their diversified events and assess geotechnical characteristics, RAMMS model was applied with the help of inputs derived using satellite remote sensing techniques. Though lots of works have already been dedicated to landslides studies in this region, application of numerical simulation to model debris flows are still in a budding stage (Bist et al. 1999; Champati ray et al. 2013 a, 2013b; Chattoraj et al. 2014; Dobhal et al. 2013; Gupta and Bist, 2004; Kimothi et al., 2005; Kumar et al., 2007; Martha and Kumar, 2013; Naithani, 2002). This study aims to fill in the knowledge gap in the field of flow modelling by addressing the variation in geophysical parameters along the flow path and thereby, helping mitigation.

RAMMS DEBRIS FLOW MODELING: AN ANSWER TO SIMULATION

Debris flow modeling, currently, is one of the most sought approach pertinent to landslides and mass wasting. The underlying principle can be applied to wide range of processes and materials has, therefore, found utmost importance and significance in the field of disaster management (Iverson et al., 1997; Cruden and Vernes, 1996). There are number of well-tested and documented empirical methods to determine dynamic and kinematic parameters of the flow, yet some numerical simulation techniques are now being preferred to predict flow paths and characterize the entrainment process (Tsai et al., 2011; Quan Luna et al., 2011). One such state of the art technique is RAMMS (Rapid Mass Movements Software) model, developed by WSL Institute for snow and Avalanche Research, SLF, Switzerland, which can be employed to simulate and forecast the geophysical mass movement very precisely right from head (release area) to base (deposition area) in three dimensions. It is incorporated with user-friendly visualization tool that allow users to easily access, display and analyses the simulation results. The software takes cues from a basic high-resolution DEM (often at resolution <10m) and other ancillary ground data including geotechnical parameters. Voellmy frictional law serves as the basic algorithm for modeling physical characteristics of entrained debris material along slope (Christen et al., 2010; Ayotte & Hunger, 2000; Rickenmann, 2005; Salm et al., 1990). The outputs include 2-D, 3-D models, kinematic parameters which can provide information on dispersion of momentum and impact pressure throughout the torrent which can be utilized for better understanding of the process and designing of suitable engineering structures (Iverson et al., 2000). Temporal variation of physical parameters along of longitudinal profiles of a particular flow can be used for time series analysis and change detection in case of repeated flows.

Algorithm

RAMMS Debris Flow physical model uses the Voellmy friction law which eventually divides the frictional resistance into two parts: a dry-Coulomb type friction (coefficient μ) that is related to the normal stress and a velocity-squared drag or viscous-turbulent friction (coefficient ζ). The frictional resistance S (Pa) is then:

$$S= \mu\rho H g\cos(\varphi) + (\rho g U^2)/\xi$$

where ρ is the density, g the gravitational acceleration, φ the slope angle, H the flow height and U the flow velocity (Salm et al., 1990). The two major frictional input parameters are μ and ξ(Salm et al., 1990). Eventually, the varied values of 'S', in the formulae, is the factor responsible for modelled outputs controlling the movement of flow.

Key Inputs to Model

The essential dataset needed to successfully execute RAMMS model are topographic data i.e. digital elevation model, release area and release mass as well as information about friction parameters. However, the DEM should have high spatial resolution and the defining the release area must be accurate for better model outputs.

Topographic Data: Source Area and Entrainment Zone

Topographic information is a vital input for debris flow simulation because slope and thus exact direction of movement will be determined by variance of elevations. In this contest a precise digital elevation model is very crucial. To fulfill this requirement, where satellite derived DEM is not available, digitized contours from respective topo sheets (53N/2 in case of Ukhimath, 53 J/6 for Uttarkashi and 53N/1-2 for Kedarnath) were interpolated to 40 m resolution DEM using the ArcGIS topo-raster function. However, in order to gain better resolution, it was resampled to 5 m from 40 m. The targeted debris flow initiation zone was identified based on satellite images and was confirmed based on field observations and geomorphological analysis.

Friction Parameters

The major difficulty in case of debris flow simulation is that it involves wide variety of materials with different frictional parameters. Therefore, calibration of the friction parameters requires several simulations with different values for each input parameters to match the expected run-out based on actual event data. Finally based on the validation of such results with field data, best simulation outputs can be used for analysis (Sosio et al., 2008).

In the present case, a number of simulations were carried out using various combinations of friction parameters. A range of values were used to find out the optimal friction value. Coefficient of dry friction values varied from 0.05 to 0.4 and for viscous turbulent flow is varied from 100-450 m/s^2. Meanwhile other parameters viz. density of materials, release height, earth pressure coefficient (lambda) and the percent of momentum were kept constant. Afterward, validation of simulation output was carried out using total run-out distance and the aerial extents of real flow path as observed on satellite images and verified by field observation.

Satellite Data Used

Indian Remote Sensing Satellite data products such as LISS-IV (Resourcesat-2) and Cartosat-2 data acquired on different dates were analyzed for pre and post disaster comparison and proper demarcation of source areas wherefrom debris flow was initiated. Cartosat-1 data of 2008 along with derived DEM and LISS IV (Resourcesat-1), Google Earth images along with ancillary terrain information from SRTM DEM Version 4 were also referred.

RESULTS AND DISCUSSION

Flow modeling using RAMMS provides four main output parameters viz. velocity, height, pressure and momentum. Other outputs like run-out longitudinal profile and point information of a specific location are also provided in 2D and 3D. Debris flow height and run out distance are very important outputs as they provided information on aerial extent of the debris flow and its possible role on blockade of the existing river course, road networks etc. Moreover, debris flow height can be used to calculate financial cost of cleaning up the debris. Velocity of materials is a crucial factor due to the presence of elements at risk that are to be affected by debris flow. Materials of higher momentum could destroy any element at risk within a short time span. On the other hand, momentum actually gives the devastation potency of debris flow. Variation of velocity and height along flow path for each of the area are discussed below:

Ukhimath

The final simulation gives the value of total release volume as 37837 m^3. Modelled result reveals that maximum velocity (12.9 m/s) is found at the initiation zone and the fan apex near Mangli village. From the longitudinal profile it is evident that velocity of 10-12m/s continues up to 400 m and then suddenly falls to 4-2 m/s between 400-600m level, due to gentler topographic slope; afterward it slightly increases to 4 m/s near Chunni village. See Figure 1.

The 2D simulated height gives maximum value of about 3.3m near Mangli village and its connecting road to Ukhimath, which is result of sudden decrease in slope near hill edge of Mangli village. The longitudinal profile indicates constant height (0.5-0.7m) between 500m level and Chunni village. Thereafter, the height is almost negligible (0.1-0.2 m) till flow reaches Mandakini River where height is approximately 0.3m.

Uttarkashi

In Uttarkashi, the value of total release volume comes out to be 27444 m^3. The maximum velocity (10.19 m/s) is found in left channel near the initiation zone.

Figure 1. Spatial variation of simulated flow parameters and corresponding longitudinal profiles (a to d: velocity, height, momentum and pressure respectively for Uttarkashi) and (e to h: pressure, momentum, height, and velocity respectively for Ukhimath)

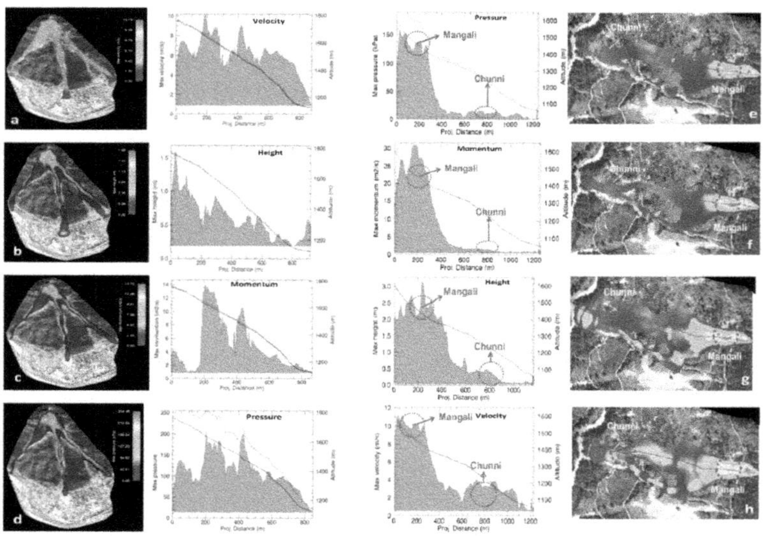

**For a more accurate representation of this figure, please see the electronic version.*

From the longitudinal profile it is evident that velocity of 6-10 m/s continues almost in every channel up to 400 m and then suddenly dwindles down to 2 m/s at a distance of 430m in left channel, which could be due to topographic flattening, afterward it increases to 4m/s and remains constant till the end. Whereas in other two channels, the average velocity of 4-9 m/s is maintained throughout the torrent zone. See Figure 1.

In all three profiles, the average height ranges between 0.56 to 0.28m. However, the maximum value (1.69m) is recorded in left channel near the release area i.e. top of Varunavat Parvat, which is validated from field information and photographs. Thereafter the height keeps on decreasing up to 800m, but it suddenly increases at the depositional zone on the bank of Bhagirathi River.

Fig. 1 Spatial variation of simulated flow parameters and corresponding longitudinal profiles: (a to d: velocity, height, momentum and pressure respectively for Uttarkashi) and (e to h: Pressure, momentum, height and velocity respectively for Ukhimath)

Kedarnath

There were 2 major debris flows in Kedarnath viz. Debris flow-1 (located upstream of the temple) and Debris flow-2 (slightly downstream of the temple). The debris flow-1 is eventually un-branched till it reached the base. The total run out length was 1.5 km. Whereas Debris flow-2 was slightly lengthier with total run out length of around 2 Km. Moreover, a small bifurcation into left and right branches is observed near the base of flow.

As for other outputs, the simulated maximum velocity at the base for debris flow-1 was 5-7m/s and for debris flow-2, was 3-5m/s. Whereas, the height at the base was observed to be 4-6m for debris flow-1 and 1-3m for debris flow-2. Both way, debris flow-1 appears to be more vigorous. It was observed that momentum was concentrated symmetrically in the middle zone of both the flows. See Figure 2.

CONCLUSION

This physical model based on numerical simulation of real event reveals both qualitative and quantitative outputs and this lend a helping hand to mitigation. Quantity wise, the modelled height, momentum and velocity of debris flow-1 at the base of major debris flow occurred at Kedarnath indicate that its contribution was high enough to provide sufficient blockade to the main course of river Sarasvati leading to changing its course and consequently, flooding the temple area. The profiles generated taking cues from modelled outputs can be readily assimilated to calculated height

Figure 2. Spatial variation of simulated flow parameters in Kedarnath (a and d: velocity; b and e: height; c and f: momentum of debris flows 1 and 2 respectively)

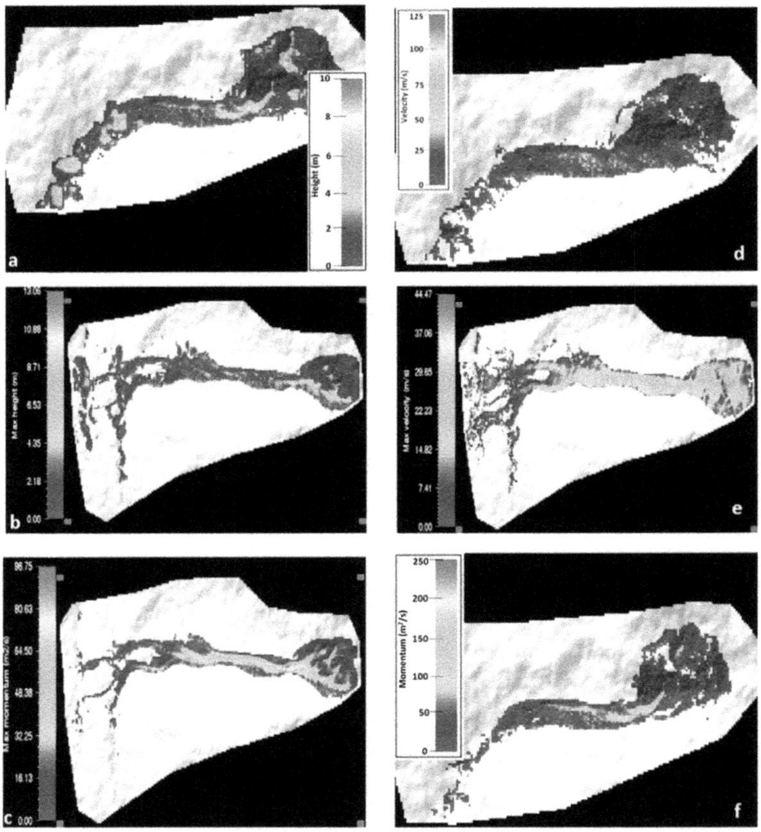

*For a more accurate representation of this figure, please see the electronic version.

and proper design of check dam to be implanted for digesting the initial trust of the heavy debris flows. It is empirically known that height of the dams depends mainly upon flow height, momentum and pressure at the point of impact with the structure.

Qualitatively the simulated results address the application of best fit Voellmy friction coefficients for simulation of other local potential flows. Thus Future potential debris flows at nearby vulnerable locations can be modelled if geology and topography is at par. Interestingly, the model, apart from satellite derived information's, also depend heavily on ground conditions. Shear strength parameters used as inputs for a good model can be validated by laboratory instrumentation techniques which is unique among this type of simulation.

As a line of caution, it is to be admitted that there were minor variations of the simulated flow compared to each the real events as no simulation technique is full proof and is based on certain scientific assumptions. In the models described here, it has been assumed that there was neither en-route erosion involved nor any contribution to main flow from its flanks. Thus, the physical parameters obtained like velocity, height etc. may be considered to be on the lower side in terms on magnitude, sensu stricto.

ACKNOWLEDGMENT

This paper has immensely benefited from contributions and consultations with several colleagues at Indian Institute of Remote Sensing and other organizations which include Dr. Ajanta Goswami, Mr. Kamal Pandey, Dr. Praveen Thakur and students/JRFs (Dr. P.K. Champati ray, Subhajit, Ashraful, Rohit, Shailja, Amit Anand, Vedika, Sashi Gaurav). Organizational support and overall guidance provided by Dr. A. Senthil Kumar, Director, IIRS and Dr. SPS Kushwaha, Dean (Academics) is also duly acknowledged. SLC is thankful to Indian Space Research Organization, Department of Space, Government of India for the financial support provided in TDP project.

REFERENCES

Ayotte, D., & Hunger, O. (2000). Calibration of a runout prediction model for debris flows and avalanches. In G.F. Wieczorek & N.D. Naeser (Eds.), *Proc. of the 2ndInternational Conference on Debris Flows*, Taipei (pp. 505-514). Balkema: Rotterdam.

Bist, K. S., & Sah, M. P. (1999). The devastating landslide of August 1998 in Ukhimath area, Rudraprayag district, Garhwal Himalaya. *Current Science*, *76*(4), 481–484.

Champati ray, P.K., Chattoraj, S.L. and Kannaujiya, S. (2013b, November 11-12). Uttarakhand Disaster 2013: Response and Mitigation measures using remote sensing and GIS. Pre workshop full publication In: National work shop on Geology and Geoheritage sites of Uttarakhand with special reference to geo-scientific development of the region organized by *Indian geological Congress (IGC), Roorkee* jointly with L.S.M. Govt. PG College (pp. 37-45).

Champati ray, P.K., Chattoraj, S.L., Chand, D.S., & Kannaujiya, S. (2013a). Aftermath of Uttarakhand disaster 2013: an appraisal on risk assessment and remedial measures for Yamunotri shrine using satellite image interpretation. *Indian Landslides*, 6(2), 61-70.

Chattoraj, S. L., Champati ray, P.K., & Bandopadhyay, S. (2014, October 28-30). Debris Flow Simulation and Modeling: A Case Study from Kedarnath Area (abstract). *Proceedings of the Geo-Environmental Hazards and Neo-Tectonic Activities in Himalaya*, HNB Garhwal University Campus Badshahi Thaul, Tehri Garhwal (p. 26).

Christen, M. K., Walski, J., & Bartelt, P. (2010). Ramms: Numerical simulation of dense snow avalanches in three-dimensional terrain. *Cold Regions Science and Technology*, *63*(1/2), 1–14. doi:10.1016/j.coldregions.2010.04.005

Cruden, D. M., & Varnes, D. J. (1996). Landslides types and processes, In A.K. Turner & R.L. Schuster (Eds.), Landslides Investigation and Mitigation (pp. 36-71). Transport Research Board Report 247.

Dai, F. C., Lee, C. F., & Ngai, Y. Y. (2002). Landslide risk assessment and management: An overview. *Engineering Geology*, *64*(1), 65–87. doi:10.1016/S0013-7952(01)00093-X

Dobhal, D. P., Gupta, A. K., Mehta, M., & Khandelwal, D. D. (2013). Kedarnath disaster: Facts and plausible causes. *Current Science*, *105*(2), 171–174.

Ghosh, J. K., & Suri, S. A. (2005). Knowledge based system for assessment of landslide hazard. *Proceedings of the Indian Geotechnical Conference(Vol. 1,* pp. 393–396). Ahmedabad: IGS.

Gupta, V., & Bist, K. S. (2004). The 23 September 2003 Varunavat Parvat landslide in Uttaranchal Township, Uttaranchal. *Current Science*, *87*, 119–131.

Iverson, R. M., Reid, M. E., & La Husen, R. G. (1997). Debris-flow mobilization from landslides. *Annual Review of Earth and Planetary Sciences*, *25*(1), 85–138. doi:10.1146/annurev.earth.25.1.85

Kimothi, M.M., Garg, J.K., Ajay, & Joshi, V. (2005). Slope Ancient religious Uttarkashi town (Garhwal Himalayas, Uttaranchal, Observation from IRS-P6 (Resourcesat-1) high resolution LISS-IV data. Map India.

Kumar, V. K., Lakhera, R. C., Martha, T. R., Chatterjee, R. S., & Bhattarcharya, A. (2007). Analysis of the 2003 Varunawat Landslide, Uttarkashi, Indian using Earth Observation data. *Environmentalgeology*, *55*(4), 789–799.

Martha, T., & Kumar, V. K. (2013). September, 2012 landslide events in Okhimath, India-an assessment of landslide consequences using very high resolution satelliete data. *Landslides*, *10*(4), 469–479. doi:10.1007/s10346-013-0420-6

Naithani, A. K. (2002). The August, 1998 Okhimath tragedy in Rudraprayag district of Garhwal Himalya, Uttranchal, India. *Gaia*, *16*, 145–156.

Quan Luna, B., Blahut, J., van Westen, C. J., Sterlacchini, S., van Asch, T. W. J., & Akbas, S. O. (2011). The application of numerical debris flow modelling for the generation of physical vulnerability curves. *Natural Hazards and Earth System Sciences*, *11*(7), 2047–2060. doi:10.5194/nhess-11-2047-2011

Rickenmann, D. (2005). Runout prediction methods. In M. Jakob & O. Hungr (Eds.), *Debris-flow Hazard and Relation Phenomena* (pp. 305–324). Chichester: Springer. doi:10.1007/3-540-27129-5_13

Salm, B., Burkhard, A., Gubler, H.U. (1990). Berechnung von Fliesslawinen: Eine Anleitungfuer Praktiker; mit Beispielen. *Mitteilungen des Eidgenoessischen Instituts fuerSchnee- und Lawinenforschung*, 47, 1–37

Sosio, R., Crosta, G. B., & Hungr, O. (2008). Complete dynamic modeling calibration for the Thurwieser rock avalanche (Italian Central Alps). *Engineering Geology*, *100*(1-2), 11–26. doi:10.1016/j.enggeo.2008.02.012

Tsai, M. P., Hsu, Y. C., Li, H. C., Shu, H. M., & Liu, K. F. (2011). Application of simulation technique on debris flow hazard zone delineation: A case study in the Daniao tribe, Eastern Taiwan. *Natural Hazards and Earth System Sciences*, *11*(11), 3053–3062. doi:10.5194/nhess-11-3053-2011

Varnes, D. J. (1978). Slope movement types and process. In Special Report 176: Landslides: Analysis and Control. In R.L. Schuster & R.J. Krizek (Eds.), Transportation and Road Research Board, National Academy of Science (pp. 11-33).

Chapter 7

Space-Time Integrated Landslide Hazard Zonation near Tehri Dam in Uttarakhand, India:

Integrated Landslide Hazard Zonation

Pooja Rana
Sphere India, India

Jeganathan Chockalingam
Birla Institute of Technology, India

Arvind Chandra Pandey
Central University of Jharkhand, India

ABSTRACT

The study aims to predict landslide hazard zones near Tehri Dam in Uttarakhand State located in the Western Himalayas in India. Four different models were analysed: Weight Factor Model (M1), Multiple Factor Model (M2), Statistical Bivariate model (M3) and Analytical Hierarchical Processes (AHP) model (M4). Five different combination of reference landslides were used for deriving weights of the classes in the factor maps: all the landslides from 1990, 2002 & 2010 (C1); landslides from

DOI: 10.4018/978-1-5225-1814-3.ch007

2010 (C2); landslides from 1990 & 2002 (C3); landslides located within 500m from roads (C4); landslides located beyond 500m from roads (C5). The accuracy resulted from each model in each combination was [Mn:C1, C2...Cn]: M1: 60,44,46,38,66%; M2:70,76,79,73,71%; M3:45,37,23,36,85%; M4:64,51,51,64,36%. Multiple Factor Model (M2) resulted in a consistently high accuracy in all the combinations. Finally, the 20 different model outputs were integrated to derive unified hazard zonation maps based on: (a) mean (85% accuracy), (b) penalisation (57% accuracy) and (c) k-means cluster (80% accuracy) approaches.

INTRODUCTION

Landslide is a major and common natural hazard in a hilly terrain. In India, natural disasters such as earthquakes, floods, and especially landslides, are frequent and invariably impact rural (and national) economy. India has about 25% of its geographical area under mountainous terrain (Rautela & Pande, 2006). Himalayas and Shivalik ranges in north are new and unstable in comparison to the southern, central and western mountains namely the Western Ghats, Satpura, Vindhyan ranges and Aravalis which are geologically very old and stable formations. The Himalayan terrain witnessed large number of major and minor landslide events causing loss of human lives, property and disruption of environment (Petley, 2013). Recently there has been a very rapid increase in the developmental activities like large scale construction of roads, mining activity, overgrazing, deforestation and opening up of steep land for agriculture in whole Himalaya which cause severe erosion resulting in extensive slope failures. The damage caused by landslides in the Himalayas estimated to cost more than one billion US dollars, besides causing 200 deaths every year, which amounts to 30% of such losses occurring world-wide (Kanungo et al. 2009). Uttarakhand is a Himalayan state that is located between Nepal and Himachal Pradesh (India). The seismic risk in the state is evaluated to be high. In the past, the state has witnessed a large number of landslide events. Four of the thirteen districts (Pithoragarh, Chamoli, Bageshwar & Rudraprayag) of the state fall completely in Zone V of the seismic risk map of India while other five (Uttarkashi, Tehri Garhwal, Pauri, Almora & Champawat) fall partially in Zone V and partially in Zone IV (Rautela & Pande, 2006). In the past decade there were few major landslide events in Uttarakhand which caused huge loss in terms of property, human life and amenities.

Apart from development activities in the Himalayan terrain, these rocks are adversely affected by tectonic movements (in the form of intense shear, faulting, fracturing), surfacial and near surfacial processes (causing joints or cracks) and precipitation. These factors make the slopes more prone to landslides. Over the years we can see many examples of major landslides during monsoon, pre-monsoon and post monsoon season in the Uttarakhand state, which depict that in addition to natural causes other causes like development activities, human intervention on the steep slopes also leads to destabilizing these slopes. Tehri Reservoir is a geologically very recent initiative in Uttarakhand which is established with a good motive but unfortunately it is located within the high landslide prone zone, where debris flow is a major threat for the population living in this area. Due to the establishment of Tehri dam (completed in the year 2006), road cuttings and changed slope conditions in the area and hence major land use and land cover changes took place resulting in frequent occurrence of landslide in the area.

During the recent decades, the use of landslide susceptibility and hazard maps for land use planning has drastically increased. The hazard zonation approach ranks different sections of land surface according to the degree of actual or potential landslide hazard. Landslide hazard refers to the possibility of occurrence of a damaging landslide within a given area which could include property damage, social-economic disruption, loss of life or injury, or environmental degradation. Remotely Sensed satellite data and GIS applications can be used together to investigate potential hazard zones and develop planning tools for particular hazard in any region. Satellite images, aerial data and digital elevation model helps us understand characteristics of terrain of any region and helps to generate different maps like aspect, slope, relief, lithology, geology, land use/land cover, soil, drainage, road network and landslide inventory, and Geographic Information system (GIS) helps in data capture, manipulation, transformation, visualization, query, analysis, modeling and output generation with its spatial data processing capability especially helps to model the relationship between different parameters and landslide inventory of a study area (Jeganathan & Chauniyal, 2000; Pandey et al. 2007). There are many techniques for modelling landslide (Anbalagan, 1992; Pachauri & Pant, 1992; Van Westen, 1993; Chung & Fabbri, 1999; Jeganathan & Chauniyal, 2000; Champati Ray et al. 2006; Sarkar et al. 2015). In most of the landslide zonation studies it is generally seen that any one of the available model is used along with available one-time landslide inventory, and hence the resultant model output would be partially reflecting the reality. In order to avoid such gaps, in this study we aim to use landslide inventory from different time periods, and variations in their spatial locations under four different models so as to derive diverse outcomes and to integrate them. Considering the potential disaster sitting at known geographical condition the current work aim to model and understand the behaviour of landslide hazard zones and finally to come out with one single representative hazard zonation map.

STUDY AREA

Uttarakhand region is one of the top destinations for spiritual tourism in India where the four most important pilgrimage locations—Badrinath, Kedarnath, Gangotri and Yamunotri—are situated along with numerous historic holy townships like Haridwar, Rishikesh and scenic locations such as valley of flowers, and landslides causes huge economic losses and also lead to loss of life in this region.

Tehri Garhwal is one of the largest districts in the hill state of Uttarakhand, India. Its administrative headquarters is at New Tehri. It is surrounded by Rudraprayag District in the east, Dehradun District in the west, Uttarkashi district in the north and Pauri Garhwal district in the south. The selected study area (longitudinal extent 78° 15' E to 78° 39' E and latitudinal extent 30° 16' N to 30° 33' N (covering 494 Sq. km area) comes under the Tehri Garhwal districts of Uttarakhand. Two major rivers are flowing in the study area, Bhagirathi and Bhilangana and confluence on Tehri dam creates reservoir of more than 45 sq. km. The dam is 260 meters high rock and earth-fill embankment dam. Its length on top is 575 meters, crest width is 20 meters, and base width is 1,128 meters. The terrain is hilly with an average slope of 40– 50 degrees. The region has high rainfall (average rainfall is 1100 mm over a period of 4-5 months) there by making it one of the highest landslide prone zone in the world. The whole study area comes under the threat of landslide and they are very frequent, especially during every monsoon Season.

DATA AND METHODS

The study utilized satellite remote sensing data from IRS P6 LISS IV satellite image (spatial resolution 5.8m) for the year 2010, LANDSAT ETM+ & TM image (spatial resolution 30m) for the year 2002 and 1990 respectively. See the study area in Figure 1 and satellite data in Figure 2. Survey of India (SOI) topographic map (53 J/7) on a scale of 1:50,000 was utilized to extract drainage features. Digital Elevation Model (DEM) at the spatial resolution of 30m derived from IRS-P5 CARTOSAT was downloaded from Bhuvan website (bhuvan.nrsc.gov.in). Geological Map of Part of Lesser Garhwal Himalaya, Tehri District was taken from Pandey (2001). After geo-referencing the three images, visual interpretation was done to delineate landslide and road cuttings in 2010, 2002 and 1990 images. LISS IV image (2010) was used for extracting land use/land cover and road network. Slope, aspect, elevations were extracted from Cartosat DEM. Distance from Drainage was calculated within Arc Map Geographical Information System (GIS) environment. All the individual layers were standardized to a common georeferencing scheme with common pixel size. Number of landslide pixels occurring in each factor classes was calculated

Figure 1. Study area

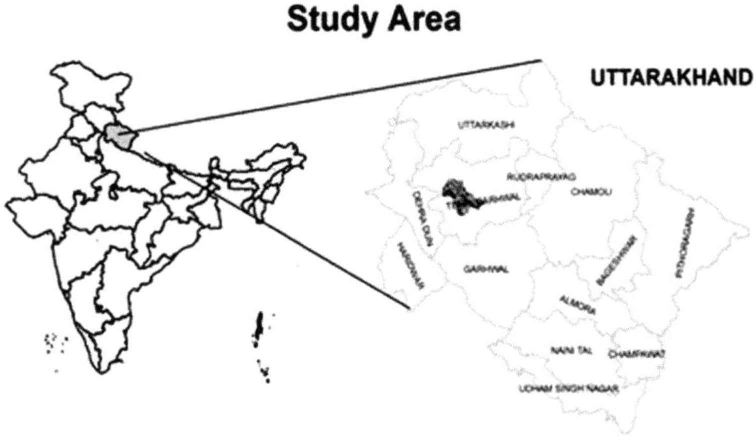

Figure 2. Multi-resolution satellite data used in the study

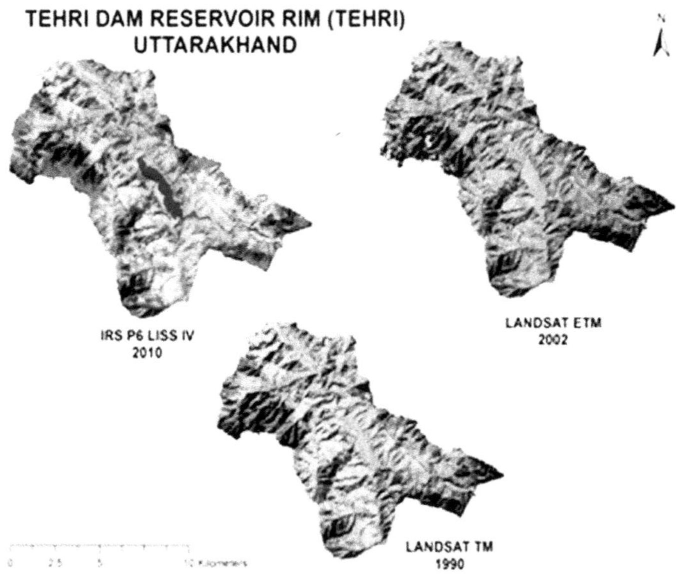

for estimating classwise weights as needed under different models. Four different models were utilized for the study: Weight Factor Model, Multiple Factor Model (Bathrelios et al. 2009), Statistical Bivariate Model and Analytical Hierarchical Processes (AHP) model.

Weight Factor Model

Weight Factor Model (WeF) is a common model in mapping landslide hazard zones (Wachal & Hudak, 2000; Ayalew et al.,2005; Bathrellos et al. 2009). See Figure 3. In WeF, classes within each factor maps are analysed for their relative contribution score towards landslide occurrence based on frequency ratio method:

Frequency Ratio = (*Landslide occupied pixels in a factor class/Total reference landslide pixels*)/(*Total pixels in a factor class/Total pixels in the study area*)

Sum of frequency ratios of all the attribute classes in a factor map is called landslide susceptibility index (LSI) of that factor map. To assign numerical weight to each input factor map, LSI was calculated for each factor maps and divided by total LSI of all the maps. Overall susceptibility index was calculated based on weighted sum of all input standardized factor maps.

Figure 3. Implementation flow for Weight Factor Model

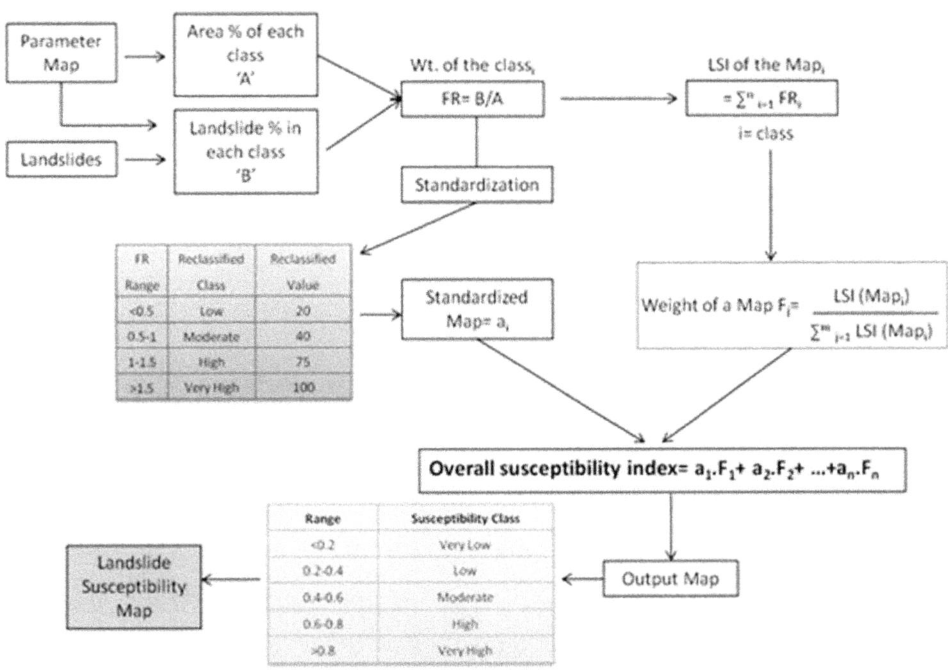

Multiple Factor Model

In Multiple Factor Model (MuF) (Batherellos et al. 2009), it is assumed that many factors can play a simultaneous role in causing landslides, and hence the additive influence of multiple factors are enhanced in this model. See Figure 4.

Statistical Bivariate Model

This method was proposed by Van Westen (1993). In this method each parameter map (geology, slope, aspect, relief, Land cover etc.) is combined with the landslide inventory. See Figure 5. The weights are derived from each attribute class within each parameter map. Mainly three types of weight estimation methods employed in this method: Information value method, weight of evidence modeling and landslide nominal susceptibility factor (LNSF) (Gupta and Joshi, 1990). In first step each attribute class within each parameter maps were provided weights. Each parameter map was separately overlayed with landslide inventory. The numbers of landslide pixels falling on each class of the parameter map are recorded and weights calculation is done by LNSF method.

Gupta and Joshi (1990) suggested LNSF value greater than 1 as high susceptibility and a value less than 1 as low susceptibility and equal to 1 refers to average susceptibility. After providing weights all the maps were reclassified into three classes (low susceptible, moderate susceptible and high susceptible). LNSF= >1

Figure 4. Implementation flow for Multiple Factor Model

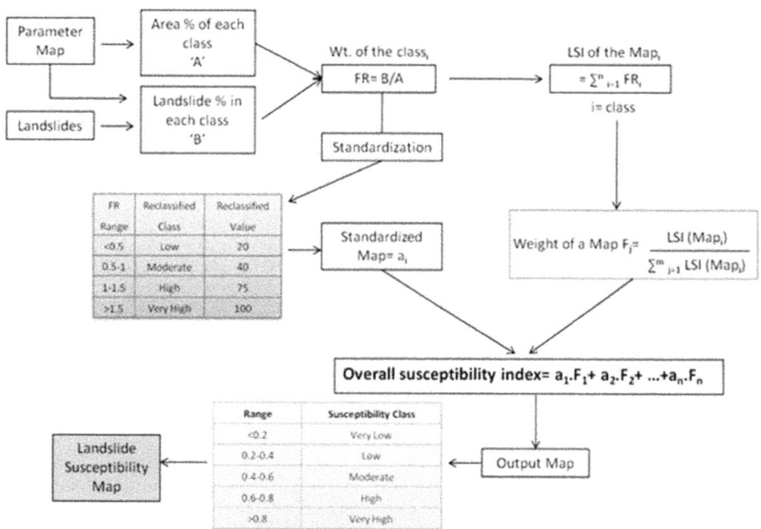

Figure 5. Implementation flow for Statistical Bivariate Model

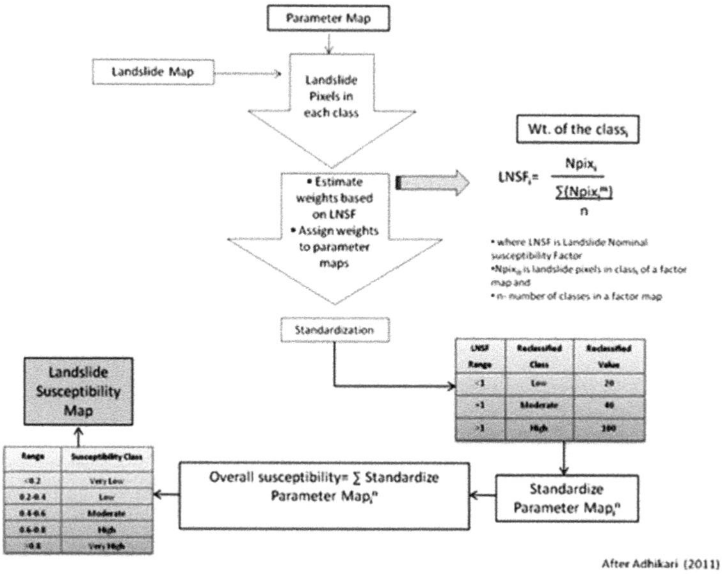

(High Susceptibility=100), LNSF= 1 (Moderate Susceptibility=40) and LNSF= <1 (Low susceptibility=20). All the weighted parameter maps were then summed up to produce the Landslide susceptibility index.

AHP Model

Analytic Hierarchy Process (AHP) is a multi-criteria decision making method which was developed by Prof. Thomas L. Saaty (Saaty, 1977, 1980 & 2008). This process uses a multi-level structure of objectives, criteria, sub criteria, and alternatives in a hierarchical order. See Figure 6. In this method the data is derived by using pair-wise comparisons matrix (PCM) (using 9 scales) which is further processed to obtain weights of each classes. Then PCM of model factors (i.e., criteria) was also created to derive importance of decision criteria. Finally, weighted sum of all the normalised factor maps would reveal landslide hazard zones.

Space-Time Integrated Landslide Hazard Zonation (LHZ)

In each of the above said models, five different landslide occurrence data were used:

1. Using new landslides from year 2010.
2. Using all landslides from 1990, 2002 and 2010.

Figure 6. Implementation flow for AHP Model

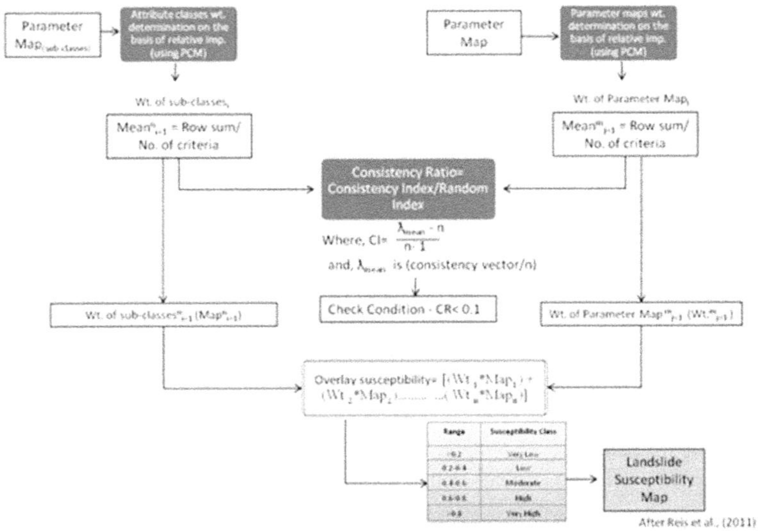

3. Using landslides near to roads.
4. Using landslides away from roads.
5. Using old landslides from 1990 and 2002.

So, overall, we have derived 20 different model outputs (i.e., 5 outputs per model x 4 models). In order to derive one representative landslide hazard map, these 20 outputs were analysed further and 3 useful metrics were extracted:

1. Mean.
2. Standard deviation [σ].
3. Max-Min difference [φ].

These three metric were used together to derive the final LHZ map. There are three ways in which one can derive final LHZ:

1. First, a simple mean of 20 models would provide a representative LHZ. But mean may have a problem if a pixel value (i.e., modelled value) in 20 outputs were widely distributed (i.e., high standard deviation). In such cases, the mean may not be a representative.
2. So, as a second case, if we penalize the mean based on its reliability (utilizing σ and φ) then we can derive a reliable representation. For this purpose, these 3 metrics were first converted into 4 classes and there were 64 combinations

if we combine these 3 classified maps. On the basis of a logic the mean value having low σ & φ is a correct representation, and on the contrary mean value with high σ & φ is a contradictive prediction, and hence high mean having high σ & φ need to be penalized to become lower landslide susceptibility class. All the 64 combinations were accordingly converted to derive a penalized mean LHZ representation.

3. Thirdly, these 3 metrics were classified using k-means clustering approach to derive classes and their clusters were assigned to a particular landslide classes based on its representative mean.

Predictive Power Estimation

In order to understand the reliability of prediction it is essential to quantify prediction power from each model. Here, we are dealing with the problem of providing measure of significance of prediction results. It confirms the validity of the results of some models over the other ones (Chung and Fabbri, 2003). To calculate the prediction power of a predictive class, we used the following equation:

Predictive Power = $Qt/r\alpha$

where, **Q** is number of actual landslide pixels within i^{th} class; **r** is total pixels in i^{th} class; **α** is total no. of actual landslide pixels in whole study area, and **t** is total pixels of study area. To make an effective prediction the ratio should either be larger than 3 (at least) or less than 0.2 (at the most) for a prediction class to be significant.

RESULTS AND DISCUSSION

First step in landslide hazard modelling is to create inventory of landslides which is usually carried out using satellite images. In the satellite image one can easily get confused between road cuttings and landslides. However, to avoid this error, the user should:

- Carefully identify and map new, under construction and old roads in the study area on satellite image to separate both the phenomena.
- Identify the differences between landslides and road cuttings by their shape as road cuttings have more lateral expansion whereas landslides are in the shape of a spoon.
- Use Google Earth (2D and 3D) as reference for better visualization. In this study we have extracted the landslides from 3 different time (1990, 2002,

2010) using different satellite data at different resolution. See Figure 7. Higher resolution data would reveal finer details and even smaller landslides. The number of landslides identified over these three different period are 24, 43 and 127 respectively.

The inventoried landslides were analysed with respect to each of the environmental factors, as seen in Figure 8 (Geology, Elevation, Slope, Aspect, Land cover and Drainage Density). The following Tables 1 through 14 provide the summary of percentage of landslides found under each of the classes.

These environmental layers and associated landslide information were fed onto different models and the resultant output from different combinations were revealed in Figures 9, 10, 11 and 12, and final integrated LHZ maps were revealed in Figure 13.

From Table 7, it could be observed that the significant predicted classes under WeF are: approach-1 (high susceptibility class), approach-2 (very high, high, very low), approach-3 (very high, very low), approach-4 (very low) and in approach-5 (very high & very low).

Figure 7. Temporal inventory of landslides and their spatial distribution

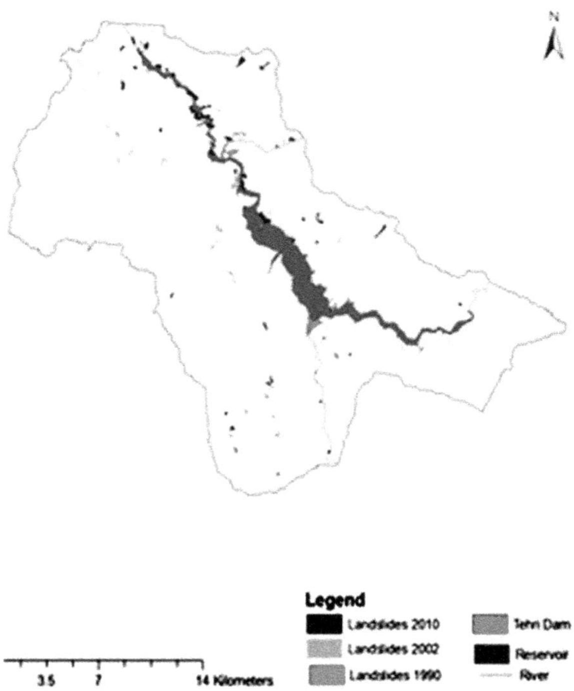

For a more accurate representation of this figure, please see the electronic version.

Figure 8. Environmental factors

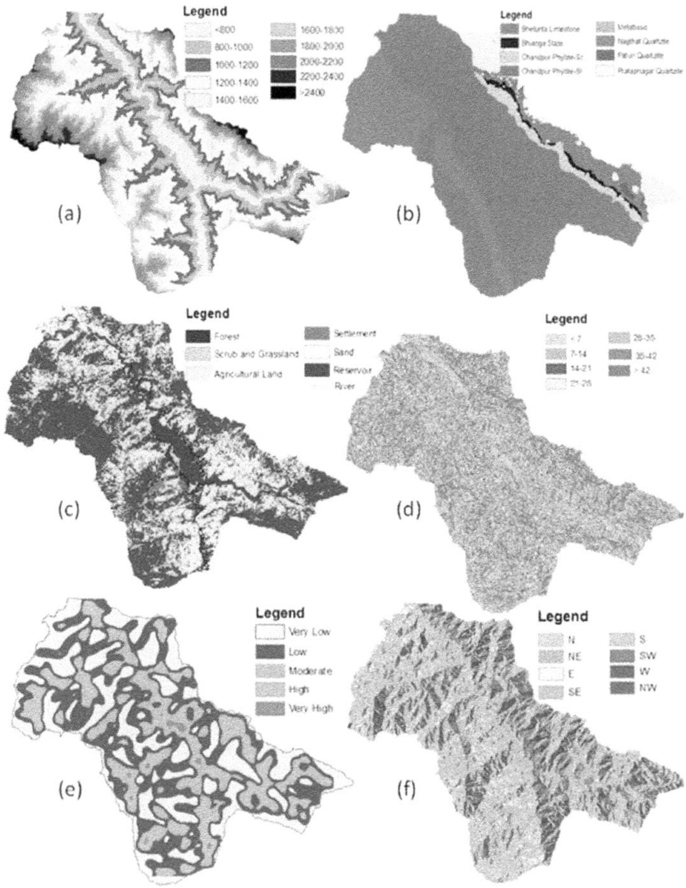

**For a more accurate representation of this figure, please see the electronic version.*

Table 1. Landslide occurrences (in %) in different slope classes

Slope Categories (degrees)	Landslide (%)				
	App-1 (New Landslides, 2010)	**App-2 (All Landslides, 1990, 2002 & 2010)**	**App-3 (Near to Roads, <500m)**	**App-4 (Away from Roads >500m)**	**App-5 (Old Landslides 1990 & 2002)**
0-7	10	10	12	10	5
7-14	12	10	12	6	10
14-21	14	13	16	10	15
21-28	16	18	17	20	19
28-35	**22**	**22**	**21**	**21**	**22**
35-42	17	16	14	19	16
>42	9	11	8	14	13

Table 2. Landslide occurrences (in %) in different aspect classes

Aspect Categories	Landslide (%)				
	App-1 (New Landslides, 2010)	**App-2 (All Landslides, 1990, 2002 & 2010)**	**App-3 (Near to Roads, <500m)**	**App-4 (Away from Roads >500m)**	**App-5 (Old Landslides 1990 & 2002)**
N	13	9	8	9	7
NE	9	10	12	4	10
E	14	15	**19**	3	14
SE	14	**16**	**18**	**22**	**18**
S	**15**	**16**	**8**	**33**	**19**
SW	**15**	12	12	13	11
W	13	12	13	9	10
NW	7	10	10	7	11

Table 3. Landslide occurrences (in %) in different land cover classes

Lu/Lc Classes	Landslide (%)				
	App-1 (New Landslides, 2010)	**App-2 (All Landslides, 1990, 2002 & 2010)**	**App-3 (Near to Roads, <500m)**	**App-4 (Away from Roads >500m)**	**App-5 (Old Landslides 1990 & 2002)**
Forest	18	8	14	8	12
Scrub & Grassland	**37**	**30**	24	**34**	**38**
Agriculture	22	20	6	24	5
Settlement	16	18	17	20	19
Sand	**4**	**11**	**5**	**11**	**8**
Water Body	3	27	**48**	18	33

Table 4. Landslide occurrences (in %) in different geological classes

Geology Categories	**Landslide (%)**				
	App-1 (New Landslides, 2010)	**App-2 (All Landslides, 1990, 2002 & 2010)**	**App-3 (Near to Roads, <500m)**	**App-4 (Away from Roads >500m)**	**App-5 (Old Landslides 1990 & 2002)**
Pratapnagar Quartzite	10	5	2	12	0.5
Bhelunta Limestone	5	6	4	10	9
Nagthat Qt.	1	2	2	3	3
Metabasic	0	0	0	0	0
Paturi	**0**	**0**	**0**	**0**	**0**
Chandpur Phyllite SC	2	4	5	0	5
Bhianga Slate	1	1	1	0	0.5
Chandpur Phyllite Sl	81	82	86	74	82

Table 5. Landslide occurrences (in %) in different drainage density classes

Drainage Density Classes	**Landslide (%)**				
	App-1 (New Landslides, 2010)	**App-2 (All Landslides, 1990, 2002 & 2010)**	**App-3 (Near to Roads, <500m)**	**App-4 (Away from Roads >500m)**	**App-5 (Old Landslides 1990 & 2002)**
Very Low	1	3	7	4	4
Low	3	10	10	8	16
Moderate	31	40	43	32	49
High	43	31	26	37	21
Very High	**22**	**16**	**14**	**19**	**10**

Table 6. Landslide percentage in various susceptibility classes from WeF

Input Types	VH		H		M		L		VL	
	Area (%)	Landslide (%)	Area (%)	Landslide (%)	Area (%)	Landslide (%)	Area (%)	Landslide (%)	Area (%)	Landslide (%)
App-1	5.49	14	9.47	46	35.52	16	39.55	17	9.99	7
App-2	15.52	4	10.21	40	34.73	39	24.14	16	15.42	1
App-3	4.96	23	9.49	23	45.46	41	30.37	12	9.74	1
App-4	17.81	28	5.73	10	20.33	32	40.90	29	15.25	1
App-5	12.83	38	15.87	28	24.96	27	26.73	6	19.63	1

Table 7. Predictive power of susceptibility classes under WeF

Input Types	VH	H	M	L	VL
App-1	2.6	5	0.3	0.3	0.7
App-2	0.2	4	1	0.7	0.1
App-3	4.8	2.5	0.8	0.3	0.02
App-4	1.5	1.5	1.5	0.5	0.1
App-5	3.5	2.1	1.3	0.3	0.1

Table 8. Landslide percentage in susceptibility classes from MuF

Input Types	VH		H		M		L		VL	
	Area (%)	Landslide (%)	Area (%)	Landslide (%)	Area (%)	Landslides (%)	Area (%)	Landslide (%)	Area (%)	Landslide (%)
App-1	4.82	8	45.86	62	42.35	29	6.86	1	0.10	0
App-2	6.16	18	45.59	58	40.72	23	7.44	1	0.09	0
App-3	5.42	5	58.43	74	34.01	20	2.06	1	0.09	0
App-4	7.11	23	35.40	50	42.07	25	14.66	2	0.76	0
App-5	2.53	5	46.29	66	45.22	27	5.82	2	0.14	0

Table 9. Predictive power of susceptibility classes under MuF

Input Types	VH	H	M	L	VL
App-1	1.7	1.3	0.7	0.1	0
App-2	3	1.2	0.5	0.2	0
App-3	0.8	1.3	0.5	0.5	0
App-4	3	1	0.5	0.1	0
App-5	1.1	1.7	0.7	0.3	0

Table 10. Landslide percentage in susceptibility classes from SB

Input types	H		M		L	
	Area (%)	Landslide (%)	Area (%)	Landslide (%)	Area (%)	Landslide (%)
App-1	19.27	45	62.86	50	17.87	5
App-2	20.31	37	67.53	57	12.16	6
App-3	18.57	23	64.05	70	17.38	7
App-4	22.07	36	62.26	60	15.67	4
App-5	64.18	85	32.86	14	2.96	1

Table 11. Predictive power of susceptibility classes under SB

Input types	H	M	L
App-1	2.3	0.7	0.3
App-2	1.8	0.8	0.5
App-3	1.3	1	0.3
App-4	1.5	0.5	0.3
App-5	1.5	0.5	0.3

Table 12. Landslide percentage in susceptibility Classes from AHP

AHP	VH		H		M		L		VL	
	Area (%)	Landslide (%)	Area (%)	Landslide (%)	Area (%)	Landslide (%)	Area (%)	Landslide (%)	Area (%)	Landslide (%)
App-1	14.18	33	21.51	31	31.82	29	19.67	6	12.39	1
App-2	8.20	20	16.80	31	32.19	32	26.50	15	15.88	2
App-3	9.21	21	17.03	30	31.84	33	25.30	15	16.18	1
App-4	10.12	33	18.19	31	32.78	26	24.57	9	13.92	1
App-5	6.98	14	16.50	22	32.34	38	26.42	22	17.33	4

Table 13. Predictive power of susceptibility classes (AHP)

AHP	VH	H	M	L	VL
App-1	2.32	1.5	0.9	0.3	0.09
App-2	3	1.9	1	0.4	0.2
App-3	9.3	1.8	1.1	0.5	0.1
App-4	3.7	1.7	0.7	0.3	0.1
App-5	3.3	1.8	1.6	0.6	0.7

Table 14. Final susceptibility area statistics from Integrated Landslide Hazard Zonation

Susceptibility Class	Total Pixel (Mean)	%	Total Pixel (Penalized)	%	Total Pixel (k-means)	%
L	18421	3.37	89645	16.39	136873	25.02
M	190379	34.80	212219	38.80	99936	18.27
H	278318	50.88	213118	38.96	185872	33.98
VH	59872	10.95	32008	5.85	124309	22.73

Figure 9. Landslide hazard zones derived using Weight Factor Model

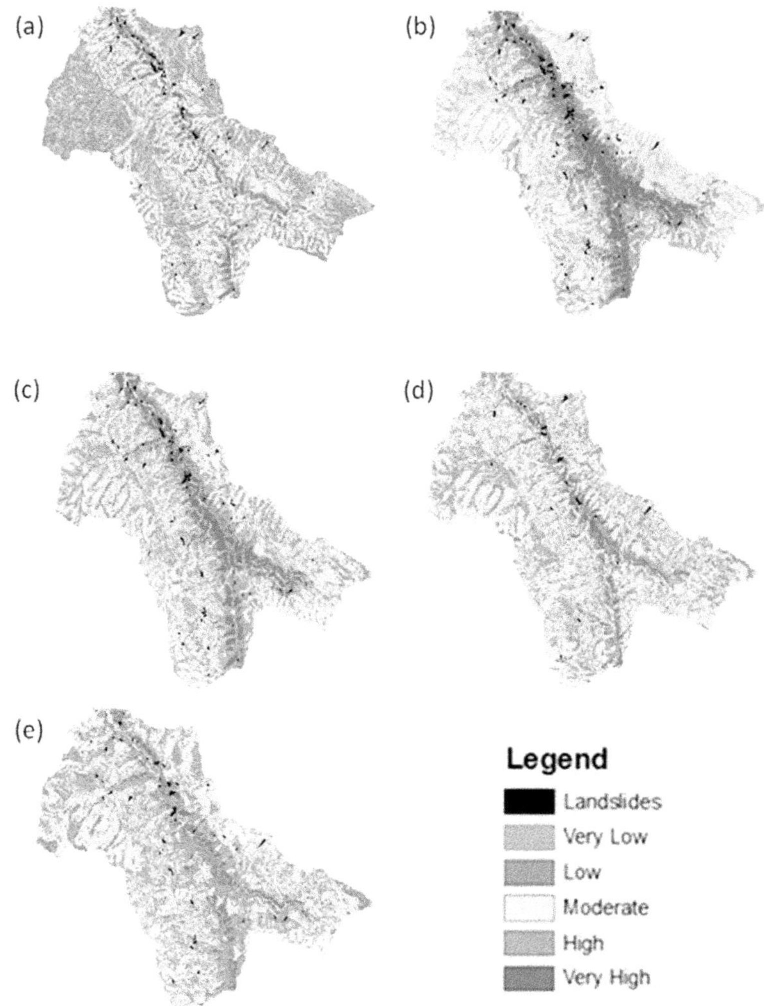

**For a more accurate representation of this figure, please see the electronic version.*

Figure 10. Landslide hazard zones derived using Multiple Factor Model

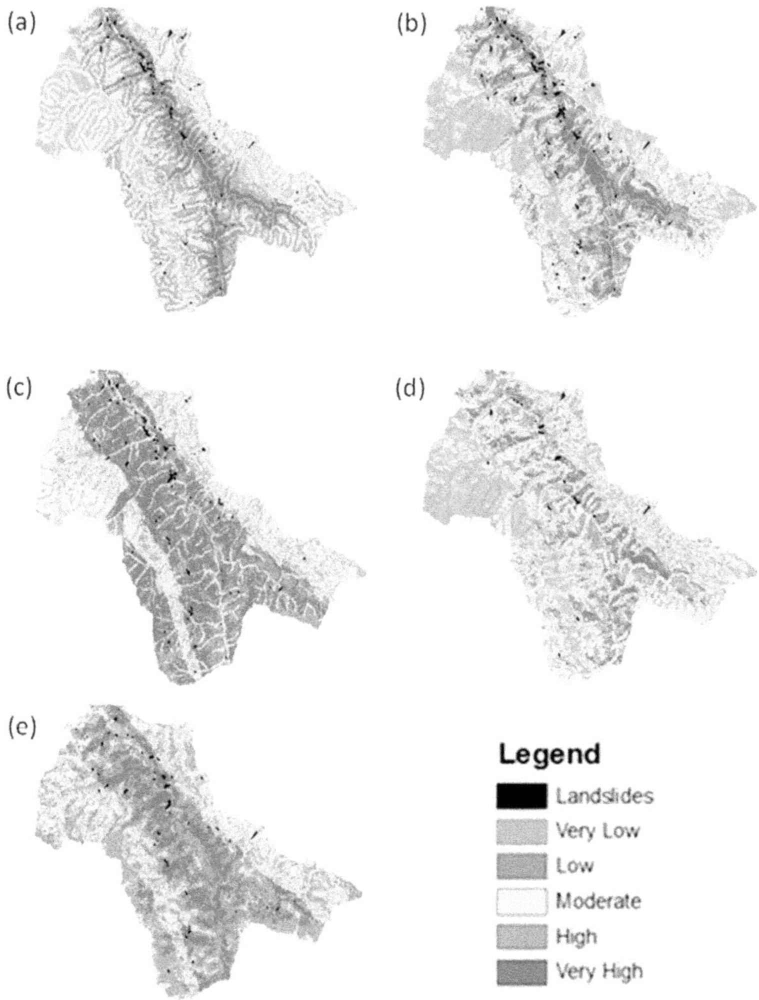

**For a more accurate representation of this figure, please see the electronic version.*

The significant predicted classes from MuF are: approach-1 (low and very low susceptibility classes), approach-2 (very low, low and very high susceptibility classes), approach-3 (very low), approach-4 (very high, low and very low) and in approach-5 (low and very low susceptibility classes).

Interestingly, there is no significant prediction made for any of the output classes from SB model.

The significant predicted classes from AHP are: approach-1 (very low class), approach-2 (very high and very low classes), approach-3 (very high and very low), approach-4 (very high and very low) and in approach-5 (very high class).

Figure 11. Landslide hazard zones derived using Statistical Bivariate Model

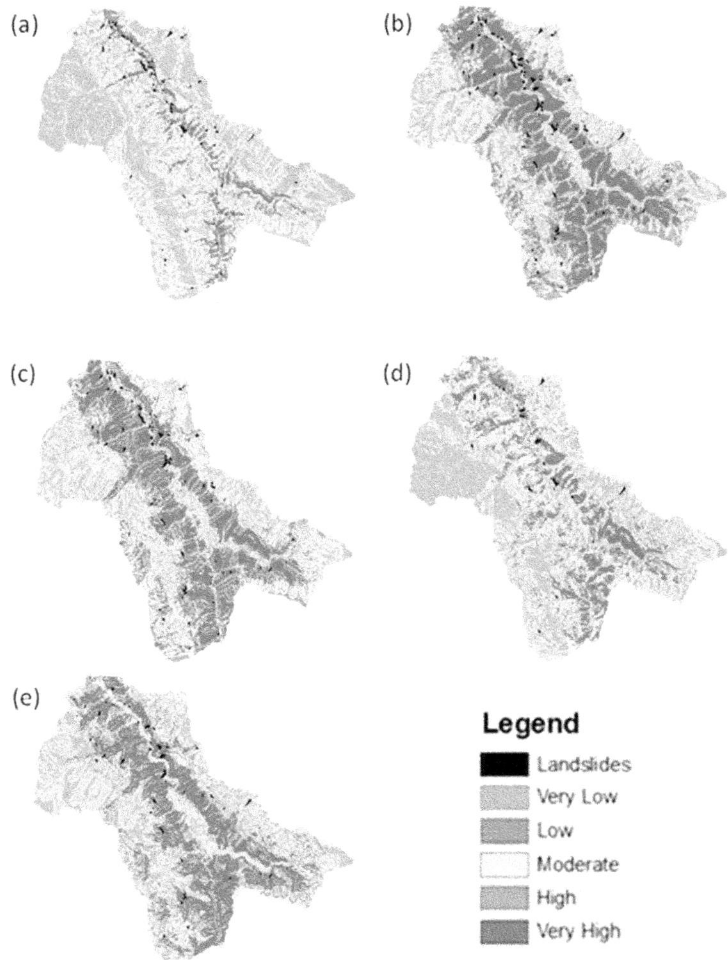

**For a more accurate representation of this figure, please see the electronic version.*

Finally, the integrated LHZ map was analyzed for its area statistics and predictive power. The landslide hazard zonation map which have maximum occurrence of landslide percentage in the high hazard prone classes is considered as more suitable. For validation of the final hazard zonation map, the landslide inventory map was overlayed and then they were statistically compared.

The spatial comparison between area and landslides in different hazard zones (of map obtained through mean) shows that about (61.83%) area is in very high & high hazard zone which constitute the maximum landslides (84.73%). See Figure 14 and Figure 15. The situation in the hazard zones comprised from penalizing the pixels is somewhat similar, where very high & high hazard zone covers the area of

Figure 12. Landslide hazard zones derived using AHP model

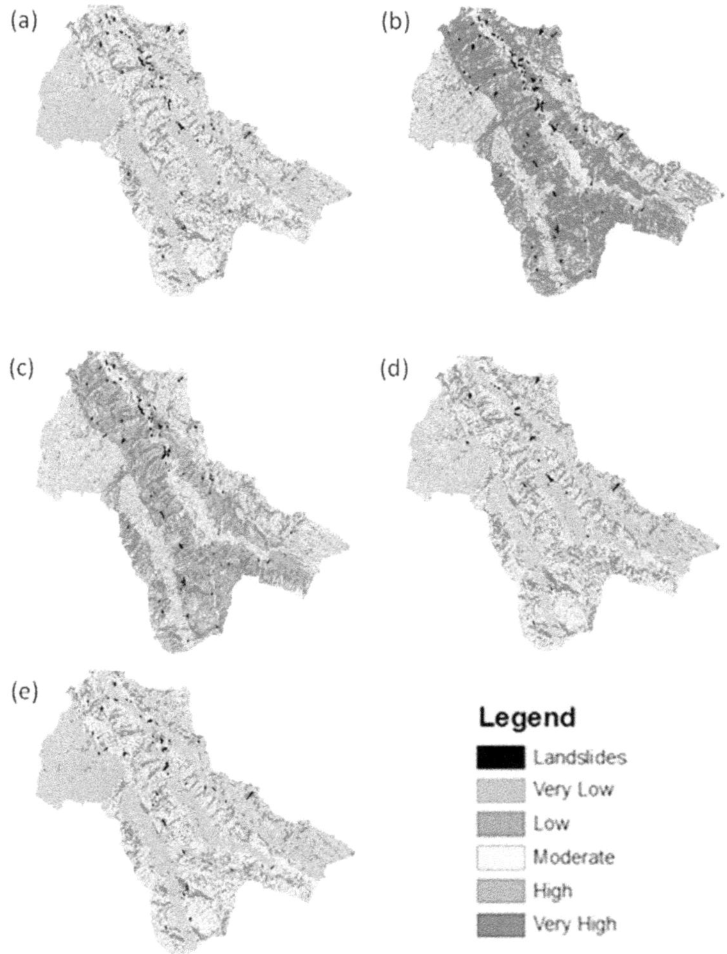

**For a more accurate representation of this figure, please see the electronic version.*

(44.81%) constituting the maximum landslides (55.52%). The susceptibility map obtained using K-means shows that almost half of the area comes under the very high & high hazard zone (56.71%) which covers the maximum percentage of landslides (80.03%) spreading in north to south direction along the water bodies (Tehri reservoir, Bhagirathi river, Bhilangana river) in the study area.

In integrated landslide hazard zonation map generated on the basis of mean and k-means cluster, the significant prediction was observed for the low hazard zones only.

Figure 13. Final integrated LHZ maps revealed

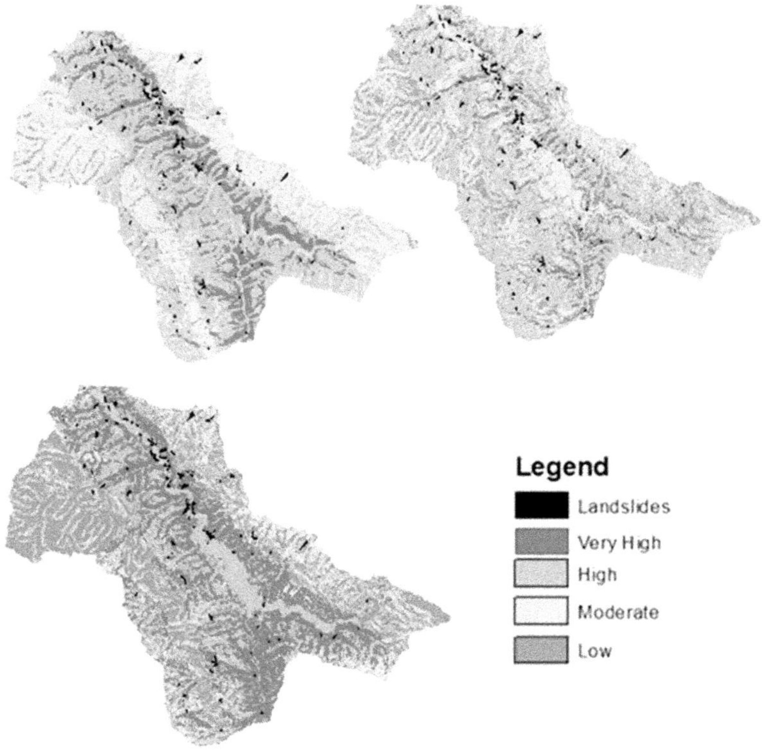

For a more accurate representation of this figure, please see the electronic version.

Figure 14. Landslide Percentage in Integrated Landslide Hazard Zones

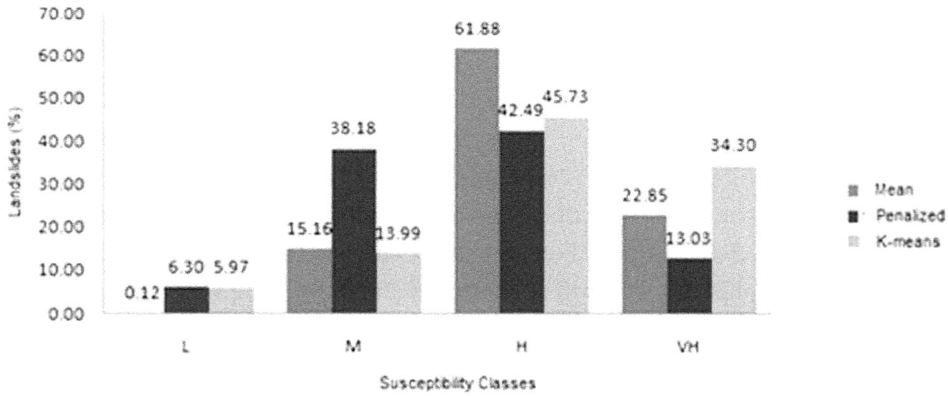

Figure 15. Area Percentage in Integrated Landslide Hazard Zone

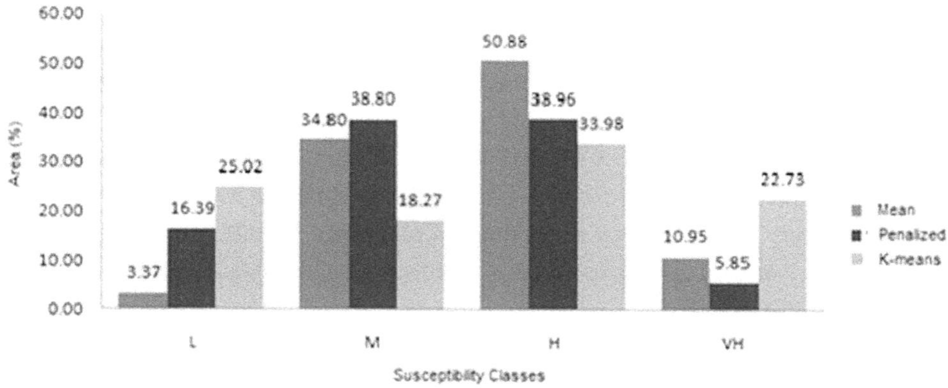

Table 15. Predictive power of susceptibility classes from integrated LHZ

Susceptibility Class	Predictive Power (Mean)	Predictive Power (Penalized)	Predictive Power (K-means cluster)
L	0.03	0.38	0.2
M	0.44	0.98	0.8
H	1.22	1.09	1.3
VH	2.09	2.23	1.5

CONCLUSION

In this study landslide hazard zonation modelling was carried out based on inventoried landslides at different time period. Four different models were analyzed:

1. Weight Factor Model (M1).
2. Multiple Factor Model (M2).
3. Statistical Bivariate model (M3).
4. Analytical Hierarchical Processes (AHP) model (M4).

Five different combination of reference landslides were used for deriving weights of the classes in the factor maps:

1. All the landslides from 1990, 2002 & 2010 (C1).
2. Landslides from 2010 (C2).
3. Landslides from 1990 & 2002 (C3).
4. Landslides located within 500m from roads (C4).

5. Landslides located beyond 500m from roads (C5).

These 5 combinations in 4 models resulted in 20 different landslide hazard zonation maps. These different combinations were tested for their accuracy using predictive power index. The accuracy resulted from each model in each combination were [M_n:C1, C2...Cn]: M1: 60,44,46,38,66%; M2:70,76,79,73,71%; M3:45,37,23,36,85%; M4:64,51,51,64,36%. Multiple Factor Model (M2) resulted in a consistently high accuracy in all the combinations. Finally, these 20 different model outputs were integrated to derive space-time integrated hazard zonation maps based on the following:

- Mean (85% accuracy).
- Penalization (57% accuracy).
- K-means cluster (80% accuracy) approaches.

For the very high hazard zone AHP was giving better results in C1, C2 and C4, whereas in C3 and C5 weight factor model predicted better results. In high hazard zone multiple factor model presented satisfied results. If we see very high and high hazard zones collectively then, multiple factor model presented the better results as 70-75% of the landslides were falling under very high and high hazard zones. Next to the multiple factor model, results obtained from AHP were more satisfied than other models as showed more than 50% of landslides in very high and high hazard zones. In the final output, the integrated landslide hazard zonation map obtained from k-means cluster showed the satisfying result as it has 80% of the landslides in very high and high hazard zones. Overall, the study hopes to help administrators and researchers in understanding various technical aspects of landslide modelling.

REFERENCES

Anbalagan, R. (1992). Landslide hazard Evaluation and zonation mapping in mountainous terrain. *Engineering Geology*, *32*(4), 269–277. doi:10.1016/0013-7952(92)90053-2

Ayalew, L., & Yamagishi, H. (2005). The application of GIS-based logistic regression for landslide susceptibility mapping in the Kakuda-Yahiko Mountains, Central Japan. *Geomorphology*, *65*(1-2), 15–31. doi:10.1016/j.geomorph.2004.06.010

Bathrellos, G., Kalivas, D., & Skilodimou, H. (2009). GIS-based landslide susceptibility mapping models applied to natural and urban planning in Trikala, Central Greece. *Estudios Geology,* 65(1), 49-65. Doi:10.3989/egeol.08642.036

Champati Ray, P. K., Dimri, S., Lakhera, R. C., & Sati, S. (2007). Fuzzy-based method for landslide hazard assessment in active seismic zone of Himalaya. *Landslides*, *4*(2), 101–111. doi:10.1007/s10346-006-0068-6

Chung, C. F., & Fabbri, A. G. (1999). Probabilistic prediction models for landslide hazard mapping. *Photogrammetric Engineering and Remote Sensing*, *65*(12), 1389–1399.

Gupta, R., & Joshi, B. (1990). Landslide hazard zoning using the GIS approach-a case study from the Ramganga catchment, Himalayas. *Engineering Geology*, *28*(1-2), 119–131. doi:10.1016/0013-7952(90)90037-2

Jeganathan, C., & Chauniyal, D. D. (2000). An evidential weighted approach for landslide hazard zonation from geo-environmental characterisation: A case study of Kelani area. *Current Science*, *79*(2), 238–243.

Kanungo, D., Arora, M., Sarkar, S., Gupta, R. (2009). Landslide Susceptibility Zonation (LSZ) Mapping - A Review. *Journal of South Asia Disaster Studies,* 2(1).

Pachauri, A. K., & Pant, M. (1992). Landslide Hazard mapping based on Geological attributed. *Engineering Geology*, *36*, 71–98.

Pandey, A. C. (2001). Geology and Morphotectonic studies in a part of lesser Garhwal Himalaya using Remote Sensing Techniques [Unpublished PhD thesis]. University of Delhi.

Petley, D. (2013). Global pattern of loss of life from landslide. *Geology*, *40*(10), 927–930. doi:10.1130/G33217.1

Rautela, P., & Pande, R. (2006). Non-monsoonal landslides in Uttaranchal Himalaya (India): implications upon disaster mitigation strategy. Disaster Prevention and Management, 15(3).

Reis, S., Yalcin, A., Atasoy, M., Nisanci, R., Bayrak, T., Erduran, M., & Ekercin, S. et al. (2012). Remote sensing and GIS-based landslide susceptibility mapping using frequency ratio and analytical hierarchy methods in Rize province (NE Turkey). *Environment and Earth Science*, *66*(7), 2063–2073. doi:10.1007/s12665-011-1432-y

Saaty, T. (1977). A scaling method for priorities in hierarchical structures. *Journal of Mathematical Psychology*, *15*(3), 234–281. doi:10.1016/0022-2496(77)90033-5

Saaty, T. (1980). *The Analytic Hierarchy Process*. New York: McGraw-Hill.

Saaty, T. L. (2008). Decision making with the analytic hierarchy process. *International Journal of Services Sciences*, *1*(1), 83–98. doi:10.1504/IJSSCI.2008.017590

Sarkar, S., Kanungo, D. P., & Sharma, S. (2015). Landslide hazard assessment in the upper Alaknanda valley of Indian Himalayas. *Geomatics Natural Hazards and Risk*, *6*(4), 308–325. doi:10.1080/19475705.2013.847501

Van Westen, C. J. (1993). Application of Geographic Information Systems to landslide hazard zonation. *ITC Publication*, *15*, 245.

Wachal, D. J., & Hudak, P. F. (2000). Mapping landslide susceptibility in Travis County, Texas, USA. *GeoJournal*, *51*(3), 245–253. doi:10.1023/A:1017524604463

Chapter 8
Land Use – Terrain Correlations in the Piedmont Tract of Eastern India:
A Case Study of the Dulung River Basin

Ashis Sarkar
Chandernagore Government College, India

Priyank Pravin Patel
Presidency University, India

ABSTRACT

The Dulung River flows across West Bengal and Jharkhand in India. The geographical variables present within the basin area are categorised into groups like Physiographic, Morphometric and Land Use-Land Cover (LULC) attributes. These facets are mapped and overlain in a GIS environment and correlations drawn between them. Factor Scores obtained through Principal Component Analysis are further compared and correlated. The different variables are fused to obtain a comprehensive grouping of the above three facets that is reflective of the overall terrain attributes and its overlying LULC classes. Through this, within the Dulung River Basin, three broad Physiographic-Soil-Land Use Units (PSLUs) are identified, which comprise of the structural ridges and residual hills, piedmont plains and floodplains. For further insight into existing LULC-landform relations, select villages across the basin landscape are examined in detail. The relations derived help in suggesting possible land management practices in this region.

DOI: 10.4018/978-1-5225-1814-3.ch008

INTRODUCTION

Land use is a dynamic phenomenon and needs continuous monitoring (Bhattacharya, 1983). Though it is mainly controlled by climatic, pedological, socioeconomic and political conditions yet geology and geomorphology too play important roles since a physical landscape is inherently constituted by its underlying lithology and structure. Depending on their occurrence in a particular climatic region, individual landforms give rise to particular soil types and certain landform-soil associations are often more suitable for specific land uses. The terrain configuration, affecting the soil and surface drainage, is borne of the sub-surface geology, surficial landforms and their slopes. Morphometric parameters, after enumeration, could be thus used in many watershed and LULC planning applications. As such, studies which integrate landform analysis with soil, hydrology and land use and land cover (LULC) facets, should ideally form the basis of the land resource evaluation of a region (Subramanyan, 1978, 1981).

LULC study of an area provides information about its present land use status and helps form a baseline study for sustainable development of that locale (Krishna, Westinga, & Huizing, 1999). This requires current and archival LULC data of the area to track temporal change patterns (Chaurasia et. al., 1996), specifically information on existing land use and the distribution of settlements, forests, agricultural and barren lands (Christian & Stewart, 1953; Natarajan, Gajbe, & Manchanda, 1986; Shankarnarayan & Sen, 1977). LULC changes are linked to the intersection of natural and human influences on environmental regimes. Biosphere changes and biogeochemical cycle alterations are triggered by heterogeneous changes in land use and manifested further through their perpetuation (Turner, 1995).

Drainage basins have always had a special relevance in geomorphology (Doornkamp & King, 1971; Strahler, 1957) and watersheds are widely adopted in most countries as natural units of ecosystem planning and development (FAO, 1977, 1987; Honore, 1999; Moore, Grayson, & Ladson, 1977). The watershed is a geo-hydrological unit of area drained to a common point and considered as an ideal unit for analysis and management of natural resources and environmental planning in any ridge-to-valley treatment, with its appropriateness in regional planning quite evident (Easter, Stow, & Jenson, 1986; Gregory & Walling, 1978; Saha & Barrow, 1981). In India too, many developmental programmes are based on watershed management (Dhruvanarayana, Sastry, & Patnaik, 1990). Furthermore, India's agrarian-cum-rural economy (on which depend over two hundred million farmers) is extremely sensitive to land and water management and even if the country's irrigation potential is maximised, about 50% of the net cultivated area will continue to depend on rainfall (ISRO/RRSSC, 2009). Much focus thus needs to be given to how the land and water management influences local vegetation and land use - land cover (LULC) (Pant & Roy, 1990).

BACKGROUND

Geographic Information Systems (GIS) technologies for data management, manipulation, analysis and integration have arguably had the greatest impact on the spatial information environment and management of land (Dunkerley, 1988; Kasturirangan, 1994). Remote sensing (RS) and GIS are effective tools for not only mapping and detecting changes in LULC conditions (Anderson, Hardy, Roach, & Wilmer, 1976; Roy et. al., 1991; Skidmore et. al., 1997), but for also predicting future trends. This technique provides much information on LULC parameters by considering tonal variations in the images (Congalton et. al., 1993; Roy et. al., 1985; Schardt, Schurek, & Winter, 1990). Sensors measure the nature and strength of multispectral solar energy reflected from each pixel in the scene in multi wavelengths, the objective being to associate LULC categories with each of these reflected energies (Shetty, Lakshman, Thockchom, & Rajesh, 2005). This characteristic spectral response of objects to radiation in different bands of the electromagnetic spectrum has been advantageously exploited for the identification of objects of interest (Ayyangar, Krishna Rao, & Rao, 1980). Synoptic viewing, continuous monitoring and change detection in the LULC classes are major advantages from such mapping (Jenson, 1986; Kachhwaha, 1985; Luong, 1993; Sharma, Jain, & Garg, 1984). Post-classification comparison is the commonly reported technique for classification-based change-detection (Burns & Joyce 1982; Estes, Stow, & Jenson, 1982; Likens & Maw, 1982; Rubec & Thie, 1978; Singh, 1989; Weismiller & Momin, 1977; Wickware & Howarth, 1981). Since the spectral reflectance varies temporally as well as spatially, it is essential to have ground based data for proper verification of the image-classified land surface features, which in turn enhances the computer-aided analysis techniques using satellite data (Sharma, Garg, & Verma, 1986).

'Morphometric methods' can be defined as the numerical systematization of landform elements measured from topographical maps and other elevation datasets (De Martonne, 1934). They are factual statements about the morphological attributes of a particular landform or landscape, helping to envision, evaluate and compare it with others, using a raft of cartographical representations to express enumerated land surface statistics. Thus morphometry delves into the measurement and mathematical analysis of the Earth's surface, and the shape and dimension of its landforms (Agarwal, 1998; Reddy, Maji, & Gajbhiye, 2002). The main purpose of such watershed-based morphometric analysis has been to evaluate basin terrain properties through the quantification of various surface elevations and slope facets and correlate their effect on the existing land cover and geomorphic processes (Krishnamurthy, Srinivas, Jayaram, & Chandrasekar, 1996; Kumar, Kumar, Lohani et al., 2000; Nag & Chakraborty, 2003; Nooka Ratnam, Srivastava, Venkateswara, Amminedu et al., 2005; Reddy, Maji, & Gajbhiye, 2002, 2004; Vijith & Satheesh, 2006). Its eventual goal is terrain characterization and the delineation of geological and landform units (Reddy & Maji, 2003), leading to terrain modelling.

RS-GIS has proved to be an effective tool in determining quantitative descriptions of landscapes (Biswas, Sudhakar, & Desai, 1999; Reddy et. al., 2002), and currently, Digital Elevation Model (DEM) analysis is the most analytical cartographic approach to represent topography by the computer manipulation of terrain heights (Maune, 2001; Tobler, 2000). High resolution remotely sensed data coupled with topographical data analysis procedures have made satellite data based morphometric analysis a highly effective tool for analysing natural resources (Srinivasan, 1988). GIS applications are thus widely used in geomorphology and hydrology (Wilson & Gallant, 2000), providing a powerful mechanism for upgrading, monitoring and integrating environmental data into the terrain database (Jain, Seth, & Nema, 1995).

The literature on utility of RS-GIS in LULC extraction and morphometric quantification is therefore a burgeoning one. Delineation of agricultural fields, fallow lands and crop differentiation through electromagnetic spectrum separation have been done by Ayyanagar, 1978; Kumar and Silva, 1974; LARS, 1938; Marshall, Thomson, & Kriegler, 1969; Nalepka, 1970; Singh et. al., 1983 and Sarma, Krishna, Malini, & Rao, 2001, among others. LULC change mapping too forms an important part of studies done in India, among others, by Jaiswal et al. (1999) in Madhya Pradesh and Brahmabhatt et al. (2000) in Gujarat. Mahajan et al. (2001), Chauhan et al. (2003) and Joshi and Gairola (2004), among others, have used RS-GIS techniques for LULC mapping in hilly watersheds. Thus it may be surmised that RS-GIS enhance the gathering, processing and presentation of data for environmental management (Carpenter, 1987). In this vein, the Department of Space (DoS), Government of India (GoI) has initiated thematic map generation and derivation of action plans at 1: 50,000 scale using RS-GIS, in 175 districts across the country under the 'Integrated Mission for Sustainable Development (IMSD)' project for planning sustainable land use practices and initiation of soil and water conservation measures depending on local land conditions and needs. (Jeyaram, Mohabey, & Krishnamurthy, 1996).

In respect of the area under focus in this particular study, Gupta and Chakraborty (1967) did pioneering research on the quantitative geomorphology of the Dulung River Basin. Mukhopadhyay (1980) did similarly for the Subarnarekha River Basin, of which the Dulung is a part. Roy (2010) has recently utilised hydrologic modelling in the upper Dulung Basin for site determination of water harvesting structures. Publications on various facets of the Dulung Basin have been done previously by the authors (e.g. Patel and Dasgupta, 2009; Patel and Sarkar, 2009, 2010; Sarkar and Patel, 2009, 2011, 2012).

THE STUDY AREA

The Dulung River Basin lies between 22⁰10′47″N - 22⁰37′4″N and 86⁰37′53″E - 87⁰08′58″E on the eastern fringe of the Chotonagpur plateau encompassing an area of 1176.38 sq.km. See Figure 1. The Dulung River is a left bank tributary of the Subarnarekha River.

Figure 1. Location of the Dulung river basin

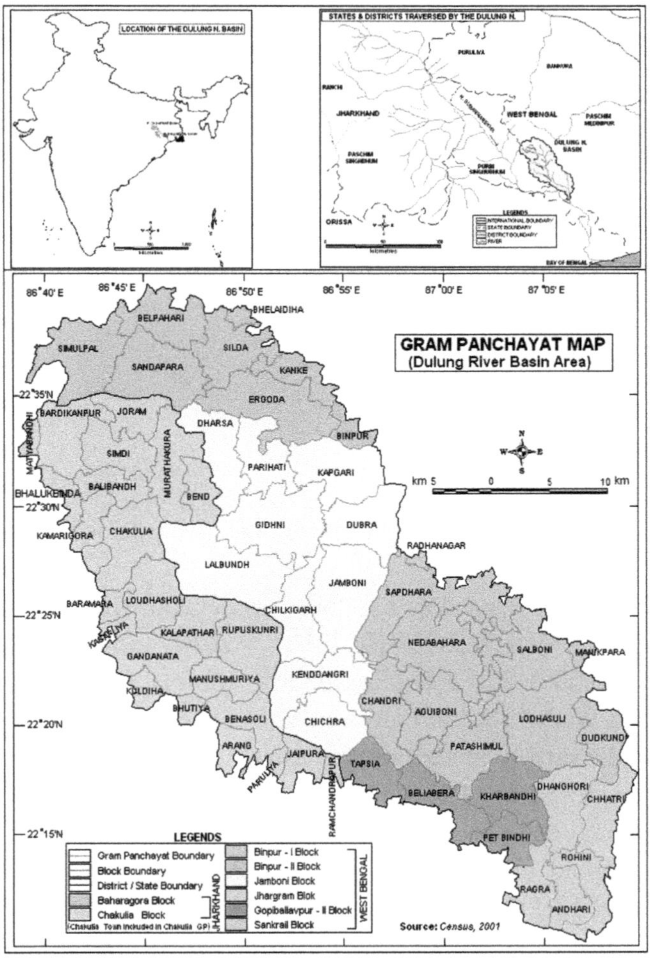

It originates from a slightly elevated area, close to Chakulia (Purbi Singhbhum district) in Jharkhand and flows southeast into the Subarnarekha near Rohini (Paschim Medinipur district) in West Bengal. Within Purbi Singhbhum, the basin covers parts of Chakulia and Baharagora blocks along the eastern fringe of the district. Within Paschim Medinipur, it covers parts of Binpur-I, Binpur-II, Jhargram, Gopiballavpur-II and Sankrail blocks while covering Jamboni block fully - all adjacent to the western margin of the district. A total of 1235 villages and one town, Chakulia, lie wholly or partly within the basin area constituting 61 gram panchayats (GPs) spread across the above 8 blocks.

OBJECTIVES

The objectives of this study are as follows:

- To spatially analyse the morphometric attributes of the Dulung basin.
- To study the LULC patterns and temporal changes therein and produce an updated LULC map for the basin.
- To classification and collate terrain and LULC information via GIS operations for distinguishing distinct land units (viz., mapping, measuring, overlaying, surface analysis, i.e., DEM, analysis).
- To identify and analyze the impact of the terrain on LULC facets, recurring basin hazards within the identified land units and outline possible mitigative measures.

METHODOLOGY

The methodology adopted to meet the above objectives are as follows:

- Planimetric map database has been generated from Survey of India Topographical Map Nos. 73J/10, 73J/11, 73J/14, 73J/15, 73J/16, 73N/3 and 73N/4 (1: 50,000). Advanced Spaceborne Thermal Emission and Reflection Radiometer Global Digital Elevation Model (ASTER GDEM) data has been used to make surface models of the basin.
- Landsat images (MSS, TM and ETM+ from United States Geological Survey Earthexplorer) have been digitally classified, enhanced and interpreted for LULC mapping along with high resolution images from Google Earth. Village level LULC data has been obtained from Census of India publications and Land and Land Reforms Department, Government of West Bengal.

Extensive field surveys have been done with a GPS for necessary ground-truthing of satellite images and land use surveys using mouza maps.

- Surface lithology was mapped from Geological Survey of India (GSI) Jamshedpur Quadrangle 73J (scale 1:250,000). The basin soil cover map was prepared by incorporating information from the National Bureau of Soil Survey and Land Use Planning (NBSS & LUP), All India Soil Survey and Land Use Planning (AIS & LUS) Reports and soil maps of Paschim Medinipur district, West Bengal and Purbi Singhbhum district, Jharkhand as prepared by National Atlas and Thematic Mapping Organisation (NATMO) and National Informatics Centre, Medinipur. Precipitation data (monthly means of overall time period 1901-2000) have been gathered from the Royal Netherlands Meteorological Institute (KNMI) for a number of stations in and around the basin along with the high-resolution (10′) global climate grids available through the Climatic Research Unit and the Tyndall Centre, University of East Anglia.
- GIS softwares have been used for the generation, integration and management of the spatial database or geo-database from the various maps, images, and field surveys.

BASIN ATTRIBUTES

Basin Relief

The study area presents a rolling topography in its northwestern part, which gradually changes to gently undulating in the central part and becomes almost flat in the south (Figure 2a).

The general appearance of the basin is of a large, open, well-cultivated plain. Towards the north and west, undulations appear, with ridges covered by thickets of *sal* trees and scrub, while the intervening depressions produce rich crops of rice. The western boundary is marked by ranges of the Chotonagpur Hills, where forests with large trees predominate.

The highest elevation of 365m above MSL is recorded near the vicinity of Chandakata Pahar ($86^0 39'04''$E, $22^0 33'43''$N) in Chakulia Block of Purbi Singhbhum District, Jharkhand in the extreme northwestern fringe of the basin area. The land falls steeply away from this northwestern fringe, with maximum elevations of just 120m and 160m above MSL being recorded along the remaining western and northern portion of the basin respectively. The elevation along the eastern fringe of the basin ranges between 120m above MSL in the northeast to about 50m above MSL in the extreme eastern portion. The eastern side of the basin is less rugged which large

Figure 2. Elevations across the Dulung river basin

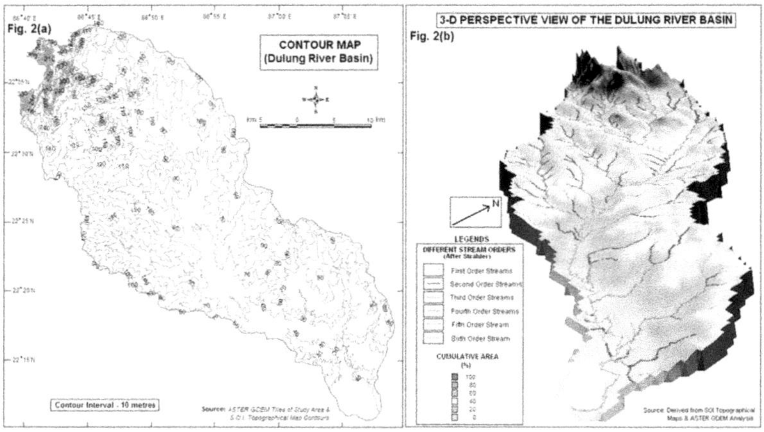

tracts of gently rolling plains grading up to from low residual hills, whose summits barely reach above 100m. The lowest elevation of 22m above MSL is recorded at the Dulung's confluence with the Subarnarekha River (87⁰04′28″E, 22⁰09′30″N) just south of Rohini town in Sankrail Block of Paschim Medinipur District, West Bengal. Thus, the relative relief of the basin is 343m.

The 3-D view of the basin (Figure 2b) clearly enunciates the above. The northern fringes of the basin are upstanding areas with a number of steeply rising hills, down whose southern flanks drains the Dulung and its upper catchment tributaries. The elevation difference between the eastern and western flanks of the basin too is apparent with the western side having a number of steep ridges radiating out from the northern plateau fringe. The eastern flank has rather more denuded and subdued water divides separating the major left bank tributaries of the Dulung, none of whose elevations are higher than their western counterparts.

Selective surface profiles have been stacked together in a series (Figure 3) keeping intact the perspective of view to show the elevation changes over the area (from Sarkar and Patel, 2011). Profiles P05 and P11 taken across the upper basin crystalline complex shows the drastic increase in height to the west in the northern part of the basin. Profile P15 cuts across the transition zone between the northern complex and the gentle, open valleys to the south.

The western half of the profile is still significantly more upstanding than its eastern counterpart but the difference is much lesser than before. To the east the narrow plain of the Dulung is spied. Profiles P20, P25, P30 and P34 traverse across the central Dulung Basin. Sections of the western ridges are prominent while portions of the lower eastern ridges and residual hills show up in these profiles. The most significant feature is the development of the broad central valley floor over

Figure 3. Serial of selected surface profiles across the Dulung river basin

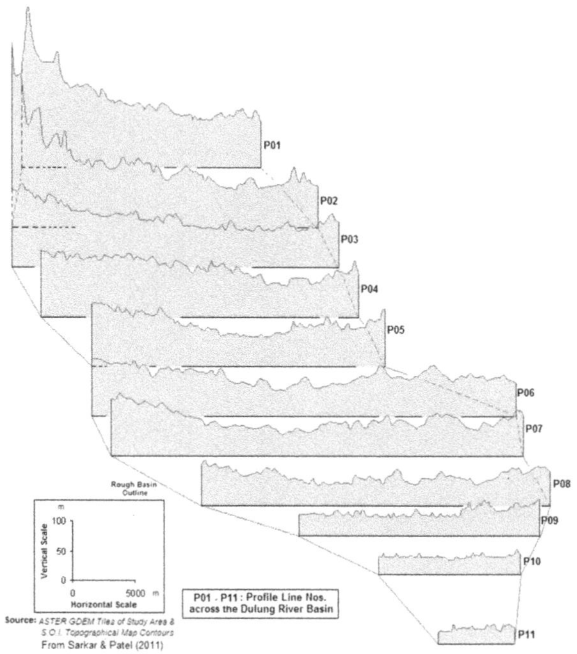

which the Dulung meanders in this stretch. Profiles P39 and P42 are taken across the lower stretch of the Dulung's course. The valley floor is much broader here and constitutes much of the profile. The last two profiles P46 and P53 are taken near the Dulung's mouth. The basin has narrowed considerably while the Dulung's valley floor has broadened and now constitutes the entire profile. It is quite an extensive floodplain with the few very minor upstanding portions often being a result of localised deposition. The narrowing of the profile lines is also indicative of the severe flooding the lower portion of the Dulung is subjected to, since all the waters gathered upstream, are channelised into this constricted, low floodplain.

Basin Lithology

Location-wise, the Dulung Basin lies on the eastern fringe of the Chotonagpur Plateau, just to the southeast of the Singhbhum Shear Zone. The area is characterized by diverse geological units ranging from hard, consolidated rocks of Precambrian age to unconsolidated alluvium of recent to sub-recent age (Figure 4a).

The area mostly consists of Cenozioc cover sediments with some basic and ultrabasic volcanics formed as intrusives in the northwestern part of the basin.

Figure 4. Distribution lithological, pedological and climatic elements across the Dulung river basin

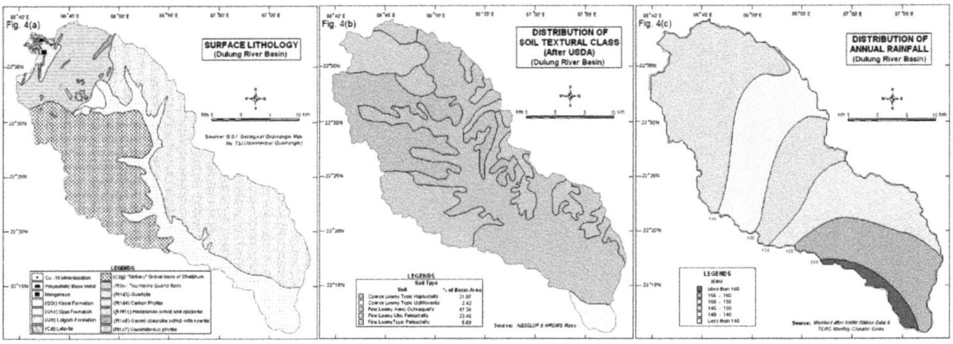

The older alluvium is often covered by laterites consisting of brown to reddish-brown sand, sandy clays and gravel, while younger alluvium occurs along river courses in the form of bar deposits. The Dulung River flows for the greater part of its course over Quaternary deposits, consisting mainly of the Sijua formation. These are the oldest alluvial terrace sediments developed above the Lalgarh formation that occupies the basin area, left of the main channel, and are comprised of hard clay and silt, impregnated with caliche concretions (GSI, 2007).

The laterite rocks cover a large area, but in the majority of cases, the only variety visible at the surface is a gravelly, pisolitic and nodular rock. Many places may be seen in pits along the roadside, where this gravelly laterite is extracted as road metal (*Moram*), for which it is admirably adapted. Often the rock becomes conglomeratic, pebbles of quartz and rounded fragments of other rocks being embedded in it. At times these pebbles are coated with oxides of iron, as in other ferruginous conglomerates.

The northwest portion is geologically more complex. The typical rock types are phyllite, mica schist, hornblende schist and quartzites. This area forms the highest elevated part and is chronically affected by droughts. The schists and phyllites are prone to getting weathered easily and consequently more easily amenable to stream incision. The quartzite bands present are resistant and form low, elongated ridges, with short streams draining off either face of these ridges. The area, especially its hard rock part, has undergone structural deformations as manifested by the presence of folds, faults, fractures and joints (GSI, 2007).

Basin Soil Cover

While red sandy soil covers the hard rock terrain, the wider pediment to the east and the floodplains to the south are covered by older alluvium soils (Figure 4b). The undulations in between are covered by lateritic soils to the east developed over the Lalgargh formation and red gravelly soils in the west developed over the Dhalbhum gravel beds. Jhargram block is composed of sandy loam to clayey loam soil while the rest of the blocks have sandy soil. Gopiballavpur-II block has 10% sandy soil and 40% of sandy loam soil (GSI, 2007).

The red soil group is part of the lateritic soil and covers two-thirds of the study area. The lateritic soil is light textured, porous and free-draining (i.e., rills and gullies can develop extensively), acidic (pH 4.8-6.5) and poor in organic content, requiring heavy application of organic manure. It is poor in phosphate content and deficient in lime, with the predominant clay mineral group being kaolinite and the colloidal fraction containing substantial quantities of sesquioxide clay. Hard ferruginous concretions are on the surface as well as in the ferruginous profiles. As these lateritic soils have formed at lower levels, they have a darker colour, probably due to a greater accumulation of humus. Their denudation has formed a sizeable pediplain in the northern portion of the basin which is subject to pronounced soil erosion and pockets of badland formation.

Basin Climate

The basin area is characterized by a sub-humid to semi-arid type of climate. The area generally experiences three seasons- summer, monsoon and winter. Winter sets in from November end till early March. December and January are the coldest months. The area receives a good amount of rainfall ranging from 130cm-170cm annually. The rainfall amount generally increases towards the eastern and southeastern part of the basin (Figure 4c). The source of this rainfall is almost entirely an offshoot of the Bay of Bengal Branch of the Southwest Summer Monsoon, occurring from June to October. Some isolated showers are received during the *Kalbaisakhi* period of March - April. Rainfall during the other months is scanty.

Basin Morphometric Attributes

Altimetric Frequency Histogram

Altimetric frequency analysis demonstrates the statistical count of the number of pixels within each altitude range. A total of 1333890 ASTER GDEM pixels, each of area 900 sq.m, have been grouped into 35 altitudinal classes of 10m interval each, with the initial class being 20m-30m and the final 360m-370m (Figure 5).

The distribution is positively skewed and multimodal, with principal mode between 80m-90m.This histogram reinforces that the basin is largely composed of

Figure 5. Altimetric frequency histogram for the Dulung river basin

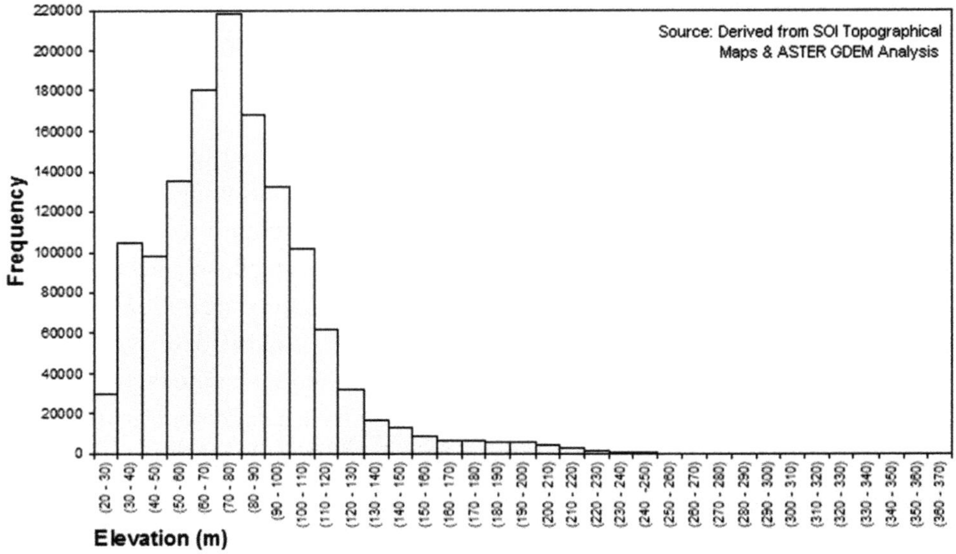

gentle lowland plains with minimal relief variations, abutting onto dissected ridges and residual hills.

Relative Relief

The relative relief (Smith, 1935) zones of interval 20m (Figure 6a) reveal that the two lowermost ranges, viz., (less than 20m) and (20m-40m) cover over 93% of the basin area. They comprise the entire lower and middle basin zones and the lion's share of the upper catchment as well. This proves that the basin has very low amplitude of relief (maximum - minimum elevation), with the topography being comprised primarily of gentle floodplains or dissected surfaces and that relief changes are gradual. The isoline trends highlight the major valley floors. Higher zones are concentrated in the northwest portion of the basin. This is the zone of steeply rising hills alternating with deep, incised valleys and dissected ridges.

Average Slope

The average slope (Wentworth, 1930) zones reveal that almost the entire basin lies in extremely gentle country (Figure 6b). The lowest average slope range (less than 2°) comprises 89% of the basin area covering the entire lower, middle and most of upper basin areas. The northern basin area comprises of slivers of higher average

Figure 6. Maps of selective morphometric attributes for the Dulung river basin

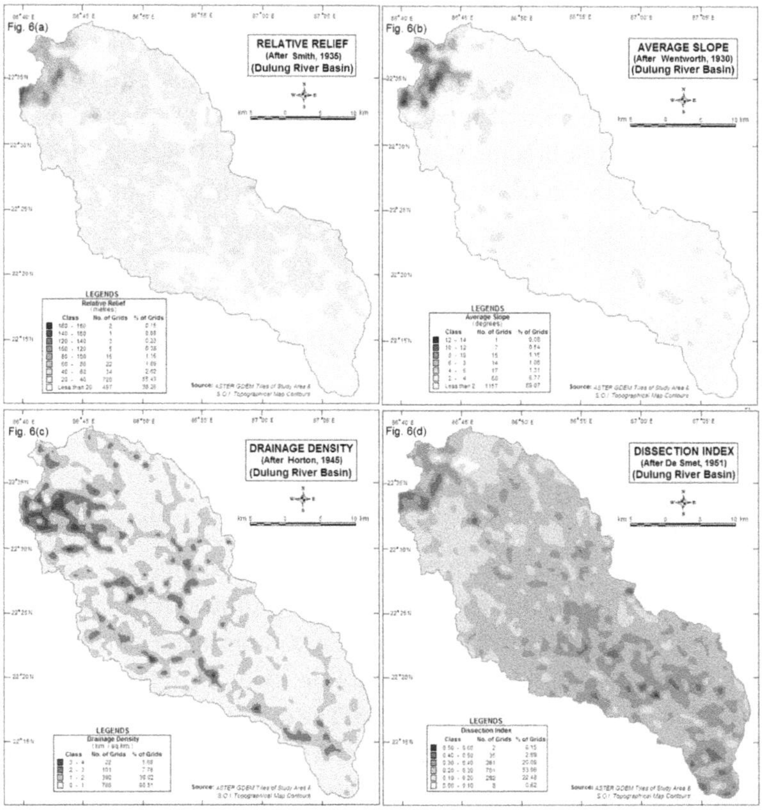

slope ranges. These ranges mark out the NE-SW trending ridges in this area and show up the higher hills therein clearly.

Drainage Density

The drainage density (length of stream per unit area - Horton, 1945) zones reveals that the greatest values (2-4 km/sq.km) are observed in the northwestern portion of the basin (Figure 6c). The rest of the basin (above 90%) has a drainage density of (0-2 km/sq.km). Small pockets of higher values are seen along the major stream courses due to confluences. Values of the right flank are more than the left flank.

Dissection Index

The interaction of the terrain matrix with the streams flowing on it sculpts the landscape and various parameters express the degree and intensity of these actions.

The dissection index (De Smet, 1951) ranges (Figure 6d) reveal that the middle basin area is relatively less dissected compared to the upper and lower portions. The northwestern part exhibits proper higher values due to greater stream incision creating high relative relief while the lower basin reaches have higher values due to a statistical anomaly wherein the amplitude of relief and the maximum grid elevation are almost similar. Following Kumar and Pandey's (1982) classification, about 76% of the basin lies in the medium dissected range while another 20% lies within the much dissected range. These two zones predominate.

MERGING TERRAIN AND PHYSIOGRAPHIC FACETS

Statistical Analysis of Morphometric Attributes

Descriptive Measures

The morphometric parameters derived from 1299 one sq.km grids across the basin comprise the variables- maximum elevation (x_1), minimum elevation (x_2), relative relief (x_3), average slope (x_4), stream frequency (x_5), drainage density (x_6), drainage texture (x_7), ruggedness number (x_8), dissection index (x_9), coefficient of relative massiveness (x_{10}), and roughness index (x_{11}). The distribution of x_4, x_5, x_6, x_7, x_8, and x_{11} is characterized by very high variability, while that of x_1, x_2 and x_3 by moderate variability, and of x_9 and x_{10} by low variability (Table 1).

Distribution is leptokurtic in cases of x_3, x_4, x_7, x_8 and x_{11}, mesokurtic in cases of x_1, x_2 and x_5 and platykurtic in cases of x_6, x_9 and x_{10}. Distribution of all the variables excepting x_{10} is positively skewed. The correlation matrix brings out the nature of statistical relationships between and among the variables (Table 2), and bears out the strongly positive correlation between the above "relief variables".

Table 1. Descriptive statistics of the morphometric variables

Parameters	x_1	x_2	x_3	x_4	x_5	x_6	x_7	x_8	x_9	x_{10}	x_{11}
Mean	95.9	80.7	24.3	1.31	1.51	0.80	2.36	0.02	0.25	0.36	1.84
Standard Deviation	41.7	34.6	15.6	1.51	1.68	0.83	4.26	0.03	0.07	0.11	2.10
Kurtosis	7.6	4.9	25.0	18.72	4.75	0.06	14.93	28.63	0.26	-0.25	17.78
Skewness	2.0	1.5	4.1	3.89	1.80	0.88	3.42	4.29	0.38	-0.28	3.68
Lowest	27.0	22.2	5.0	0.08	0	0	0	0	0.08	0.03	0
Highest	365.0	325.9	173.0	12.43	13.00	3.62	35.28	0.33	0.52	0.70	17.58

Source: Computed by the authors from the prepared topographic database

Table 2. Correlation matrix of the morphometric variables

	x_1	x_2	x_3	x_4	x_5	x_6	x_7	x_8	x_9	x_{10}	x_{11}
x_1	1.00										
x_2	*.98*	1.00									
x_3	*.83*	*.73*	1.00								
x_4	*.65*	*.59*	*.70*	1.00							
x_5	.21	.21	.13	.28	1.00						
x_6	.08	.07	.01	.19	*.83*	1.00					
x_7	.18	.18	.10	.24	*.91*	*.82*	1.00				
x_8	*.47*	.41	*.52*	*.58*	*.69*	*.71*	*.68*	1.00			
x_9	-.008	-.14	*.48*	.22	-.11	-.13	-.11	.15	1.00		
x_{10}	.22	.29	.16	.07	-.17	-.29	-.19	-.10	.05	1.00	
x_{11}	*.64*	*.59*	*.68*	*.97*	.28	.19	.24	*.57*	.20	.07	1.00

Source: Computed by the authors from the prepared topographic database

Variables x_1 and x_2 are strongly correlated due to relative relief being quite low over the basin area. As expected x_3 and x4 are strongly correlated and together, these two further correlate strongly with x_8 and x_{11}. The "drainage variables" x_5, x_6 and x_7 correlate strongly among themselves while also correlating strongly with x_8. Thus two distinct classes of morphometric variables are observed with the ruggedness index (x_8) being the linkage between them.

Factor Analysis

Factor analysis has been used to express more simply the relationships between and among variables. It uses a 'factor loadings' matrix, thus distinguishing certain 'basic' or 'abstract' variables, and expresses numerically the level of relationship between these and the original variables. It also provides a simplified data matrix, known as the 'factor score' (or weightings) matrix. If the different morphometric variables are taken to be vectors, then the angle between any two vectors reflects the degree of similarity between the variables concerned. Thus, two completely dissimilar variables would be represented by two vectors pointing in opposite directions, while identical variables would be represented by two vectors superimposed on one another. A reduction of the data to a few significant factors was achieved through the extraction and rotation of the factor-loadings matrix. The relationships between the rotated factors and the original variables are listed. These relationships are stated in terms of the percentage contributed to the variance of each variable by each rotated factor (Table 3). Thus, four factors have emerged significant explain-

Table 3. Factor analysis of the morphometric variables (total variance explained)

Compo-nents	Initial Eigenvalues			Extraction: Sums of Squared Loadings			Rotation: Sums of Squared Loadings		
	Total	% of Variance	Cumulative %	Total	% of Variance	Cumulative %	Total	% of Variance	Cumulative %
x_1	4.927	44.791	44.791	4.927	44.791	44.791	4.139	37.626	37.626
x_2	2.807	25.515	70.306	2.807	25.515	70.306	3.318	30.167	67.793
x_3	1.307	11.885	82.191	1.307	11.885	82.191	1.327	12.067	79.860
x_4	0.828	7.528	89.720	.828	7.528	89.720	1.085	9.860	89.720
x_5	0.595	5.413	95.133						
x_6	0.259	2.359	97.492						
x_7	0.126	1.146	98.638						
x_8	0.086	0.779	99.417						
x_9	0.035	0.317	99.735						
x_{10}	0.028	0.257	99.992						
x_{11}	0.001	0.008	100.000						

Extraction Method: Principal Component Analysis. Source: Computed by the authors from the prepared topographic database

ing more than 89% of the total variation. Initial loadings on Factor-1 is found to be significantly associated with the variances of $\{x_4, x_8, x_{11}, x_1, x_3$ and $x_2\}$, Factor-2 with $\{x_6, x_7$ and $x_8\}$, Factor-3 with $\{x_9\}$, and Factor-4 with $\{x_{10}\}$ (Table 4). After Varimax rotation with Kaiser normalisation the loadings on $\{x_1, x_2, x_3, x_4$ and $x_5\}$ better accounts for the variation of Factor-1, on $\{x_5, x_6, x_7$ and $x_8\}$ for the variation of Factor-2, on $\{x_9\}$ for Factor-3, and on $\{x_{10}\}$ for Factor-4.

Morphometric Classes

Factor score - 1 values have been used to identify morphometric classes or regions that are uniform with respect to the attributes of the eleven morphometric variables $\{x_1, x_2, x_3, x_4, x_5, x_6, x_7, x_8, x_9, x_{10}$ and $x_{11}\}$ (Figure 7a, Table 5).

Classification has been done with 0.5 class interval that initially produced seventeen distinct morphometric regions. Their spatial distribution values show a decrease gradually from source towards mouth of the basin. The region A mainly occupies the highest portion of the basin near its source region (4.33%) and includes nine different sub-regions (A1-A9). It includes quite *steep and moderately dissected summit areas* of the highest plateau, sharply rising *flanks of the highest hills* on upper plateau surface, moderately *dissected higher plateau surface*, moderately *dissected lower plateau surface* and relatively *steep and continuously rising flanks*

Table 4. Factor score component matrix (unrotated and rotated)

Component / Variable	Initial Component Score				Component / Variable	Rotated Component Score			
	1	2	3	4		1	2	3	4
x_4	0.836	0.277	0.170	-0.180	x_1	0.924	0.092	-0.174	0.203
x_8	0.834	-0.338	0.167	0.084	x_2	0.877	0.085	-0.312	0.276
x_{11}	0.827	0.273	0.156	-0.187	x_3	0.866	0.043	0.356	0.128
x_1	0.809	0.416	-0.315	-0.082	x_4	0.861	0.193	0.219	-0.106
x_3	0.780	0.494	0.207	0.030	x_{11}	0.854	0.190	0.203	-0.108
x_2	0.757	0.397	-0.465	-0.054	x_7	0.092	0.942	-0.067	-0.036
x_6	0.513	-0.788	0.003	0.071	x_5	0.130	0.941	-0.067	-0.021
x_7	0.587	-0.727	-0.054	0.164	x_6	0.018	0.927	-0.063	-0.160
x_5	0.618	-0.703	-0.063	0.167	x_8	0.490	0.749	0.199	-0.061
x_9	0.141	0.299	0.867	0.302	x_9	0.118	-0.082	0.964	0.035
x_{10}	0.054	0.484	-0.331	0.770	x_{10}	0.124	-0.161	0.039	0.946

Extraction Method: Principal Component Analysis. Rotation Method: Varimax with Kaiser Normalization; Rotation converged in 6 iterations. *Source: Computed by the authors from the prepared topographic database*

Figure 7. Merging terrain and morphometric units in the Dulung river basin

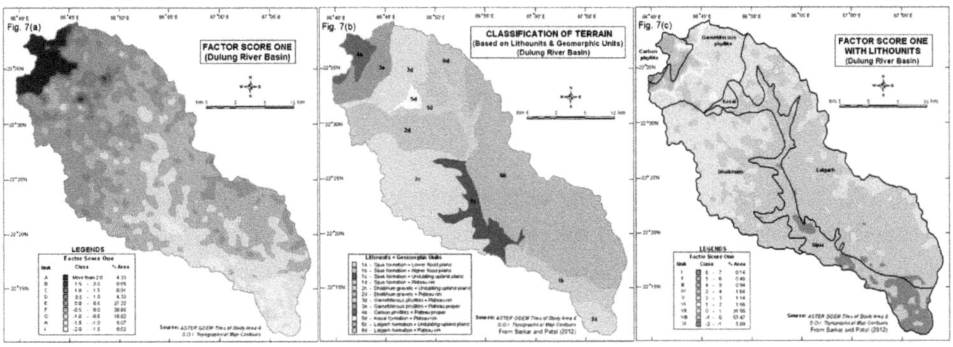

from the piedmont. It is flanked by region B that mainly comprises the *summit areas of low residual hills*. It is bounded towards southeast by the *gently rising flanks of residual hills and plateau edge* (C). The region D, covering about 4.33% of the total area is formed by *relatively steeper and dissected pediments* in the upper catchment area. It is followed by the region E, i.e. *planation surfaces with few undulations* that cover about 27.22% of the total basin area in the next lower region towards mouth. The largest proportion of area (38.65%) is the *gentle upland plains* (F) that occupy both sides away from the channel in the middle part of the basin. The *higher flood plain inundated only during the monsoon* (G) covers almost 18.82%

Table 5. Morphometric classification of the Dulung basin

Factor Score – 1	Area (sq km)	Area (%)	Description	Morphometric Classes / Regions	
6.0 - 6.5	1.61	0.14	Steep and moderately dissected summit areas of highest hills	A	A1
5.5 - 6.0	1.92	0.16	Sharply rising flanks of the highest hills on upper plateau		A2
5.0 - 5.5	3.68	0.31	- do -		A3
4.5 - 5.0	4.47	0.38	Moderately dissected upper plateau		A4
4.0 - 4.5	6.59	0.56	- do -		A5
3.5 - 4.0	9.77	0.83	Moderately dissected lower plateau		A6
3.0 - 3.5	9.52	0.81	- do -		A7
2.5 - 3.0	7.07	0.60	Steeper and continuously rising plateau rim		A8
2.0 - 2.5	6.32	0.54	- do -		A9
1.5 - 2.0	7.62	0.65	Summit areas of low residual hills	B	
1.0 - 1.5	10.75	0.91	Gently rising flanks of residual hills/plateau rim	C	
0.5 - 1.0	50.95	4.33	Relatively steeper and dissected terrain	D	
0.0 - 0.5	320.21	27.22	Undulating topography	E	
- 0.5 to 0.0	454.63	38.65	Gentle upland plains	F	
- 1.0 to - 0.5	221.41	18.82	Higher floodplain (inundated during monsoon)	G	
- 1.5 to -1.0	59.68	5.07	Lower floodplain (flooded during monsoon)	H	
- 2.0 to -1.5	0.18	0.02	Low lying areas near channel confluence	I	
	1176.38	100.00			

Source: Computed by the authors from the prepared topographic database

of the total basin area occupies the next lower region. These three regions (E, F and G) together cover about 85% of the total basin area. The *low level floodplain* region (H) covers 5.07% of the total basin area and occupies both sides of the lower course of the Dulung River. Low lying *areas near the channel confluence* is the last distinctive morphometric region, I.

Combining Lithological and Geomorphological Aspects (modified from Sarkar and Patel, 2012)

By combining the lithological units with geomorphlogical divisions, 13 terrain classes have been identified (Figure 7b, Table 6).

Each is distinguished by its own lithological matrix and geometry of the landforms produced therein. The Sijua formation includes four sub-regions, viz., 1a, 1b, 1c and 1d, respectively in the floodplains, undulating plains and plateau rim. The Dhalbhum gravels of Tertiary age have been suitably divided into two sub-regions, viz., 2c and 2d covering parts of undulating plains and plateau rims. The garnetifer-

Table 6. Terrain classes of the Dulung basin (lithology and geomorphology basis)

Terrain Classes	Area (%)	Terrain Classes	Area (%)
1a	2.26	3d	7.36
1b	5.24	3e	5.17
1c	5.16	4e	2.28
1d	4.08	5d	0.80
2c	16.75	6b	35.96
2d	9.86	6d	5.07
Method: Hierarchical & Combinatorial			

Source: Compiled by the authors from the Lithological divisions and Geomorphological divisions (from Sarkar & Patel, 2012).

ous phyllites area covers parts of the plateau proper and plateau rim divisions (3d and 3e). The carbon phyllites mainly cover the highest part of the basin with steeper slope and high dissection and ruggedness (4e). The Kasai formation is a single terrain unit distinctive with the physiographic character of the plateau rim (5d). The Lalgarh formation comprises two distinctive sub-regions, viz., 6d, and 6b respectively occupying parts of plateau rim, undulating upland plains and upper / higher floodplains.

Combining on Lithological and Morphometric Units (modified from Sarkar and Patel, 2012)

The previously identified seventeen morphometric divisions had been identified previously have been reduced to nine using a 1.0 class interval. With six lithological units superimposed onto this, 24 terrain units classes of varying dimensions, morphometric and morphologic attributes are demarcated (Figure 7c). Again, only three morphometric divisions (viz., A, B, and C) cover the bulk of the basin, i.e., more than 93% area (Table 7).

There are three terrain classes in the Sijua formation, viz., the lowest reach of the Dulung comprising the low lying confluence area and the lower flood plain, which is annually inundated during monsoon (1A), gentle upper floodplain in the middle reach (1B) and undulating plains (1C). The Dhalbhum gravel region in the right half of the Dulung River comprises two terrain classes, i.e., 2B and 2C. The garnetiferous phyllites is the most complex region with eight of the nine morphometric units, viz., 3B, 3C, 3D (lower piedmont of the dissected plateau rims), 3E (upper piedmont of the dissected plateau rims), 3F (dissected plateau), 3G (dissected plateau proper), 3H (well dissected upper plateau surface), and 3I (dissected

Table 7. Terrain classes of the Dulung basin (lithology and morphometry basis)

Factor Score – 1	Area (sq.km)	Area (%)	Description	Morphometric Classes / Regions
- 2.0 to -1.0	59.86	5.09	Lower floodplain (flooded during monsoon); Low lying areas near channel confluence	A
- 1.0 to 0	676.04	57.47	Gentle upland plains; Higher floodplain (inundated during monsoon)	B
0 - 1.0	371.16	31.55	Relatively steeper and dissected terrain; Undulating topography.	C
1.0 - 2.0	18.37	1.56	Summit areas of low residual hills; Gently rising flanks of residual hills/plateau rim.	D
2.0 - 3.0	13.39	1.14	Steeper and continuously rising plateau rim.	E
3.0 - 4.0	19.29	1.64	Moderately dissected lower plateau.	F
4.0 - 5.0	11.06	0.94	Moderately dissected upper plateau.	G
Above 5.0	7.21	0.61	Steep and moderately dissected summit areas of highest hills; Sharply rising flanks of the highest hills on upper plateau.	H

Source: From Sarkar and Patel (2012)

summits of the highest plateau). The carbon phyllites area comprise six distinctive terrain units in the most geologically complex catchment region of the Dulung, viz., 4D, 4E, 4F, 4G, 4H, and 4I. The Kasai formation covers a small area with two terrain units, viz., 5B and 5C. The Lalgarh formation in the left side of the middle reach is one of the largest geological unit with three distinctive terrain classes, viz., 6A, 6B, and 6C.

BASIN LULC COMPONENTS AND DEMARCATED PSLUs

The LULC categories derived from the different Landsat images have been extracted and compared (Figure 8a, 8b, 8c). The amount of land under crops and fallow land varies, due to the different imagery dates.

The 1973 images taken during October thus records a greater quantity of cropland and lesser fallow. The other datasets taken during November and January, during the cold, dry season, show higher amounts of fallow land with allied decrease in the land under crops. At this time of the year, except the southern portion of the basin, fields are barren due to the lack of water availability. The general LULC pattern shows the basin to be primarily rural in character with more than 56% of the basin area comprising agricultural land, viz., single cropped, single crop *boro* or more than one crop (Table 8, Figure 8d).

Due to poor availability of irrigation facility and scarcity of rainfall, almost 50% of the basin consists of just single cropped land. More than one crop land is seen mostly in parts of Sankrail block due to the year round water availability from the

Figure 8. Temporal Changes in LULC Components in the Dulung River Basin

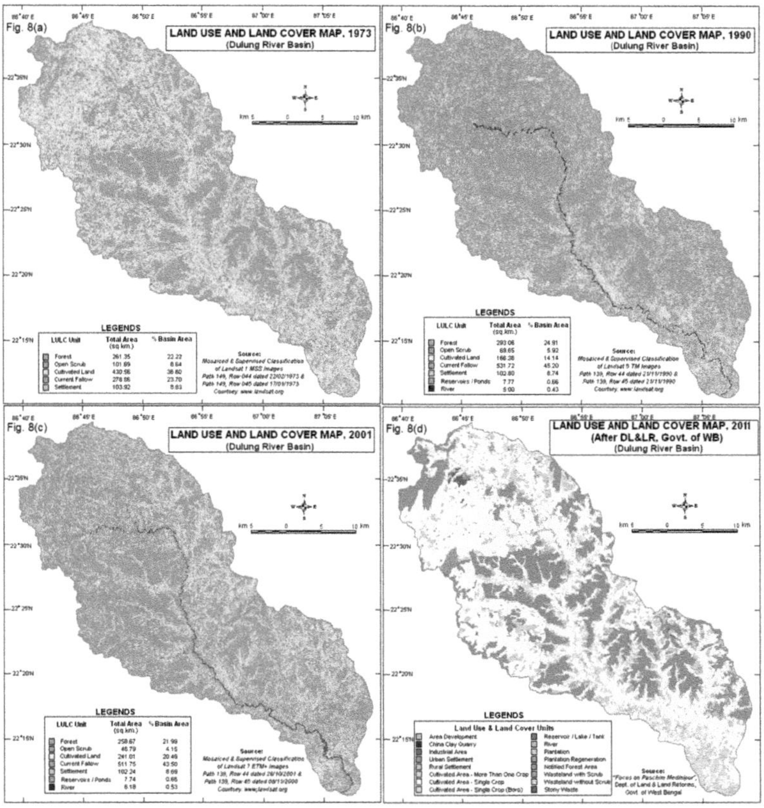

*For a more accurate representation of this figure, please see the electronic version.

Table 8. General land use and land cover composition of the Dulung river basin

LULC Class	Area (sq.km.)	Area (%)	LULC Class	Area (sq.km)	Area (%)
Area Development	0.05	0.004	River	5.64	0.47
China Clay Quarry	0.05	0.005	Rural Settlement	99.01	8.41
Industrial Area	0.36	0.031	Single Crop	587.64	49.95
More Than One Crop	79.93	6.794	Single Crop (*Boro*)	0.68	0.05
Notified Forest Area	290.36	24.698	Stony Waste	3.48	0.29
Plantation	52.54	4.467	Urban Settlement	5.94	0.50
Plantation Regeneration	3.75	0.319	Wasteland with Scrub	34.75	2.95
Reservior / Pond	7.67	0.652	Wasteland without Scrub	4.44	0.37

Source: On basis of maps prepared by L&LR Department, GoWB (2010) and NATMO (2006) and ground-truthing by the authors much enlarged Dulung River flowing through its lower course. Between 1973 and 1990, forest coverage increased from 22.22% to 24.91% but in 2001 it decreased to 21.99%. Open scrub jungle decreased steadily between 1973 and 2001 and became less than half by that time (8.64%→4.15%). Cultivable land includes both the cultivated land and the area under current fallow. These two together decreased marginally between 1973 and 1990 (60.30%→59.34%). However, in 2001 it increased by nearly 5% to 63.99%. The area under settlement does not show any significant changes; it varied between 8.69% and 8.83% (Figure 9).

The major LULC classes are enumerated for the different panchayats (Figure 10). The maximum portion of forest cover is present in the northwestern GPs of Binpur-II (Simulpal-Belpahari forests), Jambani (forests around Chilkigar) and Jhargram (Sapdhara, Aguiboni, Lodhasuli forests) and these represent the dominant LULC category in these GPs (nearly or more than 50%). Significant forest cover is also present in the Loudhasholi (Chakulia), Manushmuriya, Gandanata, Rupuskunri, Benasoli (Baharagora, the Dudhkunri Reserve Forest area), Gidhni (Jamboni) and Dudhkundi (Jhargram). Almost no forest cover is present in most of the GPs of Gopiballavpur-II and Sankrail blocks.

Un-irrigated lands are more than irrigated areas throughout except in the southern GPs of Sankrail and eastern Gopiballavpur-II, located along the lower course of the Dulung and thus having perennial water supply. Forests and un-irrigated lands form the largest LULC components. This testifies to the dry, semi-arid climate and the lack or failure of irrigation facilities. Most of the lands are mono-cropped, remaining fallow during the dry season. Irrigated lands comprising nearly or more than half the area are seen in Kharbandhi (Gopiballavpur-II), Ragra, Andhari, Ro-

Figure 9. Comparison of LULC categories for the Dulung river basin

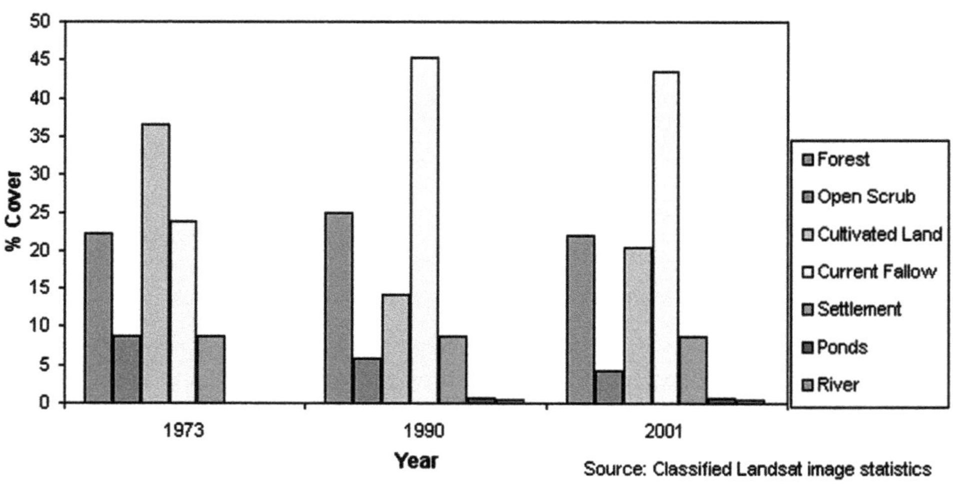

Figure 10. Land use map (Dulung river basin GPs)

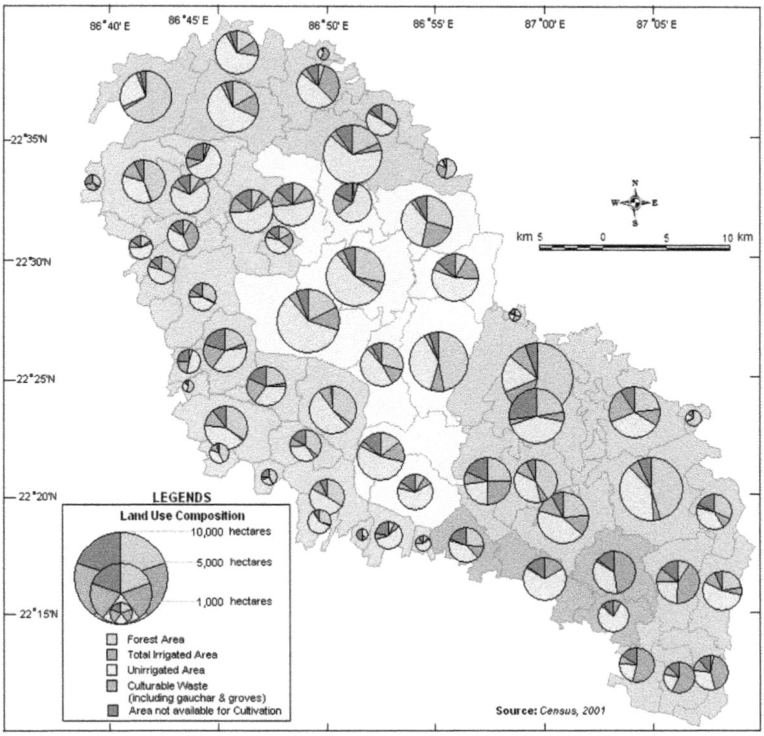

hini and Dhanghori (Sankrail) GPs, all located along the Dulung's lower course. The semi-arid nature of the area is further exhibited through the prevalence of mono-cropped land and in the indurated and leached nature of the soil profiles that have developed particularly on the higher hills and older floodplains. They are typical the *rarh* region of West Bengal. Younger, more arable soils, occupy the lower valley floors.

The most fertile croplands are obviously concentrated along the Dulung River. Other fields stretch away from the main valley, along the major tributaries and sub-tributaries, in finger-like projections of varying magnitude, often protruding into the forest areas where they have been cleared for this purpose (Figure 11). Often the small stream in its upper reaches has a small check dam.

This collects enough water to be channelled out to numerous small plots prepared subsequently on the dried streambed downstream of the check dam (Patel & Dasgupta, 2009). The streambed gets mutilated and on the ground, hardly resembles a stream at all, rather appearing as two sides of a barely undulating plain gently grading into the other. Only during the monsoon season does it assume its former nature. The dry beds can however be distinguished and delineated from high resolution

Figure 11. Human encroachment into the forests along the Dulung river's tributaries

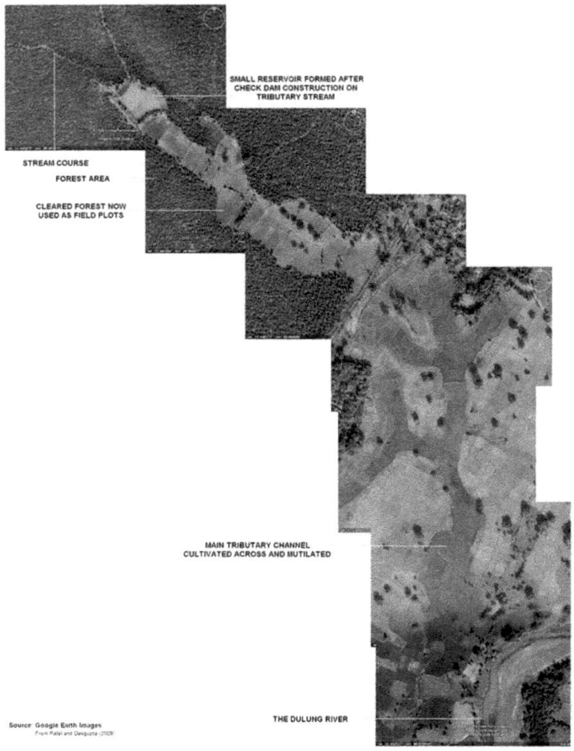

satellite images. Downstream, inundated areas are demarcated on the basis of Google Earth images and verified via field survey (Figure 12).

These tracts show up as large irregular yellow patches, i.e. crevasse splays. This residual sand, deposited over otherwise fertile arable land, greatly affects cultivation, until the sand is fully removed, a time-consuming and expensive labour-intensive process. All villages in the lower Dulung report this damage.

Based on the physiographic conditions of the area, agricultural practices have been classified into two categories- terraced cultivation and valley cultivation. Terraced cultivation is confined on the higher and intermediate slopes and is mostly rain fed at higher elevations and irrigated in low-lying areas (Table 9).

Most of the area is mono-crop land allowing harvesting of a single crop (rice, sugarcane or wheat). The southern basin area has large patches of more than one crop land primarily due to year round water availability. Mixed cropping is prevalent with paddy and sugarcane being the chief crops. Few patches of single *boro* crop land are present, these areas are usually waterlogged areas.

Figure 12. Flood affected areas in the lower course of the Dulung river

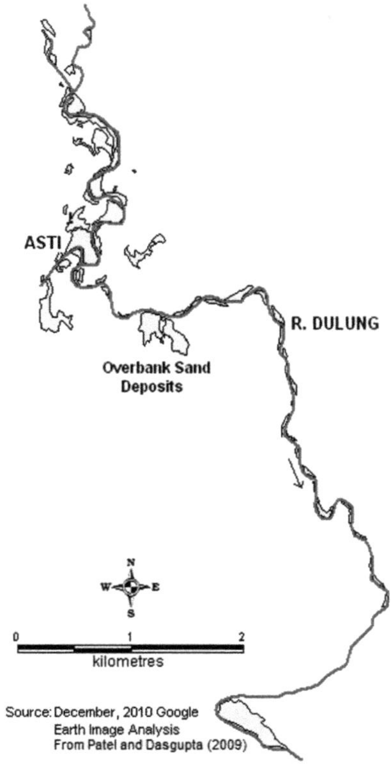

Table 9. Existent cropping pattern as per land situation

Block	Low level land	Intermediate slopes	Higher land
Jhargram	*Aus* / *Aman-Boro* / Khesari	*Aman* / *Boro* / Mustard / Wheat	*Aus* / Arhar / Maize / *Kharif* vegetable / Mustard
Jamboni	*Aus* / *Boro* / Khesari	*Aman* / Mustard / *Boro* / Wheat	*Aus* / *Kharif* vegetable / *Kharif* pulses / Mustard
Binpur-I	*Aus* / *Boro* / Khesari	*Aman* / Potato / Wheat	*Aus* / Potato / Wheat / Mustard
Binpur-II	*Aus* / *Aman-Boro* / Rabi vegetable	*Aus* / *Aman* / Rabi vegetable / Wheat	*Aus* / Maize / *Kharif* vegetable / Groundnut
Sankrail	*Aus* / *Boro* / Khesari	*Aman* / Mustard / *Boro* / Wheat	*Aus* / Arhar / Maize / *Kharif* vegetable / Mustard
Gopiballavpur-II	*Aman* / *Boro* / Watermelon / Khesari	*Aman* / Rabi pulses / mustard / watermelon	*Aus* / Arhar / Mustard / Rabi kalai
Chakulia	*Aus* / *Boro* / Khesari	*Aman* / *Boro* / Wheat	*Aus* / *Kharif* vegetable / *Kharif* pulses
Baharagora	*Aman* / *Boro*	*Aman* / Rabi pulses /	*Aus* / Arhar

Source: CGWB Technical Report D-187

LULC of Sample Villages

Plot-level distribution of the LULC classes, their association with local landforms and their variation across the basin, five villages have been surveyed in detail using cadastral maps (Table 10, Figure 13a).

Dulungdiha

Located in Simulpal gram panchayat of Binpur-II Block, Paschim Medinipur District, West Bengal, at the extreme northwest corner of the basin (Figure 13b). It is

Table 10. Surveyed villages across the Dulung basin

Panchayat	Block	J.L. No.	Mouza	Area (hectares)	Location
Simulpal	Binpur – II	100	Dulungdiha	111.11	Source of River Dulung at head of the Basin; quartzite-phyllite zone of dissected ridges
Simulpal	Binpur – II	148	Basudebpur	66.66	Quartzite-phyllite zone in upper basin; Residual hills with intervening plains
Dharsa	Jamboni	36	Nagdi	89.44	Pediment area in upper central basin; Confluence of Dulung & Kupan Rivers
Andhari	Sankrail	38	Asti	38.89	Upper floodplain surface; most ravaged by floods
Andhari	Sankrail	71	Ramanandapur	72.88	Lower floodplain surface; mouth of River Dulung

Source: Selected by the authors through sampling

Figure 13. LULC components of select villages in the Dulung river basin

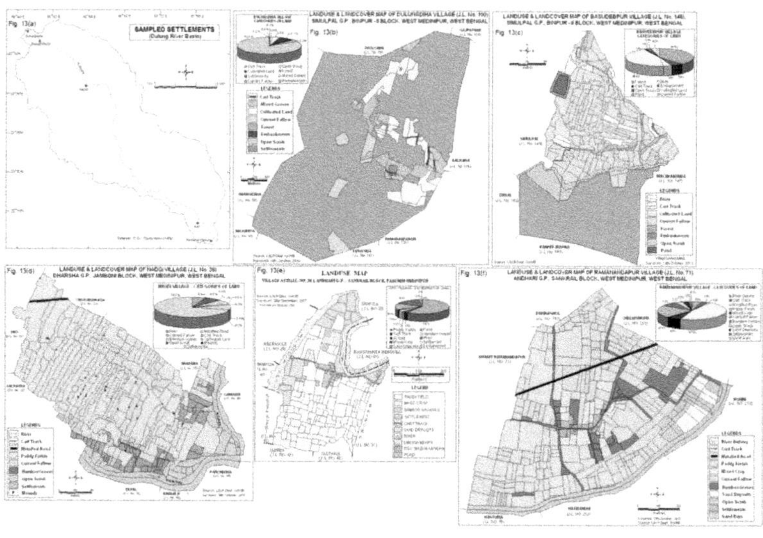

the source of the River Dulung. The village lies nestled between the linear ridges of the Chotonagpur plateau and is thus densely forested. More than four-fifths of the mouza area is covered by open mixed and dense jungles. The main channel of the upper Dulung arises from this mouza. Its channel along with those of its small tributaries provides the only cultivable patches in this mouza. Only a few huts are present, aligned along the narrow streams. Due to the scarcity of water, most of the cultivable fields remain as fallow for most of the year. Square patches, 5m x 5m, are often cut into the hard lateritic crust to allow percolation of whatever rainwater may assimilate during brief showers. This forms the basis for all cultivation here.

Basudebpur

Located in Simulpal gram panchayat of Binpur-II block of Paschim Medinipur district, West Bengal (Figure 13c). It is uninhabited. The village lies in the transition zone between the cultivated pediplains to the north and the forested ridges and residual hills to the south, markedly reflected in the map, with the northern half divided into small arable plots while the southern half is forested. The village area covers the confluence of two small streams that course through it. Another small tributary stream emanating from the large pond near the northern apex has been converted into tilled plots. The west-east flowing stream in the south, marks the boundary between the cultivated lands and the forests. South of this, the land rises in a series of low mounds which are densely wooded. Alignment of the field plots indicates the local slope.

Nagdi

Situated on the right bank of the Dulung River in Dharsha gram panchayat of Jambani block of Paschim Medinipur district, West Bengal (Figure 13d). It lies at the confluence of the Dulung and its most important tributary, the Kupan. The Dulung flows in along the eastern edge of the village while the Kupan flows in along the south, converging at the southeastern extremity of the village. The village also lies in the lithological transition zone between the red lateritic zone to the north and the unconsolidated sediments to the south, with the soil within the village area showing an admixture of red gravels to fine yellow sands and clays. The rectangular field plots align themselves roughly perpendicular to both the major streams. Small mounds of laterite dot the village landscape, breaking out above the level of the fields. Settlements are clustered in two groups, along the right bank of the Dulung and other along the left bank of the Kupan.

Asti

Situated on the right bank (along the cliff slope) of the River Dulung's meander in Andhari gram panchayat of Sankrail block, Paschim Medinipur district, West Bengal (Figure 13e). The village lies in the nook between the confluence of the Dulung River to its east and the Hudia Khal flowing in from its west and north. With its apex to the north, the village spreads out laterally southwards. Settlements are positioned for protection behind the long embankment along the Dulung, to the southwest, with fields encircling them on all other sides. Floods are recurrent and the village has often borne the brunt of the damage, with embankments being breached almost every monsoon. The floods of 2007 caused heavy damage. 14 out of 238 village plots were rendered un-arable by over-bank residual sand deposits. These covered 15.3% of the total village area, quite valuable agricultural land (Patel & Dasgupta, 2009).

Ramanandapur

Located in Andhari gram panchayat of Sankrail Block, Paschim Medinipur District, West Bengal (Figure 13f). It is the last village on the Dulung's right bank, before the river sweeps westward and debouches into the River Subarnarekha. The village radiates outward from top of a large, low mound like a sector of a circle. Due to this, the village narrows towards the north at its apex, broadening towards its south and east, along which flows the Dulung, continuously curving westward. The fertile fields too reflect the above, there being a mixture of paddy and sugarcane cultivation in this area. The village is criss-crossed by a number of cart tracks and an important metalled road that leads from Rohini, across the Dulung on the other bank, to Ragra, further west. Bank heights range from 20m-25m and are frequently undercut, especially along the eastern section which was eroded away in the 2007 floods. Over bank sand deposits, remnant of those floods, still cover some the field plots and scrub land, rendering them unusable for cultivation.

Physiographic - Soil - Landuse Units (PSLU) within the Dulung River Basin

Land system mapping units are areas or groups of areas of a recurring pattern of topography (landform), soil and vegetation (Christian & Stewart, 1969). This recurrence since the landforms develop on similar parent rocks, under identical climatic conditions and geomorphic processes, thus showing similarities in soil-vegetation associations (Raghavswamy, 1982). The physiographic, terrain and LULC maps of the Dulung Basin reveal striking correlations among the above facets which exhibit distinct controls on the land utilisation pattern. The control is direct where the relationship is obvious and indirect where there is an intermediate control (Kumar, Chennaiah & Gautam, 1983). The direct controls observed are:

- The dissected plateau region, elongated ridges and the residual hillocks, due to their rugged terrain offer best possibilities for natural vegetation (forests and scrubs).
- The bare rock exposures and lateritic caps, unfit for profitable use, are left as wastelands or fallows.
- The northern pediplains and the valley fills are almost level surfaces without undulations of great amplitude and with relatively nutrient rich soils. They are thus preferred for agriculture.

The indirect controls are mostly lineaments and morphological facets which influence water availability, in turn controlling agriculture and settlement patterns. These are:

- In the northern pediplain area, there is a greater expanse of agriculturally suitable land whereas cultivated lands in the central area of the basin tends to be aligned along the relatively narrow valley floors, encroaching at times onto the lateritic uplands that form low water-divides between adjacent basins.
- Few settlements are present in the dissected plateau area due to its poor agricultural productivity. Settlements increase in the pediplain and along the valley floors of the major streams, forming linear trends. Settlements numbers are highest in the southern alluvial plains where most lands support more than one crop. Agriculture is facilitated here by groundwater aquifers and surface lift irrigation.

Thus, within the Dulung River Basin, the four broad PSLUs identified (Figure 14) are:

1. **Structural Ridges and Residual Hills:** This unit consists of high hills and elongated ridges with rocky, narrow and broad crests to sharp conical and rounded summits having an elevation range of about 150m-365m. Situated in the north-western portion, it covers about 8% of the basin area. The debris slopes are steep, with above 50% gradient in places, while escarpments are quite vertical. Thin gravelly soils (often containing plinthite) occurring on the slopes, get eroded due to severe runoff and the narrow valleys are the accretion zones of sandy valley-fills and washed down lateritic gravel. Escarpments are usually forested though some are bare. Their present LULC pertains to forests

Figure 14. Demarcated PSLUs in the Dulung river basin

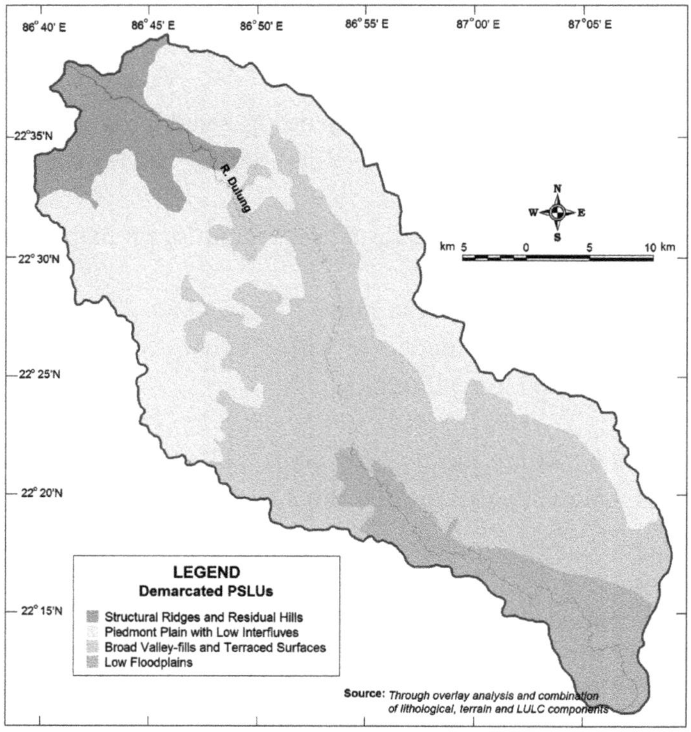

and *kharif* cultivation of paddy, millets, potato and wheat in valleys and on debris slopes. This is the least inhabited zone in the basin landscape.

2. **Piedmont Plain with Low Interfluves:** This unit (covering about 39% of the basin area) consists of gently sloping to undulating plains having 4% to 6% slope with an elevation range of about 90m-150m and the gently rising interfluves that alternate between the small streams draining this surface. The interfluve crests are often laterite-capped, being remnants of leached old alluvium deposits. The valleys cut into this surface comprise of finer material brought and deposited from the foothills by the meandering streams. The fine sandy soil is often subjected to sheet and gully erosion. This area covers the northern portion of the basin and parts of its eastern and western flanks. The present LULC includes scattered bamboo and sub-tropical forests and cultivation of paddy, wheat, potato and millets in the valley floors and on terraced upper plains. The land is usually mono-cropped with rare pockets of double cropped land where water availability is assured.
3. **Broad Valley-Fills and Terraced Surfaces:** Abutting onto the piedmont zone is this tract (covering 41% of basin area), where the initially constricted

valleys of streams broaden to manifold times the channel width, allowing them to shift gradually across the surface and meander extensively. The soils are sandy with admixtures of clay, underlain by weathered lateritic surfaces, which crop out occasionally along valley up-slopes. Numerous ponds dot this PSLU with small rural communities arranged around them. The low terraces of the Dulung and its principal tributaries have been modified into small plots, principally for paddy. Bank erosion is a recurring hazard here, with topsoil loss from field runoff further augmenting this.

4. **Low Floodplains:** This unit (covering 12% of the basin area) consists primarily of the double cropped lands near the mouth of the basin, in its southern tracts, having an elevation range of about 22m-50m. These lands, though fertile due to the thick deposits of alluvium and sandy loams, are also prone to submergence during floods and can be covered by thick crevasse-splay deposits. This PSLU is most densely populated and has year-round access to water, allowing multiple crops to be harvested. The settlements are located linearly following the trend of the main Dulung River channel and field plots radiate perpendicularly away from the bank edges. Notionally higher areas within the villages serve to site households.

LAND MANAGEMENT ISSUES AND FEASIBLE RECOMMENDATIONS IN THE PSLUs

Watershed development involves continuous interaction and exchange between various sectors e.g. the livestock maintained is dependent on fodder availability, which is related to soil and water management. Firewood availability is related to the extent of forest cover and land productivity. Emphasis is given here on land development aspects since it is the underlying component of all other aspects. Enhancing agricultural productivity is thus paramount. The land, however, is afflicted by certain problems. Some hindrances against improving agriculture in this area are outlined below.

- **Slow Permeability:** Except along sand or loam based valley floors, the area has slow percolation rates, particularly where stoniness exists in underdeveloped soil, causing heavier resulting in sheet and rill erosion. This also adversely affects groundwater recharge.
- **Plinthite in Soil Profile or Presence of Gravel:** Stoniness and gravels, especially in the soils of the north-western portion of the basin area makes them difficult to cultivate. The stoniness affects root development downward as well as water seepage adversely.

- **Poor Solum Depth and Medium to Poor Soil Texture:** In the northwestern tracts, soils are relatively shallower, bereft of proper horizon development. The only pockets of cultivation are along narrow channels whose beds are composed of an admixture of sand-loam-clay.
- **Formation of Hard Pans:** Seen where laterite layers exist, making cultivation difficult and such *moram* are usually only useful for laying roads due to their tough nature. Being less permeable, where massive, they also aid higher surface runoff.
- **Moderate to Poor Moisture Holding Capacity and Prolonged Dryness:** The extended dry season desiccates soils, particularly if they contain clay. Slower permeability rates do not allow the moisture to penetrate deeper into the soil layers.
- **High Runoff Along with Seasonal Inundation:** The lower basin area regularly gets inundated by the over-spilling Dulung River, submerging field plots and leaving thick deposits of unconsolidated sand via crevasse splays. The land cannot be used unless these are cleared totally.
- **Poor Soil Tillage and Weed Infestation:** This often occurs due to improper farming practices or the unavailability of requisite implements to plough the land properly.
- **Low Soil pH and Poor Fertility:** This occurs due to improper development of the soil profile in places, poor soil texture and structure and leaching over the lateritic landscapes.
- **Steeper Land Surface Slope and Soil Erosion Susceptibility:** This is seen particularly in the more upland portions of the basin.

In light of the above issues, some recommendations have been put forward, which could to a certain extent help alleviate the nature and severity of the problems. These are as follows:

- Application of soil-testing based recommended doses of fertilizers which are crop–specific, along with the application of organic and green manures as well as mulching to conserve soil moisture.
- Crop rotation and mixed cropping, along with agronomic practices like eradication of weeds and application of lime to the soil to reduce the soil acidity.
- Terrace and field bunding along with levelling of land in the steeper sloped areas. Strip–cropping for erosion control and moisture conservation.
- Gully plugging and check dam construction with suitable vegetation cover to retard badland expansion, especially where *moram* quarrying occurs or brick kilns source soil from. Development of areas considered to be wasteland and reclamation of degraded lands needs to be given priority.

- Adoption of proper dry-land agricultural practices and cultivation of less-water-requiring variety of crops.
- Plantation of more economically beneficial tree species within suitable enclosures for protection and afforestation through farm and social forestry with fire protection.
- Restricted or controlled livestock grazing with restricted tree-felling.

Work on modifying the land towards greater soil conservation, runoff preservation and improving agricultural output should ideally be executed in a "ridge-to-valley" perspective, i.e., starting from the top and coming downwards (NABARD, 2007). This would mean that the treatment of the hilltops and hill slopes would be completed first and the lands along the valley bottom be treated. This ensures that soil erosion is reduced, helping water conservation and ground water recharge. The treatment must be on area basis, i.e., on both public and private lands. If these measures are carried out well, the whole watershed basin will act as a large reservoir with recharged ground water. Therefore, check dams need only be constructed at the mouth of the watershed to impound the excess water after allowing water to seep into the aquifer through the entire course of the drainage line of the watershed (NABARD, 2007). Some of the above points are already being tended to, under the aegis of the Mahatma Gandhi Rural Employment Guarantee Act (MNREGA) (Table 11). Some of the various sectors which it has focused on are:

Table 11. Number of completed works under various activities of the MNREGA in the study area

Gram Panchayat	A	B	C	D	E	F	G	H	I	J	K
da	9	0	7	7	0	3	12	3	0	0	41
Belpahari	13	0	2	14	2	1	0	5	1	0	38
Bhelaidiha	25	0	17	21	10	0	1	1	1	0	76
Ergoda	0	0	0	0	0	0	0	0	0	0	0
Kanke	29	0	18	13	2	0	0	1	1	0	64
Sandapara	0	0	0	0	0	0	0	0	0	0	0
Silda	0	0	0	0	0	0	0	0	0	0	0
Shimulpal	23	0	0	0	0	23	0	0	0	0	46
Beliabera	58	5	16	455	21	1	59	122	2	0	739
Kharbandhi	11	0	7	0	0	0	0	6	0	0	24
Petbindhi	13	0	1	5	0	0	0	0	0	0	19
Tapsia	9	0	2	0	0	0	0	0	0	0	11

continued on next page

Table 11. Continued

Gram Panchayat	A	B	C	D	E	F	G	H	I	J	K
Chilkigarh	11	0	2	0	1	3	0	1	0	0	18
Chichra	2	0	0	0	0	0	0	10	2	0	14
Dharsa	3	0	35	0	0	0	0	0	0	0	38
Dubra	14	2	17	0	0	5	0	2	0	0	40
Gidhni	6	0	5	6	4	0	0	1	0	0	22
Jamboni	2	0	0	10	1	5	0	13	0	0	31
Kapgari	19	0	40	9	9	22	0	7	1	0	107
Keddangri	0	0	1	0	0	0	0	0	0	0	1
Lalbundh	4	0	3	13	0	2	0	0	0	0	22
Parihati	9	0	7	0	3	2	1	8	0	0	30
Aguiboni	0	0	0	0	0	0	0	0	0	0	0
Chandri	9	0	7	2	0	1	0	15	0	0	34
Dudhkundi	2	0	7	12	7	0	0	18	0	0	46
Lodhasuli	2	0	0	0	0	0	0	3	0	0	5
Manikpara	13	0	8	4	1	0	0	0	0	0	26
Nedabahara	1	0	4	0	6	6	0	0	0	0	17
Patashimul	0	0	0	0	0	0	0	0	0	0	0
Radhanagar	9	0	13	1	0	0	0	6	0	0	29
Sapdhara	4	1	7	4	6	0	0	28	0	0	50
Salboni	0	0	0	0	0	0	0	0	0	0	0
Andhari	0	1	1	0	0	0	0	38	0	0	40
Chhatri	0	0	13	0	0	0	0	4	0	0	17
Dhanghori	0	0	0	0	0	0	0	0	0	0	0
Ragra	1	0	1	5	0	0	0	2	0	0	9
Rohini	6	0	4	3	0	0	0	2	0	0	15
Benasoli	0	0	0	0	0	0	0	0	0	0	0
Bhutia	0	0	0	0	0	0	0	5	0	0	5
Paruliya	1	0	0	1	0	0	0	0	0	0	2
Ramchandrapur	0	0	0	0	0	0	0	0	0	0	0
Manushmuriya	1	0	0	1	0	0	0	0	0	0	2
Gandanata	0	0	0	3	0	2	0	2	0	0	7
Jaipura	0	0	0	0	0	0	0	0	0	0	0
Kuldiha	0	0	0	0	0	0	0	2	0	0	2
Arang	0	0	1	0	0	0	0	0	0	0	1
Baramara	3	0	0	0	0	0	0	0	0	0	3

continued on next page

Table 11. Continued

Gram Panchayat	A	B	C	D	E	F	G	H	I	J	K
Bardikanpur	0	0	0	0	0	0	0	0	0	0	0
Bend	2	0	0	0	0	0	0	0	0	0	2
Bhalukbinda	3	0	0	0	0	0	0	0	0	0	3
Joram	0	0	0	0	0	0	0	0	0	0	0
Kamarigora	0	0	2	0	0	0	0	0	0	0	2
Kalapathar	1	0	1	0	0	0	0	0	0	0	2
Balibandh	3	0	1	0	0	0	0	0	0	0	4
Lodhasholi	0	0	0	0	0	0	0	0	0	0	0
Matiyabandhi	1	0	1	0	0	0	0	0	0	0	2
Rupuskunri	1	0	0	0	0	0	0	0	0	0	1
Simdi	0	0	3	0	0	0	0	0	0	0	3

Source: Website of the Mahatma Gandhi National Rural Employment Guarantee Act (MNREGA) (www.nrega.nic.in)

Notations: A: Rural Connectivity; B: Flood Control; C: Water Conservation and Water Harvesting; D: Renovation of Traditional Water Bodies; E: Drought Proofing; F: Irrigation Canals; G: Irrigation Facilities to SC/ST/IAY/LR; H: Land Development; I: Other Works; J: Rajiv Gandhi Seva Kendra; K: Total Works Accomplished. Data not available for Chakulia, Murathakura and Kasafliya.

- **Flood Control:** It is only the southern portion of the basin area, near its mouths that gets affected by floods. Therefore, flood control works, mostly involving repairing of embankments, has been mainly undertaken in the gram panchayats of Gopiballavpur-II and Sankrail blocks.
- **Water Conservation and Water Harvesting:** These have involved excavation of small ponds, de-silting of existing tanks, construction of few bunds or check dams across streams. Ponds often dry up due to the truancy of the monsoonal rains and rapid evaporation rates in the region. The raised embankments around the ponds have instead suffered from rilling, becoming quite breached in some places.
- **Renovation of Traditional Water Bodies:** Through excavation of private ponds, primarily for fishing and also an attempted restoration of some wetlands near Parihati and Rohini. A large number of such works have been done in Beliabera gram panchayat of Gopiballavpur-II block.
- **Drought Proofing:** Measures such as creation of minor irrigation structures for tapping of ground water have been implemented in certain places.
- **Irrigation Canals:** Renovation of some of the canals within the study area has been undertaken. However, with the river regime of the Dulung River and its principal tributaries being much influenced by the monsoon, there still remains a sizeable water scarcity in the area during the dry months.

- **Land Development:** Certain land development works like land levelling, afforestation and creation of plantations have been undertaken.

Overall, the MNREGA impact is seen to be more profound in the blocks which are part of Paschim Medinipur in West Bengal than in those which are part of Purbi Singhbhum in Jharkhand.

FUTURE RESEARCH DIRECTIONS

Small test farms to monitor the suitability of particular crops and proper soil nutrient management could greatly increase productivity, with timely irrigation. Within the basin area, a few mineral localities could be used for setting up small scale industries. The Katachua clay deposit is of good quality (Basu and Pal, 1988) and this along locally found clay could serve small scale ceramic industries making tiles or sanitary wares. Talc or stelite occurring here could be used for manufacturing utensils. Black stone, suitable for road construction is found within the basin (Das, 1972) and crushing units could be set up to utilise these. Some stone crushing plants operate already near Parihati, and scrap iron based steel plants near Lodhashuli, which has the locational advantage of being along Highway NH6. Expansion of their capacities could help develop the surrounding area, though care must be taken to hot harm the local environment.

CONCLUSION

The philosophical aim underlying this study has been the formulation of a methodology for the terrain assessment of watersheds for land management purposes.

Morphometric analysis has a rich history within Geomorphology but the quality of analyses done can now be increased manifold using DEM datasets and RS-GIS techniques. An in-depth analysis of the terrain has been achieved via spatial analysis of the morphometric attributes of the Dulung River Basin for a comprehensive multivariate classification. Combinations of the lithology, geomorphology and morphometry coupled with the LULC units demarcated, help identify distinct landscape units (PSLUs) within the basin. From the structural and morphological standpoint, the basin shows clear divisions between the granite-phyllite hill ranges in the northwestern part, the laterite-older alluvium piedmont zone in the middle section and the low floodplains in the south composed of younger alluvium and unconsolidated sand deposits. The majority of the basin is quite low lying and gently sloped. Surface profile analysis across the area reinforces this further. This corresponds with the geological and geomorphological maps produced for the area and also with the terrain classes derived for the basin via factor score analysis of the various morphometric variables.

Studies on the LULC highlight spatio-temporal changes in those attributes. Areas afflicted by deforestation are identified and variations in the village-level LULC composition from basin head to basin mouth are showcased. Each distinctive unit suffers from its own problems related to the terrain and thus needs to be looked at separately while planning for any sustainable development measures. These natural hazards are quite strongly correlated to the physiography, with rapid runoff across the gentle piedmont zone causing much rill and sheet erosion while bank erosion and flooding are a recurrent problem in the middle and lower reaches of the basin comprised of low floodplains.

In formulating suitable plans for proper land and water resource management, the various obstacles have been outlined. Problems afflicting the land in different portions of the basin are looked into and certain feasible ameliorative measures have been suggested through a raft of recommended measures. The present status of various developmental works undertaken across the basin area has also been highlighted.

REFERENCES

Agarwal, C. S. (1998). Study of drainage pattern through aerial data in Naugarh area of Varanasi district, U.P. *Journal of the Indian Society of Remote Sensing*, *24*(4), 169–175. doi:10.1007/BF02990795

Anderson, J. R., Hardy, E. E., Roach, J. T., & Wirmer, R. E. (1976). A land use land cover classification system for use with remote sensing data. United States Geological Survey (USGS) Professional Paper 964.

Ayyangar, R. S. (1978). Experiment to evolve methods for separation and identification of agricultural crops from multispectral information.*Proceedings of 12th International Symposium on Remote Sensing of Environment*, Environmental Research Institute of Michigan, Ann Arbor, Michigan.

Ayyangar, R. S., Krishna Rao, M. V., & Rao, K. R. (1980). Interpretation and analysis of multispectral data for agricultural crop cover types- role of crop spectral response and crop spectral signature. *Journal of the Indian Society of Remote Sensing*, *8*(2), 39–48.

Basu, S. N., & Pal, P. K. (1986). Mineral resources of Midnapore District: Scope of some mineral based industries. *Proceedings of the National Seminar on Small Mineral Deposits: their development and industrial possibilities, New Delhi*. The Mining and Metallurgical Institute of India, Kolkata.

Bhattacharya, A. (1983). Geologic and geomorphic controls in land use patterns. *Journal of the Indian Society of Remote Sensing*, *11*(3), 57–66.

Biswas, S., Sudhakar, S., & Desai, V. R. (1999). Prioritization of sub-watersheds based on morphometric analysis and drainage basin. a Remote Sensing and GIS approach. *Journal of Indian Society of Remote Sensing*, *27*(3), 155–166. doi:10.1007/BF02991569

Brahmabhatt, V. S., Dalwadi, G. B., Chhabra, S. B., Ray, S. S., & Dadhwal, V. K. (2000). Land use / land cover change mapping in Mahi Canal Command area, Gujarat, using multi-temporal satellite data. *Journal of the Indian Society of Remote Sensing*, *28*(4), 221–232. doi:10.1007/BF02990813

Burns, G. S., & Joyce, A. T. (1982). Evaluation of land cover change detection techniques using Landsat MSS data. *Proceedings of Conference on Remote Sensing with Special Emphasis on Output to Geographic Information Systems in the 1980s.*, 7th PECORA Symposium, Sioux Falls, South Dakota, USA (pp. 252-257).

Carpenter, R. A. (1987). What to do While Waiting for an Environmental Ethics. *Environment and Progress*, *9*(4), 327–335.

CGWB (Central Ground Water Board). (2006). *Report on Reappraisal Hydrological Survey in Parts of West Midnapur District, West Bengal* (Technical Report No. D-187). CGWB-Eastern Region (ER), Ministry of Water Resources, Govt. of India, Kolkata.

Chauhan, E. S., Porwal, M. C., Sharma, L., & Negi, J. D. S. (2003). Change detection in sal forest in Dehradun forest division using Remote Sensing and Geographical Information System. *Journal of the Indian Society of Remote Sensing*, *31*(3), 211–218. doi:10.1007/BF03030827

Chaurasia, R., Loshali, D. C., Dhaliwal, S. S., Minakshi, , Sharma, P. K., Kudrat, M., & Tiwari, A. K. (1996). Land use change analysis for agricultural management-A case study of Tehsil Talwandi Sabo. Punjab. *Journal of the Indian Society of Remote Sensing*, *24*(2), 115–123. doi:10.1007/BF03016124

Christian, C. S., & Stewart, G. A. (1953). General Report on Survey of Katherine Darwin Region. The Commonwealth Scientific and Industrial Research Organization (CSIRO).

Christian, C. S., & Stewart, G. A. (1969). Lands of the Oueandeyan-Shoalhaven area. In R.H. Gunn, R. Story, R.W. Galloway, P.J.B. Duffy, G.A. Yapp, & J.R. McAlpine (Eds.). Lands of the Oueandeyan – Shoalhaven Area, A.C.T. and N.S.W. The Commonwealth Scientific and Industrial Research Organization (CSIRO) Publication.

Congalton, R., Green, K., & Teply, J. (1993). Mapping old growth forests on national forest and park lands in Pacific Northwest from remotely sensed data. *Photogrammetric Engineering and Remote Sensing*, *59*(4), 529–535.

Das, D. K. (1972). Report on some of the stone deposit in Jhargram Sub-Division: District Midnapore. Kolkata: Geological Survey of India, Report No. G.P. 30/72.

De Martonne, E. (1934). Les régions arides du nord argentin et chilien. *Bulletin de lAssociation de Geographes Francais*, *79*(79), 58–62. doi:10.3406/bagf.1934.6686

De Smet, R. (1951). Principles elementaires de morphometrie. *Rev. Cercle des Sciences*, *1*(4), 13–16.

Dhruvanarayana, V. V., Sastry, G., & Patnaik, U. S. (1990). Watershed Management, New Delhi: Indian Council for Agricultural Research (ICAR).

Doornkamp, J. C., & King, C. A. M. (1971). *Numerical Analysis in Geomorphology-An Introduction*. London: Edward Arnold.

Dunkerley, D. L. (1988). LEVPLOT. A BASIC program for processing and plotting survey sections (Technical Memorandum). Dept. of Geography, Monash University, Melbourne.

Easter, K. W., Dixon, J. A., & Hufschmidt, M. M. (1986). Watershed Resources Management. An integrated framework with studies from Asia and the Pacific. Boulder, Colorado: Westview Press Studies in Water Policy and Management.

Estes, J. E., Stow, D., & Jenson, J. R. (1982). Monitoring land use and land cover changes. In C. J. Johannsen & J. L. Sanders (Eds.), *Remote Sensing for Resources Management* (pp. 100–110). Ankeny, Iowa: Soil Conservation Society of America.

FAO (Food and Agricultural Organization of the United Nations). (1977). *Guidelines for Watershed Management. 1*. Rome: FAO.

FAO (Food and Agricultural Organization of the United Nations). (1987). *Soil and Water Conservation in Semi-Arid Areas. Soil Bulletin 57*. Rome: FAO.

Gregory, K. J., & Walling, D. E. (1978). *Drainage Basin Form and Processes*. London: Edward Arnold.

GSI (Geological Survey of India). (1998). *Geological Quadrangle Map - 73J (Jamshedpur Quadrangle)*. Kolkata: GSI.

GSI (Geological Survey of India). (2007). *District Resource Map - Medinipur, West Bengal*. Kolkata: GSI.

Gupta, A., & Chakraborty, S. (1967). Quantitative analysis of landforms-A case study on the Dulung drainage basin.*Proceedings of Seminar on Geomorphological Studies inIndia*, University of Saugar (pp. 190-199).

Honore, G. (1999). Our Land, Ourselves-A Guide to Watershed Management in India. New Delhi: Govt. of India.

Horton, R. E. (1945). Erosional development of streams and their drainage basins: Hydrophysical approach to quantitative morphology. *Bulletin of the Geological Society of America*, *56*(3), 275–370. doi:10.1130/0016-7606(1945)56[275:EDOSAT]2.0.CO;2

Indian Space Research Organisation / Regional Remote Sensing Centre / Indian Society of Remote Sensing. (2009, September 15-16). Pre-Symposium Tutorial of National Symposium on Advances in Geo-Spatial Technologies with special emphasis on Sustainable Rainfed Agriculture, Nagpur (Vol. 1 & 2).

Jain, S. K., Seth, S. M., & Nema, R. K. (1995). Morphometric Analysis of Sei Dam Catchment using GIS.*Proceedings of the national Symposium on Hydrology*, Jaipur (pp. 5-7).

Jaiswal, R. K., Saxena, R., & Mukherjee, S. (1999). Application of Remote Sensing technology for land use / land cover change analysis. *Journal of the Indian Society of Remote Sensing*, *27*(2), 123–128. doi:10.1007/BF02990808

Jensen, J. R. (1986). *Introductory Image Processing. A remote sensing perspective.* USA: Prentice-Hall.

Jeyaram, A., Mohabey, N. K., & Krishanamurthy, Y. V. N. (1996). Groundwater potential using Remote Sensing and Geographical Information System. In N.K. Mohabey, L.G. Gwalani, & D.M. Mohabey (Eds.), *Mineral and Groundwater Resources of Vidarbha,* Golden Jubilee Symposium, Nagpur University, Nagpur (pp. 257-267).

Joshi, P. K., & Gairola, S. (2004). Land cover dynamics in Garhwal Himalayas-A case study of Balkhila sub-watershed. *Journal of the Indian Society of Remote Sensing*, *32*(2), 199–208. doi:10.1007/BF03030876

Kachhwaha, T. S. (1985, November 21-26). Temporal monitoring of forest land for change detection and forest cover mapping through satellite remote sensing.*Proceedings of the Sixth Asian Conference on Remote Sensing*, Hyderabad (pp. 77-83).

Kasturirangan, K., Aravamudan, R., Deekshatulu, B. L., Joseph, G., & Chandrasekhar, M. G. (1996). Indian Remote Sensing satellite (IRS) IC-The beginning of a new era. *Current Science, 70*(7), 495–500.

Krishna, N. D. R., Westinga, E., & Huizing, H. (1999, March 9-11). Monitoring land cover changes using Geoinformatics in some communal lands of Zimbabwe. *Proceedings of "Geoinformatics. Beyond 2000"International Conference on Geoinformatics for Natural Resource Assessment, Monitoring and Management*, Dehradun (pp. 92-101).

Krishnamurthy, J., Srinivas, G., Jayaram, V., & Chandrasekhar, M. G. (1996). Influence of rock type and structure in the development of drainage networks in typical hard rock terrain. *ITC Journal, 3*(4), 252–259.

Kumar, A., & Pandey, R. H. (1982). Morphometric properties of the polygenetic drainage basin: A case study of the Burha Nadi. *Geographical Review of India, 37*, 329–338.

Kumar, R., Kumar, S., Lohani, A. K., Nema, R. K., & Singh, R. D. (2000). Evaluation of geomorphological characteristics of a catchment using GIS. *GIS India, 9*(3), 13–17.

Kumar, R., & Silva, L. P. (1974). Statistical separability of agricultural cover types in subsets of one to twelve spectral channels. *Proceedings of the Ninth International Symposium on Remote Sensing of Environment*, Environmental Research Institute of Michigan (ERIM), Ann Arbor, Michigan.

Kumar, S. S., Chennaiah, G. Ch., & Gautam, N. C. (1983). Study of landform control over land utilisation pattern in parts of southern U.P.- A remote sensing approach. *Journal of the Indian Society of Remote Sensing, 11*(1), 49–53.

Laboratory for Application of Remote Sensing (LARS). (1968). *Remote Multi-spectral Sensing in Agriculture, 3-Annual Report*. West Lafayette, Indiana: Purdue University.

Land and Land Reforms Department, Govt. of West Bengal. (2010). Focus on Paschim Medinipur. Kolkata: Dept. of L&LR, Govt. of West Bengal.

Likens, W., & Maw, K. (1982). Hierarchical modelling for image classification. *Proceedings of Conference on Remote Sensing with Special Emphasis on Output to Geographic Information Systems in the1980s, 7th PECORA Symposium*, Sioux Falls, South Dakota, USA (pp. 290-300).

Luong, P. T. (1993). The detection of land use / land cover changes using Remote Sensing and GIS in Vietnam. *Asian-Pacific Remote Sensing Journal*, *5*(2), 63–66.

Mahajan, S., Panwar, P., & Kaundal, D. (2001). Land use changes in Ashwani Khad watershed using GIS techniques. *Journal of the Indian Society of Remote Sensing*, *33*(2), 243–248. doi:10.1007/BF02995730

Marshall, R. E., Thomson, F., & Kriegler, F. (1969). Use of multispectral recognition techniques for conducting rapid, wide wheat surveys. *Proceedings of the Sixth International Symposium on Remote Sensing of Environment*, Environmental Research Institute of Michigan (ERIM), Ann Arbor, Michigan.

Maune, D. F. (Ed.). (2001). *Digital Elevation Model Technologies and Applications. The DEM Users Manual*. Bethesda, Maryland: American Society of Photogrammetry and Remote Sensing.

Moore, I. D., Grayson, R. B., & Ladson, A. R. (1991). Digital terrain modelling. A review of hydrological, geomorphological and biological applications. *Hydrological Processes*, *5*(1), 3–30. doi:10.1002/hyp.3360050103

Mukhopadhyay, S. C. (1980). *Geomorphology of the Subarnarekha Basin*. Burdwan: University of Burdwan Publication.

NABARD (National Bank for Agricultural and Rural Development). (2007). *District Agricultural Plan-East Singhbhum (2008-2009 to 20011-2012)*. Ranchi: NABARD Consultancy Services, Jharkhand Regional Office.

Nag, S. K., & Chakraborty, S. (2003). Influence of rock types and structures in the development of drainage network in hard rock area. *Journal of the Indian Society of Remote Sensing*, *31*(1), 25–35. doi:10.1007/BF03030749

Nalepka, R. (1970). Investigation of multispectral discrimination techniques (Final Report 2264-12-1). Willow Run Laboratories, University of Michigan, Michigan, USA.

Natarajan, S., Gajbe, M. Y., & Manchanda, M. L. (1986). Landuse study in Mewat area, Haryana using Remote Sensing techniques. *Journal of the Indian Society of Remote Sensing*, *14*(1), 17–24. doi:10.1007/BF03007219

NATMO (National Atlas and Thematic Mapping Organisation). (2000). *District Planning Map Series - Purbi Singhbhum, Jharkhand.* Kolkata: NATMO.

NATMO (National Atlas and Thematic Mapping Organisation). (2006). *District Planning Map Series - Paschim Medinipur, West Bengal.* Kolkata: NATMO.

Nooka Ratnam, K., Srivastava, Y. K., Venkateswara Rao, V., Amminedu, E., & Murthy, K. S. R. (2005). Check dam positioning by prioritization of microwatersheds using SYI model and morphometric analysis - Remote Sensing and GIS perspective. *Journal of the Indian Society of Remote Sensing*, *33*(1), 25–38. doi:10.1007/BF02989988

Pant, D. N., & Roy, P. S. (1990). Vegetation and land use analysis of Aglar watershed using satellite remote sensing technique. *Journal of the Indian Society of Remote Sensing*, *18*(4), 1–14. doi:10.1007/BF02997068

Patel, P. P., & Dasgupta, R. (2009). Flood induced land use change in the Dulung River valley, West Bengal. In Geoinformatics for Monitoring and Modelling Land-Use, Bio-diversity and Climate Change (pp. 103-123). NMCC Publication.

Patel, P. P., & Sarkar, A. (2009). Application of SRTM data in evaluating the morphometric attributes: A Case Study of the Dulung River Basin. *Practising Geographer*, *13*(2), 249–265.

Patel, P. P., & Sarkar, A. (2010). Terrain Characterization Using SRTM Data. *Journal of the Indian Society of Remote Sensing*, *38*(1), 11–24. doi:10.1007/s12524-010-0008-8

Raghavswamy, V. (1982). Role of satellite remote sensing in land system mapping, land resources inventory and land use planning: A sample study of Kemang River basin, Arunachal Pradesh. *Journal of the Indian Society of Remote Sensing*, *10*(3), 31–40.

Reddy, G. P. O., & Maji, A. K. (2003). Delineation and characterzation of geomorphological features in a part of lower Maharashtra metamorphic plateau, using IRS-1D, LISS-III data. *Journal of the Indian Society of Remote Sensing*, *31*(4), 241–250. doi:10.1007/BF03007344

Reddy, G. P. O., Maji, A. K., & Gajbhiye, K. S. (2002). GIS for morphometric analysis of drainage basins. *GIS India*, *11*(4), 9–14.

Reddy, G. P. O., Maji, A. K., & Gajbhiye, K. S. (2004). Drainage morphometry and its influence on landform characteristics in basaltic terrain-A Remote Sensing and GIS approach. *International Journal of Applied Earth Observation and Geoinformatics*, *6*(1), 1–16. doi:10.1016/j.jag.2004.06.003

Roy, D. (2010). Selection and Design of Appropriate Structure for Watershed Management around Upper Reach of Dulung Nala in the Subarnarekha Basin at Belpahari, West Midnapore, West Bengal (Research Project). School of Water Resources Engineering, Jadavpur University, Kolkata, sponsored by Dept. of Science and Technology, Govt. of West Bengal.

Roy, P. S., Kaul, R. N., Roy, M. R., & Gabryal, S. S. (1985). Forest type stratification and delineation of shifting cultivation in the eastern part of Arunachal Pradesh using Landsat MSS data. *International Journal of Remote Sensing*, *6*(3&4), 411–418. doi:10.1080/01431168508948463

Roy, P. S., Ranganath, B. K., Diwakar, P. G., Vohra, T. P. S., Bhan, S. K., Singh, I. H., & Pandian, V. C. (1991). Tropical forest type mapping and monitoring using remote sensing. *International Journal of Remote Sensing*, *12*(11), 2205–2225. doi:10.1080/01431169108955253

Rubec, C. D., & Thie, J. (1978). Land use monitoring with Landsat digital data in southwestern Manitoba.*Proceedings of the Fifth Canadian Symposium on Remote Sensing of Environment*, Victoria, British Columbia (pp. 136-149).

Saha, S. K., & Barrow, C. J. (1981). *River Basin Planning. Theory and Practice.* Chichester: Wiley.

Sarkar, A., & Patel, P. P. (2009). Drainage analysis of the Dulung Basin. In H. S. Sharma & V. S. Kale (Eds.), *Geomorphology in India (Prof. Savindra Singh Felicitation Volume)* (pp. 133–154). Allahabad: Prayag Pustak Bhavan.

Sarkar, A., & Patel, P. P. (2011). Topographic analysis of the Dulung R. Basin. *eTraverse - The Indian Journal of Spatial Science*, 2(1).

Sarkar, A., & Patel, P. P. (2012). Terrain Classification of the Dulung Drainage Basin. *eTraverse - The Indian Journal of Spatial Science*, 3(1).

Sarma, V. V. L. N., Krishna, G. M., Malini, B. H., & Rao, K. N. (2001). Land use / land cover change detection through Remote Sensing and its climatic implications in the Godavari delta region. *Journal of the Indian Society of Remote Sensing*, *29*(1&2), 85–91. doi:10.1007/BF02989918

Schardt, M. K., Schurek, A., & Winter, R. (1990). Forest mapping using satellite imagery - The Riegenburg map sheet 1.2,000,000 as example. *International Journal for Photogrammetry and Remote Sensing (ISPRS). Journal of Photogrammetry and Remote Sensing*, *45*, 33–46. doi:10.1016/0924-2716(90)90004-U

Shankarnarayan, K. A., & Sen, A. K. (1977). The methodology of mapping land resources for rational utilization. In P.S. Jaiswal (Ed.), Desertification and its Control (Ch. 9, pp. 77-80). New Delhi: Indian Council for Agricultural Research (ICAR).

Sharma, K. P., Garg, P. K., & Verma, S. K. (1986). Spectral reflectance pattern study of some Indian land use / land cover features. *Journal of the Indian Society of Remote Sensing*, *14*(1), 11–16. doi:10.1007/BF03007218

Sharma, K. P., Jain, S. C., & Garg, P. K. (1984). Monitoring land use and land cover changes using Landsat images. *Journal of the Indian Society of Remote Sensing*, *12*(2), 65–70.

Shetty, A., Lakshman, N., Thokchom, S., & Rajesh, M. V. S. (2005). Land use - land cover mapping using satellite data for a forested watershed, Udupi district, Karnataka state, India. *Journal of the Indian Society of Remote Sensing*, *33*(2), 233–238. doi:10.1007/BF02990040

Singh, A. (1989). Digital change detection techniques using remotely sensed data. *International Journal of Remote Sensing*, *10*(6), 989–1003. doi:10.1080/01431168908903939

Singh, B. M., Manchanda, M. L., Iyer, H. S., Pande, L. M., Prasad, J., & Rao, B. R. M. (1983). Use of remote sensing techniques for agricultural land use survey, classification and planning. *Journal of the Indian Society of Remote Sensing*, *11*(3), 1–13.

Skidmore, A. K., Bijker, W., Schmidt, K., & Kumar, L. (1997). Use of Remote Sensing and GIS for sustainable land management. *ITC Journal*, *3*(4), 302–315.

Smith, G. H. (1935). The Relative Relief of Ohio. *Geographical Review*, *25*, 272–284.

Srinivasan, T. N. (1988). Population Growth and Economic Development. *Journal of Policy Modeling*, *10*(1), 7–27.

Strahler, A. N. (1957). Quantitative analysis of watershed geomorphology. *Transactions - American Geophysical Union*, *38*(6), 913–920.

Subramanyan, V. (1978). Land system studies for resource evaluation: A demonstration around Sagar. *Madhya Pradesh. Journal of the Indian Society of Remote Sensing*, *6*(1), 31–42.

Subramanyan, V. (1981). Geomorphology of the Deccan Volcanic Province, India. In K. V. Subbarao & R. N. Sukheswala (Eds.), *Deccan Volcanism and related Basalt Provinces in other parts of the World* (pp. 101–116). Bangalore: Geological Society of India.

Tobler, W. (2000). The development of analytical cartography: A personal note. *Cartography and Geographic Information Science*, *27*(2), 189–194.

Turner II, B.L. (1995). Linking the Natural and Social Sciences. *The Land Use / Cover Change Core Project of IGBP, IGBP Newsletter*, 22.

Vijith, H., & Satheesh, R. (2006). GIS based morphometric analysis of two major upland sub-watersheds of Meenachil River in Kerala. *Journal of the Indian Society of Remote Sensing*, *34*(2), 181–185.

Weismiller, R. A., & Momin, S. J. (1977). Change detection in coastal zone environments. *Journal of Photogrammetric Engineering and Remote Sensing*, *43*(12), 1533–1539.

Wentworth, C. K. (1930). A Simplified Method of Determining the Average Slope of Land Surfaces. *American Journal of Science*, *20*, 184–194.

Wickware, G. M., & Howarth, P. J. (1981). Change detection in the Peace-Athabasca delta using digital Landsat data. *Remote Sensing of Environment*, *11*(1), 1, 9–25.

Wilson, J. P., & Gallant, J. C. (Eds.). (2000). *Terrain Analysis: Principles and Applications*. New York: Wiley.

KEY TERMS AND DEFINITIONS

Land Cover: Generally, denotes the type of features covering some area of the surface of the Earth.

Land Use: Primarily indicates the degree of human activity which in turn reflects inhabitants' lifestyle.

Morphometry: The examination of the surface form through quantitative techniques involving mapping and statistical analysis.

Terrain: The physical framework of the Earth surface, comprising of lithological, pedological and landform elements.

Chapter 9
Remote Sensing Techniques and Its Application in Arid Zones of India

Lokesh Kumar Jain
College of Agriculture, India

ABSTRACT

Remote sensing technologies offer the potential for contributing the security to human existence on arid zones in the country in variety of ways. Remote Sensing in agriculture particularly for natural resource management. It provides important coverage, mapping and classification of land cover features. The remote view of the sensor and the ability to store, analyze, and display the sensed data on field maps are make remote sensing a potentially important tool for agriculture. The aerial photography gives two main advantages viz., speedy survey in very large area or remote area and precise description and recording of resources status. Remotely sensed images provide a means to assess field conditions and gave valuable insights into agronomic management. It led to understanding of leaf reflectance and leaf emittance changes in response to leaf thickness, species, canopy shape, leaf age, nutrient status, and water status. Understanding of leaf reflectance has led to quantify various agronomic parameters, e.g., leaf area, crop cover, biomass, crop type, nutrient status, and yield.

DOI: 10.4018/978-1-5225-1814-3.ch009

INTRODUCTION

The ancient peoples were totally dependent on the astronomy i.e. the behavior of planets and stars and their movement to follow the field practices starting from field preparation to harvesting along with daily routine activities. The predictions were almost perfect and still dominant and active particularly for rural in Indian context. It was similar to modern remote sensing techniques in terms of no physical contact between object and planets. The present remote sensing technique is only varying in terms of more and more realistic and comprehensive study as it is science and technology based. The agriculture in ancient times was subsistence types so farming community is limited to their locality only. As the agriculture is changed and it became a business so wide spread and informative study is needed in present time of technology. Comprehensive, reliable and timely information on agricultural resources are very much necessary for a country like India whose mainstay of the economy is agriculture (Mittal, 2013). In fact, agricultural survey is a backbone of planning and allocation of the limited resources to different sectors of the economy. To make the agriculture from subsistence to sustainable, the adaptation of new research tools particularly from aerospace is one of the prime objectives and ensures food security without affecting the agro-ecological balance. This area is still relatively new compared with other agronomic fields; however, the information content is providing valuable insights into improved management decisions.

In this chapter I tried to elaborate the efficient use of natural resources using remote sensing techniques so sustainability may be maintaining in the dry tracts of India defined by very difficult normal life under extremes of climatic conditions.

DEFINITION AND IMPORTANCE OF REMOTE SENSING

Remote sensing is defined as the art and science of gathering information about objects or areas from a distance without having physical contact with objects area being investigated. The word remote sensing was coined by Fisher in 1960AD. Remote sensing has been used in detection and analysis of resources. There are several types of remote sensing systems used in agriculture but the most common is a passive system that senses the electromagnetic energy reflected from plants. The sun is the most common source of energy for passive systems. Passive system sensors can be mounted on satellites, manned or unmanned aircraft, or directly on farm equipment. Interactions between reflected, absorbed, and transmitted energy can be detected by remote sensing. The relationship between reflected, absorbed and transmitted energy is used to determine spectral signatures of individual plants as spectral signatures are unique to plant species. Interactions between reflected, absorbed, and transmitted energy can be detected by remote sensing (Navalgaud, Jayraman, & Roy, 2007).

The wavelengths used in most agricultural remote sensing applications cover only a small region of the electromagnetic spectrum. Both the visible and infrared regions are used in agricultural remote sensing. When electromagnetic energy from the sun strikes plants, three things i.e. reflection, absorption, or transmission can happen. The differences in leaf colors, textures, shapes or even how the leaves are attached to plants, determine how much energy will be reflected, absorbed or transmitted. There are several factors considered while choosing a remote sensing system for a particular application, including spatial resolution, spectral resolution, radiometric resolution, and temporal resolution. Geo-stationary satellites can provide continuous sensing while normal orbiting satellites can only provide data each time they pass over an area.

REMOTE SENSING PLATFORMS

Three platforms are generally used for remote sensing techniques. They are ground based, air based and satellite based. Infrared thermometer, Spectral radiometer, Pilot-Balloons and Radars are some of the ground based remote sensing tools while aircrafts air based remote sensing tools. Since the ground based and air based platforms are very costly and have limited use, space based satellite technology has become handy for wider application of remote sensing techniques. The digital image processing, using powerful computers, is the key tool for analyzing and interpretation of remotely sensed data (Patel & Singh, 2015). The advantages of satellite remote sensing are:

1. **Synoptic View:** Wide area can be covered by a single image/photo (One scene of Indian Remote Sensing Satellite IRS series cover about 148 x 178 sq.km area).
2. **Receptivity:** Can get the data of any area repeatedly (IRS series cover the same area every 16- 22 days).
3. **Coverage:** Inaccessible areas like mountains, swampy areas and thick forests are easily covered. Space based remote sensing is the process of obtaining information about the earth from the instruments mounted on the Earth Observation Satellites.

The satellites are subdivided into two classes and the types of satellite are as follows:

- **Polar orbiting Satellites:** These operate at an altitude between 550 and 1,600 km along an inclined circular plane over the poles. These satellites are used for remote sensing purposes. LANDSAT (USA), SPOT (FRANCE), and IRS (INDIA) are some of the Remote Sensing Satellites.
- **Geostationary Satellites:** These have orbits around the equator at an altitude of 36,000 km and move with the same speed as the earth so as to view the same area on the earth continuously. They are used for telecommunication and weather forecasting purposes. INSAT series are launched from India for the above purposes. All these satellites have sensors on board operating in the visible and near infrared regions of the electromagnetic spectrum.

IMPORTANCE OF REMOTE SENSING

Sustainable agricultural development/ sustainable increase in crop production could be achieved by adopting below mentioned variety of agricultural technologies and it can be achieved more efficiently by timely availability of information and its diffusion up to the farmers at ground level and policy makers at the top end. These objectives only achieved by using remote sensing technology to avoid human error also (Levin, 1999). Crop stress, soil problems and natural disasters, these problems can be effectively tackled by using remote sensing and GIS applications. The export potential of precision farming can be tilted in favour of our country which can be motivated in large tracts of land (Shukla, Patel, & Radadiya, 2014).

- Improved crop management technology through the use of high yielding, input responsive and soil, climatic and biotic stresses - tolerant crop varieties;
- Suitable cropping systems for different agro-ecological regions based on soil, terrain and climatic suitability;
- Integrated nutrient management for improving soil productivity and minimization of the risk of pollution of soil, water and environment;
- Integrated pest management for effective pests control as well as to reduce the adverse effects of pesticides on environment;
- Soil and water conservation for controlling soil degradation and improving moisture availability.
- Input use efficiency maximization in terms of economic return with minimal input.

FIELD OF APPLICATION OF REMOTE SENSING

- The remote sensing will facilitate farmers in improving the efficiency and productivity of agriculture and allied activities by providing timely and quality information inputs for decision making.
- It will helpful to extension workers to access the latest information and improve their ability to serve the farming community effectively.
- Remote sensing contributes to agricultural productivity and empowering farmers to take informed and quality decisions.
- This approach is capital intensive and useful where large tracts of land are Involved particularly sand dunes of Thar desert.
- This technique will useful for arid farmers to remain competitive in changing environment.
- Remote sensing continues to evolve as a valuable agronomic tool that provides information to scientists, consultants, and producers about the status of their crops and resources management.

CHARACTERISTICS OF ARID REGIONS

The arid regions of the country comprise of the ecosystem of Thar Desert which is very fragile. The rainfall is being the major source of sustaining entire production system and recurrent droughts causes land degradation frequently. Irrespective of the irrigation facilities created, the successful agriculture is still governed by the rains. The district has low food and fodder productivity and peoples are characterized by low literacy and less awareness, poor socio economic status and low risk bearing capacity. The intensive cultivation of land without conservation of soil fertility and soil structure, would lead ultimately to the springing up of the deserts and threats to sustainable agriculture. In discriminative use of synthetic pesticides fungicides and herbicides could cause adverse change in biological balance as well as leads to an increase in the incidence of diseases of cancer and many others through the toxic residues present in the grains. The vigorous exploitation of natural resources like water will pose serious threat to the sustainability of the environment. The lives and livelihoods of the small and marginal famers can be sustainable if their basic necessities are fulfilled from their farm itself and their dependency on the market is reduced. The main aims to develop sustainability by accelerating income, utilizing surroundings environmental resources and reducing risk from farm. The drought prone areas of arid areas are facing serious livelihood threats due to changing environment. As a part of environment the hydrology of arid areas of country is quite complex and defy any simplistic description. There is great variability in terms of temporal and spatial in pattern of water in flow, inundation, depth, duration, submergence, water regime, water receding pattern, sand/ silt casting and soil and land erosion etc. this variability give rise to a complex, dynamic, heterogeneous and high risk arid ecosystem. These areas are rich in terms of natural resources, bio-diversity and traditional knowledge but due to changing environment and lack of adaptation strategies, agricultural productivity of this zone is low as well as unstable. Farmers are pooling their livelihood on their resilience capacity and indigenous traditional knowledge backup. Green revolution (high external input commodity based) technology generated thinking perhaps may not suite to these particular areas. The small and marginal and especially women farmers are often kept aloof from the mainstream of knowledge and information which in- turn hinders their growth in their own field for raising crops (Jain, Hada, & Bhaduria, 2014).

APPLICATION OF REMOTE SENSING IN ARID AREAS

The drought prone areas of country are rich in terms of natural resources, bio-diversity and traditional knowledge but due to changing environment and lack of adaptation strategies, agricultural productivity of this zone is low as well as unstable. Low agricultural productivity in arid areas continues baffle researchers, development administrators and policy makers. Remotely sensed images can be used to identify nutrient deficiencies, diseases, water deficiency or surplus, weed infestations, insect damage, hail damage, wind damage, herbicide damage, and plant populations. The remote sensing is applicable for biomass estimation, grassland type identification, trees and shrubs identification and chemical composition estimation in arid zones.

The country needs to use this technology more and more to address various agricultural/environmental problems like water management system, in the estimation of soil nutrient profile, water profiling, biotic and abiotic plant stress, crop coverage, pest conditions and surveillance, deforestation, biodiversity assessment, marine biology, stabilization of sand dunes and so on. Remote Sensing combining with conventional as well as frontier technologies like Geographic Information Systems (GIS) will results in sustainable agriculture and provide food security for ever increasing population. The broad objective of sustainable agriculture is to balance the inherent land resource with crop requirements, paying special attention to optimization of resource use towards achievement of sustained productivity over a long period. Remote sensing will reflect the economy of farmers and reduce the cost of cultivation without affecting soil environment adversely.

Remote sensing techniques are used in agricultural and allied fields in arid zones particularly are discussed under following heads.

Alternate Land Use Systems

A number of diversified farming systems have been evolved for low-rainfall areas, which include agro-forestry, agri-horticulture and agri-silvi-pasture, to sustain livelihood during crop failure and to maintain livestock during drought. Improved practices for pasture and rangeland management, especially through silvi-horti-pastoral systems and rotational grazing, and rehabilitation of mine spoils through vegetative means have been developed and are being propagated by R&D institutions. Remote sensing in arid land/rangeland regions has developed to meet the need for low cost management information over large expanses of land. Applications include rangeland management, watershed analysis, anti-desertification, wildlife habitat management, mine waste reclamation, management of the arid land-irrigated agriculture interface, and outdoor recreation (Jain & Gupta, 2013).

Assessment and Improvement of Degradation Status of Rangelands

With the help of remote sensing farmers or ranchers observe their fields or pastures to assess their condition without physically touching them. Observing the colors of leaves or the overall appearances of plants can determine the plant's condition. Ranchers use remote sensing to identify prime grazing areas, overgrazed areas or areas of weed infestations. Lending institutions use remote sensing data to evaluate the relative values of land by comparing archived images with those of surrounding fields. For water erosion control on arable lands, contour cultivation, bunding, graded bunding and bench terracing are adopted in conjunction with minimum tillage, cover crops, inter-cropping, strip cropping, contour vegetative barriers, etc. For non-arable lands check dams, gully plugging, stabilization of gully heads and vegetative measures are advocated. These measures and appropriate land uses are integrated on catchments basis with due regard to capability of the land (Rao, Gautam, Nagaraja, & Mohan, 1996).

Moisture Stress

Emittance from crop canopies is a measure of leaf temperature and infrared thermometers have fostered crop stress indices currently used to quantify water requirements. Remote sensing is used to identify stressed areas in fields by first establishing the spectral signatures of healthy plants. The spectral signatures of stressed plants appear altered from those of healthy plants. The vegetative index i.e. comparison of the reflectance values at different wavelengths is commonly used to determine plant vigor. The most common vegetative index is the normalized difference vegetative index compares the reflectance values of the red and NIR regions of the electromagnetic spectrum. The NDVI value of each area on an image helps identify areas of varying levels of plant vigor within fields. This helps in water management or mid-season corrections for saving the vegetation. Remote sensing assisted to farmers to determine how much and when to irrigate, to minimize the amount of water and fertilizer reaching the groundwater beneath their crops. Changes in water content in the upper 2 cm of soil causes the largest variations in reflectance. Water has low reflectance and the addition of water films around the soil aggregates causes an increased absorption of the incident radiation. As the soil is wetted there is darkening in colour and lighten as the soils dries. This variation in the reflectance from soils due to changing soil water content helps in water management to the plants (Chopada & Sharma, 2014).

Rain Water Management

The vigorous exploitation of natural resources like water will pose serious threat to the sustainability of the environment. The drought prone areas of western Rajasthan are facing serious livelihood threats due to changing environment. As a part of environment the hydrology of arid areas is quite complex and defy any simplistic description. There is great variability –temporal and spatial in pattern of water in flow, inundation, depth, duration, submergence, water regime, water receding pattern, sand/ silt casting and soil and land erosion etc. this variability give rise to a complex, dynamic, heterogeneous and high risk arid ecosystem. Rain water conservation, its harvesting and efficient utilization are in-built in watershed management programmes. Combating desertification through land care and enhancing agricultural productivity is the underlying principle for sustainable land management in the desert area. Thus, remote sensing will help in improving runoff water harvesting and in enhancing water use efficiency of water harvesting structures like pons, tanks and khadins. Mapping or locating low land areas in arid zones provide opportunities for better and efficient use of rain water or planning watersheds.

Weed Management

Combined use of remote sensing, GIS (Geographic Information System) and GPS (Global Positioning System) is a powerful tool in detecting, mapping and monitoring the spread of weeds over inaccessible areas. Identification of weed species using remote sensing in range or wild range lands can be achieved. Site specific weed management (SSWM) may be undertaken using these tools. Computer-analyzed video images of the digital data can provide the area estimates of weeds (Das, Sharma, & Pathak, 2012). Utilization of satellite positioning technology and advanced application controls allow growers to apply herbicides with pinpoint accuracy and to vary application rate based on the history of weed pressure or on observations of high-density weed patches.

Desertification

The remotely sensed satellite imagery is one of the widely used primary sources of data for land use/land cover classification and change detection analyses. It provides the spatial distribution of features and qualitative and quantitative information of features changes. Monitoring techniques based on multispectral satellite acquired data have demonstrated potential as a means to detect, identify and map changes in land use/land cover. It helps in identifying change between two or more dates that is uncharacterized of normal variation. Change detection is useful in many applications such as land use changes, rate of deforestation/afforestation etc. through spatial and temporal analysis techniques such as remote sensing along with digital image processing techniques (Ramesh, 2012)

The ecosystem of arid region is very fragile and land degradation is more during drought because rainfall is being the major source of sustaining entire production system as they suffered recurrent droughts. Within the scope of the combat against desertification, remote sensing allows to follow-up and monitor risk areas in the long term, to determine desertification factors, to support decision-makers in defining relevant measures of environmental management, and to assess their impacts

The intensive cultivation of land without conservation of soil fertility and soil structure, would lead ultimately to the springing up of the deserts and threats to sustainable agriculture. Remote sensing allows to follow-up and monitor risk areas in the long term, to determine desertification factors, to support decision-makers in defining relevant measures of environmental management, and to assess their impacts.

Nutrient Management

Soil is one of the important renewable natural resource that is the centre of all the activities controlling agricultural production. The production potential of soils varies with their fertility and inherent limitations. Therefore, reliable and accurate information on soil is of paramount importance for putting it to the best use. Information from remote sensing can be used as base maps in variable rate applications of fertilizers and pesticides so précised application and sustainable use of non- renewable resources may be taken. Growers need to apply nitrogen-based fertilizer in sufficient quantities to achieve the highest possible crop yields without over-applying - a situation that could lead to serious environmental effects. Current methods for determining the optimum timing of nitrogen fertilizer application can be costly, time consuming, and difficult. It was sometimes faces the human error also. Growers need to apply nitrogen-based fertilizer in sufficient quantities to achieve the highest possible crop yields without over-applying- a situation that could lead to serious environmental effects. In wheat, a critical factor comes down to timing in order to determine how efficiently plants will use nitrogen fertilizer. Current methods for determining the optimum timing of nitrogen fertilizer application can be costly, time consuming, and difficult. The remote sensing technique was found to accurately time nitrogen fertilizer applications 86% of the time across all field locations. Similarly, the development of different soil color and other texture modification may be distinguished by remote sensing and will help in necessary ameliorative measure.

Canopies of plants are grown in fields with varying soil and soil also has some unique reflective properties. The variation across wavelength is less for soil than for leaves; however, the reflectance changes in response to modifications in the surface. The addition of organic matter as residue on the surface reduces the reflectance. Soils vary in reflectance due to mineral composition and weathering of minerals (Chopada & Sharma, 2014).

The NDVI value of each area on an image helps identify areas of varying levels of plant nutrients status as the color of foliage is correlated with N supply within fields (Kumbar, Hatti, Sanjay, & Saunshi, 2014). This is based on the principle that growing plants absorb photo synthetically active radiation and reflect near infrared radiation. This helps in nitrogen management in terms of dose and N use efficiency. Remote sensing assisted to farmers to determine how much and when to supply nitrogen to minimize the loss through leaching to contribute groundwater for eutrophication and volatilization o pollute aerial environment.

Soil Salinity

Soil salinity caused by natural or human-induced processes is a major environmental hazard. Salinity reduces growth and productivity of vegetable crops in many salt-affected areas. It is estimated that about 20% of cultivated lands and 33% of irrigated agricultural lands worldwide are afflicted by high salinity (Foolad, 2004; Ghassemi, Jakeman, & Nix, 1995). In addition, the salinized areas are increasing at a rate of 10% annually; low precipitation, high surface evaporation, weathering of native rocks, irrigation with saline water, and poor cultural practices are the major contributors to the increasing soil salinity. In spite of the physiological cause ion toxicity, water deficit, and/or nutritional imbalance, high salinity in the root area sternly inhibits normal plant growth and development, resulting in reduced crop productivity or total crop failure (Ghassemi, et al., 1995).

Soil reclamation requires careful monitoring of the soil salinity status and variation to curb degradation trends, and secure sustainable land use and management.

Multi-temporal optical and microwave remote sensing can significantly contribute to detecting temporal changes of salt-related surface features. Various sensors (e.g. aerial photographs, satellite- and airborne multispectral sensors, microwave sensors, video imagery, airborne geophysics, hyper-spectral sensors, and electromagnetic induction meters) and approaches used for remote sensing helps identification and mapping of salt-affected areas (Metternicht & Zinck, 2003).

The addition of high salt content of soil or irrigation water even in large amounts causes the foliage temperature to be warmer than those plants irrigated with the same volume of salt free water or in neutral soil. The variation in temperature and its accurate measurement is possible by portable and hand held infrared thermometer or remote sensing (Chopada & Sharma, 2014).

Similarly, the development of different soil color and other texture modification may be distinguished by remote sensing and will help in necessary ameliorative measure. The temporal and spatial changes of salinity using combined approaches that incorporate different data fusion and data integration techniques are prime requirement for successful reclamation.

ASSESSMENT OF SOIL AND WATER RESOURCES

Remote sensing is very useful for assessing water availability in reservoirs, to determine irrigation water demand over space and time, water logging and salinity problems in irrigated lands and to evaluate the performance of irrigation systems (Kumar, 1999). Remote sensing has proven applications in various fields like, land use/land cover, land degradation, geology, soil conservation planning, water resources, forest and natural vegetation, agriculture outcrops, human settlement, etc. Farmers are pooling their livelihood on their resilience capacity and indigenous traditional knowledge backup in arid zones. Green revolution (high external input commodity based) technology generated thinking perhaps may not suite to these particular areas. In such circumstances the life and development is only based on integration and efficient utilization of available resources on sustained basis. The remote sensing will have helped in identification of locations for drilling wells and tube wells for effective exploitation of groundwater resources in arid zone under the circumstances of dark zones.

Insect Pest Management

Implementation of remote sensing techniques in the field of plant protection is necessary to decrease infestation and increase in sustainability yield. Excessive and indiscriminative use of insecticides in majority of crops leads to problems of insecticides resurgence, accumulation of harmful residues and toxicity to non-target organisms (Umbarkar & Wajire, 2014). The main aim of remote sensing for plants is to record as well as correlate radiation emitted by plants with various biotic and abiotic factors. This technique may also use for surveying and monitoring the insect pests (pest detections). As the remote sensing is very fast scouting system and cover a large area at a time is required for a sound economic and ecological pest management and especially in outbreak situations. The honey dew secretions from insect invite sooty mould fungus which leads to blackening of the foliage provides an evidence of insect and pathogens incidence that is readily visible by aerial photography. So the remote sensing can utilize on the integrated pest management programmes.

Precision Crop Management

Precision crop management is an agricultural management system designed to target crop and soil inputs according to within, field requirements to optimize profitability and protect the environment. The image-based remote sensing information provide spatially and temporally distribution of resources for precision crop management (PCM) as it has been hampered by a lack of timely distributed information on crop and soil conditions. We found that both aircraft- and satellite-based remote sensing could provide valuable information for PCM applications. Images from aircraft-based sensors have a unique role for monitoring seasonally variable crop/soil conditions and for time specific and time-critical crop management; current satellite based sensors have limited, but important, applications; and upcoming commercial Earth observation satellites may provide the resolution, timeliness, and high quality required for many PCM operations. Advising growers of the chance of night time frost, both early in the day and as the evening develops (Moran, Inoue, & Barnes, 1997).

Crop Marketing and Trade

In India, we still predominantly use traditional techniques such as field based crop cutting experiments (CCE) to assess the crop yield and acreage. Remote sensing helps in collecting basic data for monitoring of crop growth and helps in estimating the cropped area along with forecasting the crop production. It is worthwhile to note that in India all crop exports and import decisions are still based on historical production data (previous year's production records), as against the growing international trend of basing these decisions on more scientific and accurate methods such as assessing the current year's yield and acreage much in advance of the actual production by using remote sensing and GIS techniques. The ramifications of taking crucial export and import decisions based on historical data are that there could be a perceived shortage or surplus. The rainfall distribution in spatial and temporal seriously affects the production in most of agriculture commodities while remaining may or may not be affected but overall results into hike in prices in local market. The imports of such commodity without estimation of actual shortage will lower down the prices of local produce of the same commodity.

Remote sensing is increasingly being used to take vital decisions for crop marketing and export, crop inventory and commodity trading. It helps in crop acreage and yield estimation; health monitoring by agriculture information system through supply chain management, etc. by trimming the cost and human effort.

Agricultural data is currently generated by multiple agencies in multifarious ways by multiple agencies; both conventional field surveys based as well as advance information technology based. However, a prudent examination reveals that all the above data varies drastically. Aiming to resolve such issues of vagaries in the AI data, RMSI has established a geospatial approach to rationalize a competent methodology for SCM (Supply Chain Management) which can benefit farmers, traders, exporters, industrial, government, and federal agencies to combat the exports. Thus, remote sensing helps in estimating crop acreage estimation, crop yield estimation and production estimation for various crops.

Alternative Income Generation Activities

Mostly arid areas receive very less rainfall with improper distribution forced the farmers to take monocropping. Under such situation the residents of arid regions get employment during only four or five months and also unsecure. The remote sensing helps in selection of best sites for alternate land use systems easily assessable to peoples to get timber, fruit, fodder etc. throughout the year and generate additional employment and gave additional income to them. If sufficient employment is available in local vicinity and earning of income for livelihood, in such circumstances it stops migration towards urban areas and employment in riskier occupations.

Policy Relevant Analysis

The population in arid areas is very scared and development of them depends on the policies created by government for their development. Before formulating any policy and its successful implementation, the knowledge regarding inventory of infrastructure and available resources is pre-requisite. The reliable and accurate detailed information about resources and infrastructure will only be possible if provided by remote sensing. The information will also available in very short periods and lack of human errors so necessary actions may be taken very soon to avoid any causality.

Animal Productivity

The more than 50% of rural livelihood is almost dependent on livestock. The productivity of livestock is largely dependent on availability of quality fodder and drinking water. The water is scare and the quality of fodder is adversely affected in terms of digestibility and palatability. The quality of fodder in terms of accumulation of harmful ingredients under shortage of water is more and affects the production from these qualitatively and quantitatively. Similarly, the fertility or sometimes cause death in animal hits the normal life of the peoples. With the use of remote sensing images and information regarding type of livestock dominant in particular areas, useful vegetation of quality fodder matching the availability of water may be recommended. Similarly, necessary remedial measures against any disorders in animals and sometimes warning may be forecasted.

Urban and Regional Planning

As discussed in above paragraph under policy relevant analysis, the government or any local bodies may create infrastructures like construction of roads, e-kiosks, community service centers, custom hiring centers etc. easily approachable to local community is only possible by assessing the resources and available infrastructures though remote sensing only. Thus remote sensing helps in urban and rural developments and planning for better hood in arid areas.

Environmental Resource Monitoring and Generations

There is a vast scope of mining petroleum from arid area. These mining helps in generation of income and helps in increasing the opportunities for employment. Remote sensing helps in identifying the cost effective and locations rich area for establishment of refineries and allied industries. These mining helps in generating non-renewal resource within the country and save a lot of money expended on import of these materials ultimately will be utilized on other beneficiary projects. The CSR activities also helped in upliftment of rural economy. Environmentally-sound agriculture increasingly is information-driven. Implementation of new management paradigms such as Integrated Pest Management and Integrated Crop Management requires ready access, by the manager, to environmental data.

Plant Diseases Control

The monocropping is prevalent in arid regions to grow food grains for fulfilling the dietary requirement. The monocropping year after year increase the intensity of diseases regularly and sometimes takes epidemics particularly soil borne pathogens. Similarly, the plant pathologist is adopting new remote sensing techniques to study a variety of vector-borne diseases specially caused by arthropod like mosquitoes, ticks, blackflies, tsetse flies, and sandflies and are incurable in standing vegetation which are responsible for causing the majority of vector-borne diseases. Remote sensing makes it possible to merge more accurately ground observations of vector demographics and disease incidence with satellite data. Similarly, remote sensing helps in interpreting complex set of biotic and abiotic factors influence the emergence and spread of vector-borne diseases (Kalluri, Gilruth, Rogers, & Szczur, 2007). Remote sensing techniques can aid in determining the influence of abiotic environmental factors on their spread. Anthropogenic factors such as land cover conversion and increase in greenhouse gases are resulting in global climate change, which could expand the ecosystem boundaries of disease-carrying vectors, resulting in an increase in infectious diseases. There are impacting the routine use of remote sensing in epidemiology include accessibility to high resolution and low-cost imagery, as well as issues with data continuity in terms of consistencies in spatial, spectral, and temporal resolutions among satellite sensors during different years. Plant pathologist have used remote sensing tools for a number of years and were among the first use of color infrared photography to assess the presence of diseases in trees (Chopada & Sharma, 2014). Predicting development of important and endemic diseases of different crops, allow the growers for timely application of plant protective measures like fungicide sprays only when necessary.

Maps showing seasonal risks of vector-borne diseases will be necessary to monitor the impacts of global change on vector ecologies will help us to improve forecasts of epidemics. Automation of satellite data processing could lead to the generation of risk maps in near real-time to warn local health care professionals. Associations between satellite-derived environmental variables such as temperature, humidity, and land cover type and vector density are used to identify and characterize vector habitats. The use of remote sensing techniques to map vector-borne diseases has evolved significantly over the past 25 years. The status of remote sensing studies with vector habitats, as well as complex statistical models that link satellite-derived multi-temporal meteorological observations with vector biology and abundance, are discussed here. It assesses environmental variables important for epidemics such as air temperature, humidity, and rainfall in real-time and use as inputs for modeling and modifications Simple statistical models could help deduce the epidemiological processes from an analysis of the observed spatial patterns of the disease and the environment. The availability of resources for gathering, processing, and modeling geospatial data, training of personnel on the proper interpretation of results, cost effectiveness of these surveillance techniques, and the continuous availability of remote sensing data in a timely manner. It should also be emphasized that the allocation of resources for these novel monitoring techniques should not come at the cost of basic disease prevention and management activities at the community level.

LIMITATIONS AND CONSTRAINTS FOR SUCCESSFUL REMOTE SENSING

Unique remote sensing problems in arid regions are related to sparse vegetation, multiple species, and considerable bare ground. Thus, spectral interpretations must consider: multiple intermingled green and senescent species; considerable bare ground which includes cryptogamic soil crusts and powdery, endurated, or salinized surfaces; standing dead vegetation; litter; and shadows. Pixel modeling will be required in these heterogeneous environments. In particular, the lack of greenness tends to preclude the application of vegetation indices based on infrared/red ratios. New interpretation approaches to scene understanding, such as those included in this issue, should lead to useful procedures for arid lands:

- High cost involved.
- Bias and errors.
- The Paucity of secondary data.
- Lacks appropriate tools for advanced spatial analysis.
- Cloud free satellite data especially during Kharif season. Cloud cover can interfere with the data from a scheduled remotely sensed data system.
- Remote sensors located in fields or attached to agricultural equipment can provide the most frequent temporal resolution.
- The small and marginal and especially women farmers are often kept aloof from the mainstream of knowledge and information which in turn hinders their growth in their own field for raising crops.
- It does not help in prediction to evolve the new vector-borne pathogens.
- The failure of satellites and delays in the launch of new earth observing satellites.
- The restricted spectral range, coarse spatial resolution, slow turnaround time, and inadequate repeat coverage.
- Remote sensing taken from cameras mounted on airplanes is often used to provide data for applications requiring more frequent sensing.
- The interpretation of the data from instruments build on aircrafts or satellite through remote sensing needs some special training for accuracy.
- The low bandwidth is one of the major limitations in providing effective e-services to farmers.

HOW TO OVERCOME

- However, if the industrialists and federal government jointly produce such data, future generations will have easy utility of this supply chain management model.
- Establishment of effective and efficient kiosk or seva Kendras at rural and remotely situated farmers.
- Mass deployment of information kiosks for effective use of the Internet based content and services in economically feasible and handles important agricultural services for the people living in rural areas. These will further strengthen through distance education, training, insurance, rural post office, e-Governance etc. for timely dissemination of information to local rural and poor resources farmers.
- The initiatives have already been made to provide IT based services to rural community but most of the services revolve around limited subjects. It is necessary to form a coordination mechanism to strive for a concerted effort to support farming community in the country.
- The rural peoples generally like pictorial or graphic presentation because they are illiterate. For easy use and fast understanding, graphics based presentation must be prepared to encourage greater participation.
- The regional language fonts and mechanisms for synchronization of the content provide a challenge that needs to be met with careful planning.
- The timely alerts to the farmers and improved the efficiency of administration have major impact on the farmers and help them to appreciate the potential of information technology.
- The power supply must be insuring for usefulness of the intended services by explore solar power.
- Reliable connectivity is a prerequisite for a successful penetration of IT into rural areas.
- The networks with high bandwidth are being set up by several companies passing through rural segments which can be utilized by rural. The dynamic information which changes more frequently can be accessed from remote locations to obtain the latest status.

REFERENCES

Chopada, G., & Sharma, S. (2014). Integrating Remote sensing into Plant Pathology. *Agrobios Newsletter*, *12*(11), 42–43.

Das, T. K., Sharma, A. R., & Pathak, H. (2012). *Crop-Weed Balance under Climate Change. Climate Change Impact, Adaptation and Mitigation in Agriculture: Methodology for Assessment and Application*. New Delhi: Indian Agricultural Research Institute.

Foolad, M. R. (2004). Recent Advances in Genetics of Salt Tolerance in Tomato. [journal article]. *Plant Cell, Tissue and Organ Culture*, *76*(2), 101–119. doi:10.1023/B:TICU.0000007308.47608.88

Ghassemi, F., Jakeman, A. J., & Nix, H. A. (1995). *Salinisation of Land and Water Resources: Human Causes, Extent Management and Case Studies*. Wallingford, England: CAB International.

Jain, L. K., & Gupta, S. (2013). Sustainable Crop Production in Arid and Semi Arid Areas of Rajasthan (Conservation Agriculture: The New Paradigm). In S. G. N. Hada (Ed.), *Biotech* (pp. 27–51). Books.

Jain, L. K., Hada, N., & Bhaduria, S. S. (2014). Alternate use of Land for Better Livelihood in Western Rajasthan. In S. G. S. S. Tomar (Ed.), *Sustainable Rural Development through agriculture* (pp. 345–349). Biotech Books.

Kalluri, S., Gilruth, P., Rogers, D., & Szczur, M. (2007). Surveillance of Arthropod Vector-Borne Infectious Diseases Using Remote Sensing Techniques: A Review. *PLoS Pathog, 3*(10), e116. doi: 10.1371/journal.ppat.0030116

Kumar, D. N. (1999). Satellite Remote Sensing & GIS for Irrigation Management. Retrieved from civil.iisc.ernet.in/~nagesh/pubs/B01_RS_Irr_Manak_1999.pdf

Kumbar, B., Hatti, V., Sanjay, M. T., & Saunshi, S. (2014). Strategies to Enhance Nitrogen Use Efficiency in Agriculture. *Agrobios Newsletter*, *XII*(11), 37–40.

Levin, N. (1999). *Fundamentals of Remote Sensing* (L. Noam, Ed.).

Metternicht, G. I., & Zinck, J. A. (2003). Remote sensing of soil salinity: Potentials and constraints. *Remote Sensing of Environment*, *85*(1), 1–20. doi:10.1016/S0034-4257(02)00188-8

Mittal, S. C. (2013). Role of Information Technology in Agriculture and its Scope in India. Retrieved from http://125.19.12.220/applications/Brihaspat.nsf/6dca49b7264f71ce65256a81003ad1cb/82f2c15ccd4dd9a065256b37001af3fe/$FILE/it_fai.pdf

Moran, M. S., Inoue, Y., & Barnes, E. M. (1997). Opportunities and limitations for image-based remote sensing in precision crop management. *Remote Sensing of Environment*, *61*(3), 319–346. doi:10.1016/S0034-4257(97)00045-X

Navalgaud, R. R., Jayraman, V., & Roy, P. S. (2007). Remote sensing applications: An overview. *Current Science*, *93*(12), 1747–1766.

Patel, A. N., & Singh, S. (2015). Remote Sensing Principles and Applications (2nd rev. ed.). Scientific Publishers: India.

Ramesh, H. (2012). Change Detection in Harangi Watershed using Multi-temporal Satellite Imagery. *Paper presented at theInternational Symposium on Mountain resource management in a changing environment (MRMCE '12).*

Rao, D. P., Gautam, N. C., Nagaraja, R., & Mohan, P. (1996). IRS-1C applications in land use mapping and planning. *Current Science*, *7*(3), 390–395.

Shukla, K. K., Patel, D. J., & Radadiya, B. L. (2014). Role of Information Technology in Improvement of Current Scenario in Agriculture. *Oriental Journal of Computer Science and Technology*, *7*(3), 390–395.

Umbarkar, P. S., & Wajire, N. S. (2014). Remote sensing: A novel approach for insect pests management. *Agrobios Newsletter*, *XII*(11), 61–62.

Chapter 10
Applications of Active Remote Sensing Technologies for Natural Disaster Damage Assessments

Muhammad Tauhidur Rahman
King Fahd University of Petroleum and Minerals, Saudi Arabia

ABSTRACT

Immediately following a natural disaster, it is imperative to accurately assess the damages caused by the disaster for effective rescue and relief operations. Passive remote sensing imageries have been analyzed and used for over four decades for such assessments. However, they do have their limitations including inability to collect data during violent weather conditions, medium to low spatial resolution, and assessing areas and pixels on a damages/no damage basis. Recent advances in active remote sensing data collection methods can resolve some of these limitations. In this chapter, the basic theories and processing techniques of active remote sensing data is first discussed. It then provides some of the advantages and limitations of using active remote sensing data for disaster damage assessments. Finally, the chapter concludes by discussing how data from active sensors are used to assess damages from various types of natural disasters.

DOI: 10.4018/978-1-5225-1814-3.ch010

INTRODUCTION

Every year, natural hazards such as hurricanes and floods cause immediate destruction and loss of human lives and economic assets both in the United States and around the world. They also cause long-term impacts on ecological conditions and the bio-physical environment. These impacts are a compound result of the disaster itself and existing environmental management practices, but they also shape the management of risks associated with future disasters, including the response and recovery processes. As shown by Hurricane Katrina and Asian Tsunami, post-disaster environmental conditions often exert even more long-lasting adverse multidimensional impacts on human welfare than the disaster itself (Dolfman & Bergman, 2007; Slattery, Willett, Cobb, & Benson, 2010).

In 1995, recognizing the range and similarity between environmental and ecological problems that can be induced by disasters, the Federal Emergency Management Agency (FEMA) of the United States has adopted an all- hazards approach to manage and handle disasters and crisis events. The goal of this approach is not only to reduce the risk of loss of life and injuries, but also reduce the economic losses of man-made buildings and infrastructures, and of natural environmental losses including trees and crops from all major natural hazards. To achieve this goal, the U.S. National Academies recommend that researchers as well as emergency management teams use geospatial technologies (GIS and Remote Sensing (RS)) to monitor and assess damages to the natural and built environment immediately following a disaster (Mileti, 1999). Utilizing them will not only help in saving lives but will also allow rescue and relief operations to be carried out smoothly, and societies to get back to normal.

Review of literature show that remote sensing technology has been used extensively to monitor changes in the landscape (Araya & Cabral, 2010; Fichera, 2012; Hassan, Mahmud-Ul-Islam, & Rahman, 2015; Jamali & Rahman, 2016; M. T. Rahman, 2016). However, most of the optical sensors observe change (or damage) in change/no change basis. Due to the medium to large spatial resolution of these sensors, their usability is also limited when damage needs to be assessed at an individual neighborhood basis. For such cases, active remote sensing data such as Light Detection and Ranging (LiDAR) can be used to resolve some of the limitations. They have been used extensively within the last several years to create highly accurate 3D-Maps of the features on earth's surface. They have also been used on moving vehicles to create 360-degrees vertical profiles of buildings, dams, levees, and other structures and assess how much damage they sustained from storms and floods (Zhou, Gong, & Guo, 2015).

In this chapter, the basic theories behind active remote sensing will be first discussed. This will be followed by a discussion on the advantages and limitations of using active remote sensing technologies for disaster damage assessments. Finally, keeping the theme of the book in mind, it will conclude by providing description of how active remote sensing data and technologies can be used to assess damages from various natural disasters.

FUNDAMENTALS OF ACTIVE REMOTE SENSING

Active remote sensing technology primarily includes two types of sensors: RAdio Detection and Ranging (RADAR) and Light Detection and Ranging (LiDAR) sensors. RADAR has been widely used for various purposes including assessing soil moisture, planning transportation routes, analyzing geological features, estimating canopy bulk density and foliage moisture content, exploring minerals, planning land-use, navigating aircraft, modeling hydrologic features, and mapping rivers and large volcanic cones (Kornelsen & Coulibaly, 2013; Soja, Persson, & Ulander, 2015; Zajc, Pogačnik, & Gosar, 2014). It primarily works by having side-looking airborne RADAR (SLAR) apparatus attached to a space shuttle or flying aircraft (Figure 1). The SLAR unit consists of a pulse-generating device, a transmitter, a duplexer, an antenna, a receiver, a recording device, and a monitor or display showing the data being collected from the unit (Campbell & Wynne, 2011). The sensor transmits a microwave (radio) signal at a given frequency towards the target and detects the time it takes to receive the echo back from the surface of the earth. The strength of the backscattered signal is measured to discriminate between different targets and the time delay between the transmitted and reflected signals determines the distance (or range) from the unit to the target. Finally, the recorder displays and records the signal as an image for the analyst to view the surface of the earth (Campbell & Wynne, 2011). Table 1 provides a summary of the basic characteristics of some of the commonly used data available from RADAR sensors.

Figure 1. Direction of data capturing by RADAR (left) and LiDAR (right)

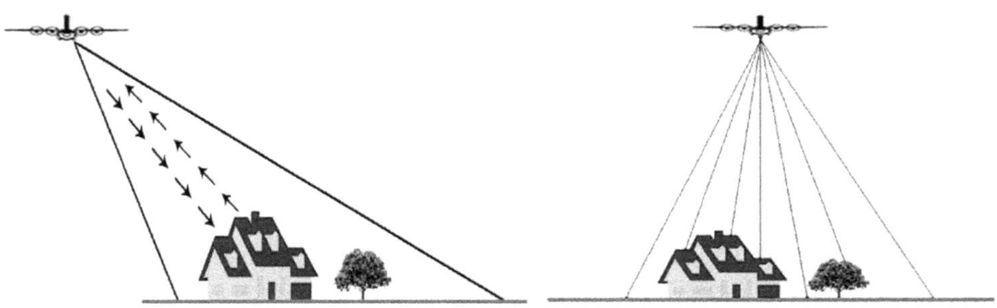

Table 1. Characteristics of commonly used RADAR sensors for disaster damage assessments

Satellite	Sensor	Swath (km)	Spatial Resolution	Repeat Cycle
ALOS	PALSAR (Fine)	40-70	10	44
	PALSAR (ScanSAR)	250-350	100	
Radarsat-1/-2	Fine	50	8	Every few days
Radarsat-2	Quad-pol fine	25	8	
Radarsat-1/-2	Standard	100	25	
Radarsat-2	Quad-pol standard	25	25	
Radarsat-1	Wide	150	30	
Radarsat-1/-2	ScanSAR narrow	300	50	
Radarsat-1/-2	ScanSAR wide	500	100	
Radarsat-1/-2	Extended high	75	25	
Radarsat-1	Extended low	170	35	
Envisat	ASAR standard	100	30	36 days
	ASAR ScanSAR	405	1000	
TerraSAR-X	Spotlight	10	1	11 days
	Stripmap	30	3	2.5-days
	ScanSAR	100	18	
Cosmo-Skymed	ScanSAR	100-200	30-100	37 hours
Sentinel-1		5-40	12	6 days

In each RADAR data set, many elements with varying brightness values are present and beginning interpreters often face difficulties extracting or analyzing features from the dataset. In RADAR data, the image is usually presented as an analog product and grey level of the scenes corresponds to the relative strength of the microwave energy backscattered from the surface. First step in data analysis includes calibrating the RADAR data and it can be a challenge for new interpreters to do so. Typically, data brightness values that are very dark and/or very bright landscape can be confused for shadows or solar illumination. Tone on the RADAR imagery is usually inversely related with the average intensity of the backscattered signal. High intensity returns as light tones while low intensity returns as dark tones on the image. Features such as streets, bridges, highways, airports, and racetracks can then be identified by their shape in the RADAR image and algorithms has been developed to extract and transfer them in remote sensing or GIS-compatible software. The size of these features can also be used as a ratio of the scale and dimensions for estimating other features on the data imageries.

Similar to RADAR sensors, data from LiDAR sensors are collected by a sensor attached to a small airplane flying at a low altitude (<3,000 m) (Figure 1). The sensor emits laser energy pulses from the airplane, which reach buildings, tree-tops, its understory branches, roof-tops, and the ground surface and then reflect back to the sensor (Arp, Greisbach, & Burns, 1982; Lim, Treitz, Baldwin, Morrison, & Green, 2003; Mata-Lima, Alvino-Borba, Pinheiro, Mata-Lima, & Almeida, 2013). The LiDAR system records the amount of time it takes for the laser pulses to return back to the sensor and the time measurement is converted into distance using the speed of light. The distance measurements are used to derive a three-dimensional view of the surface using the Differential Global Positioning Systems (DGPS). Data from the LiDAR sensors usually are delivered as a text file containing the latitude, longitude, and the elevation for every laser point that struck the earth's surface. When plotted, the continuous wave of laser reflections forms a three-dimensional model with x, y, and z coordinates. Once the data is pre-processed, the analyst has to use a GIS or remote sensing package to open and view the file. Using the text file, the data is interpolated based on the topmost laser points in order to create a raster grid showing (referred as digital surface model (DSM)) the highest elevation points for the entire area of interest (AOI). Kriging, Inverse Distance Weight (IDW), and Nearest Neighbor (NN) are the most common interpolation techniques that have been used. While Kriging has produced the most accurate interpolated results, due to calculation time constraints and computer memory limitations, IDW and NN is often preferred over Kriging (M. T. Rahman & Rashed, 2015). The user will also need to have a digital elevation model (DEM) of the entire AOI and calculate the difference between the pixel values of DSM and DEM in order to calculate the height of each individual object (i.e. trees, buildings, etc.).

ADVANTAGES AND LIMITATIONS OF ACTIVE REMOTE SENSING DATA

Compared to passive remote sensors, active remote sensing technology has several general advantages. First, passive sensors are sensitive and dependent on the solar illumination and their capabilities are limited by the time of day and the weather conditions. On the contrary, active sensors use their own energy to illuminate an area on the earth's surface. This allows data collection during day or night and in normal or severe weather conditions (Plank, 2014). Often data from active sensors are available before cloud free data from passive sensors are collected. Also, because the energy is emitted by the sensor itself, the operator or the analyst can know the complete properties and limitations of the sensor. Furthermore, active remote sensors provide unique information (i.e. air flow and sea wave direction) based on frequency characteristics, Doppler Effect, polarization, back scattering etc. This kind of information cannot be observed with visible and infrared sensors. When used for disaster damage assessment, pre- and post-storm disaster active remotely sensed data allows the emergency managers, researchers, and government organizations to quantify the damage, something which is impossible to do with passive remote sensing technology (Dash, Steinle, Singh, & Bahr, 2004; M. Li et al., 2008; M. T. Rahman & Rashed, 2015). Also, damage assessment can be done at very high (meter or even sub-meter) spatial resolution and this allows assessing damages at an individual item-by-item basis. Finally, the active sensors such as SAR have larger wavelength and are very sensitive to surface roughness and geometric configurations (Arciniegas, Bijker, Kerle, & Tolpekin, 2007). This property is essential for disaster damage assessments.

In terms of disadvantages, the high cost of obtaining the data (i.e. LiDAR) is a major setback of active sensors. Special aircrafts are often needed to collect the data. Sometimes these aircrafts are not available and even if they are, they have to be flown from a different part of the country or world. In poor developing countries of Asia and Africa, the aircrafts and/or sensors are lacking and data from passive remote sensors may be the only option for detecting damage or changes on the surface of the earth. Also, although RADAR signals are capable of penetrating clouds and other solid features, the analyst often face difficulties interpreting the amount of RADAR penetration and interpreting specific images. The data processing sometimes can be complicated, time consuming (depending on the speed and power of computers), and subjective on the interpreter experience and knowledge.

USE OF ACTIVE REMOTE SENSING DATA FOR DISASTER DAMAGE ASSESSMENT

Active remote sensing has been used extensively for assessing damages in different natural disasters in various parts of the world. In the following sections, a description of how active remote sensing data and technologies are used to assess damages from various natural disasters is summarized.

Infrastructure Damage Assessment from Earthquakes

Resulting from the sudden release of energy, earthquakes are well known for annually causing massive damage to buildings, roads, and infrastructure and injuring and killing hundreds if not thousands of people within a short period of time. To rescue and save human lives, techniques utilizing remote sensing data from SAR sensors have been found to be highly effective and accurate in assessing building and infrastructure damages from recent earthquakes. One of the most popular techniques for detecting small changes and deformation on the earth's surface due to earthquakes is the Differential SAR Interferometry (DInSAR). It calculates the damages causes by an earthquake by subtracting and calculating the differences in topography between SAR images (i.e. TerrraSAR-X, PALSAR) obtained before and after an earthquake event. Details of the methods and algorithms that are used for DInSAR are given in (Strozzi, Wegmuller, Tosi, Bitelli, & Spreckels, 2001). The technology has been used effectively in recent earthquakes of Landers (using interferometric C-band ERS SAR data), Wenchuan (from L-band SAR interferometric data), Bam and L'Aquila (utilizing ENVISAT ASAR data), Lushan (from RADARSAT-2 sensor), and Solomon Islands (from ALOS PALSAR Data) (Hashimoto, Enomoto, & Fukushima, 2010; Massonnet et al., 1993; Miyagi, Ozawa, & Shimada, 2009; Stramondo, Moro, Tolomei, Cinti, & Doumaz, 2005; Zebker, Rosen, Goldstein, Gabriel, & Werner, 1994). While the quality of the two compared SAR data are factors affecting accuracy, change accuracy level of about 0.5-1 cm for X-Band, 1-2 cm for C-band, and 2-3 cm for the L-band of the SAR data are obtainable for medium to highly damaged areas (Solano & Scavuzzo, 2013).

Earthquakes often lead to other types of disasters including landslides, avalanches, soil liquefaction, and tsunamis and it is often necessary to assess the damages from these events as well. The 8.0 magnitude Wenchuan earthquake of 2008 caused numerous landslides and completely destroyed the buildings, roads, and bridges in the county seats of Yingxiu town and Beichuan along the Yingxiu-Beichuan fault in the southwestern province of Sichuan, China (Dong, Li, Dou, & Wang, 2011; Fu et al., 2011). The landslides blocked the rivers in Longmenshan mountain region, which resulted in the formation of dozen barrier lakes. Using a combination of Synthetic Aperture Radars (SAR) at different spatial resolutions, researchers from the China Earthquake Administration and the Chinese Academy of Sciences were able to assess the damages in the cities. They found that the SAR sensor ALOS PALSAR (with 10 m spatial resolution) showed areas of landslides with strong backscattering and barrier lakes with lower backscattering. Using the 1-m high resolution TerraSAR-X data, they were able to identify individual building damages since collapsed buildings had lower backscattering compared undamaged buildings (Z. Li, Chen, Zhou, & Tian, 2009). With a combination of the SAR data and pre-earthquake GIS ancillary data along with the high resolution optical remote sensing data, it is now possible to asses and map the damage of individual buildings quite accurately.

Flood Inundation Assessments

Flooding are the most frequent, devastating, and costly natural hazard affecting the urban residents and destroying agricultural lands in the tropical and subtropical regions of the world (Creutin et al., 2013; Gruntfest, Showalter, & Ruin, 2008; Nahiduzzaman, Aldosary, & Rahman, 2015; Ruin, Creutin, Anquetin, Gruntfest, & Lutoff, 2009; Špitalar et al., 2014). Flooding can generally be categorized into flash flooding and riverine/regional flooding (Muhammad Tauhidur Rahman, Aldosary, Nahiduzzaman, & Reza, 2016). Regardless of the category, it is critical (and often the first step) to identify and map the areas vulnerable to flooding for creating an effective flood management strategy. For creating such maps and identifying the flood prone areas, remote sensing and GIS has played a crucial role for the past two decades. The use of passive remote sensors is quite limited for flood monitoring due to cloud cover and severe weather conditions during the flooding events. In such conditions, RADAR imageries are quite beneficial since the RADAR pulses can penetrate the cloud covers. The RADAR imageries also have an added benefit since the RADAR pulses can accurately distinguish between land and water.

In identifying the flooding areas from RADAR imageries, thresholding is possibly the most popular and frequently used technique (Brivio, Colombo, Maggi, & Tomasoni, 2002; Matgen, Schumann, Henry, Hoffmann, & Pfister, 2007; Townsend & Walsh, 1998). It works by first determining and setting a threshold of RADAR back scatter in decibel values. Based on that value, a binary algorithm is used to identify pixels that are below the threshold value to be classified as flooded zones. The remaining areas are then classified as dry areas (Martinis, Twele, & Voigt, 2009). The method is computationally simple and produces reliable and accurate maps quickly and efficiently. Once the areas are identified using thresholding method and if multi-temporal RADAR imageries are available, it is also possible to examine the progress of the flooding event between certain time periods. However, the method does have some problems and limitations. First, the results are dependent on the roughness of the terrain and water body. Rough waters produced by heavy winds and bad weather conditions during flooding appear to be brighter in the SAR imageries when compared to calm waters (Sanyal & Lu, 2004). This can confuse the interpreter in determining the threshold value for different flooding events. Second, forest coverage can also be quite problematic as flooded forest areas are brighter than non-flooded forest areas in RADAR imageries (Hess, Melack, & Simonett, 1990; Kundus, Karszenbaum, Pultz, Paramuchi, & Bava, 2001). Similarly, in urban settlement areas, the backscattering of the buildings can be similar to the backscattering of the flood water (Oberstadler, Honsch, & Huth, 1997). Third, the identification of flood inundation zones is subjective as it depends on the threshold values set by the operator (Martinis et al., 2009). Therefore, the accuracy of the results depends on the background and experience of the operator, the geography of the area, and climatic conditions during the flooding events. Newer SAR sensors (i.e. RADARsat) have advanced maneuvering capabilities that can reduce some of the above-mentioned limitations and identify flood affected areas in different parts of the world (Sanyal & Lu, 2004).

Hurricanes and Coastal Damage Assessments

Every year, hurricanes cause extensive damage to homes, trees, and vegetation in the eastern coastal areas of the United States. They are also responsible for causing hundreds of fatalities and billions of dollars in damages. For quick response, recovery, and reconstruction of the damaged areas and restoration of public services, it is imperative to assess the damages quickly and accurately within hours and days after the hurricanes make landfall. While traditional assessments involved house by house field surveys and inspection of damages, rapid developments in LiDAR data collection systems over the past two decades allows disaster managers and city planners to quickly fly/drive over the damaged areas and quantify the amount of damage to buildings and infrastructures. After the Hurricane Sandy made landfall on the north eastern coastal city of Fire Island, New York on October 29th of 2012, the United States Geological Survey (USGS) used LiDAR sensors to collect and create 3-D models of the coastal beaches and urban areas within three days before and after the hurricane made landfall. Results showed that the hurricane caused extensive damage to the homes and eroded the sands in the coastal beaches. In some areas, winds along with the high waves moved and washed-over sands almost 100 meters inland (Gong & Maher, 2014).

In addition to the traditional LiDAR sensors, mobile Terrestrial LiDAR" (or T-LiDAR that allows collection of vertical topographic profiles) sensors are quickly being deployed to create 3-D models of buildings and infrastructures. After Hurricane Sandy, also in 2012, Hurricane Isaac made landfall in 4 southern coastal states of the United States. After making landfall, T-LiDAR were deployed in the coastal cities and within a very short time, it allowed researchers from the USGS to collect over millions of topographic data points for buildings, dams, homes, and highways in a full 360-degree angle. By comparing to pre-storm data, interactive flood inundation and damage assessment maps were able to be created (Gong & Maher, 2014). Accuracy assessment results show that the damage estimations of buildings and infrastructures obtained by the LiDAR sensors are highly accurate and are not affected by environmental factors such as shadows, color and shape of buildings, and damage size (Kashani & Graettinger, 2015).

Assessment of Trees and Vegetation after Ice Storms and Hurricanes

In addition to damaging infrastructures and agriculture, hurricanes and ice storms also damage trees and vegetation in urban areas and coastal forests and marshlands. In urban areas, these trees cause millions of dollar damages due to destruction to power lines, residential houses, and agriculture. To assess such damages, data from Airborne Laser Scanners (ALS) collected before and after the storm can be useful in quantifying the damages sustained by each tree. However, quantifying such damages can be a time consuming process depending on the resolution and the areal coverage of the ALS data. Damage assessment typically involves three major steps. First, taking the differences between the first and last return ALS points produces a model (known as Canopy Height Model (CHM)) that represents the height of individual trees. Based on the model, the location and outline of individual tree boundaries are extracted using various tree extracting algorithms (Clark, Clark, & Roberts, 2004; Koukoulas & Blackburn, 2005; Maltamo, Mustonen, Hyyppä, Pitkänen, & Yu, 2004; Popescu, Wynne, & Nelson, 2002, 2003). Finally, the pre- and post-storm canopy height models are compared within the tree boundaries to calculate the differences in height and canopy due to the storm. Results from an ice storm of December, 2007 show that tree canopy and height damage estimates obtained for single standing trees were highly accurate (>81%) (M. T. Rahman & Rashed, 2015). However, the accuracy rate drops to 42% for detecting canopy damage of grouped trees since it is difficult to precisely extract the boundaries of grouped trees. The methodology can be used to assess tree damages from hurricanes in coastal areas as well.

CONCLUSION

The chapter attempted to review and discuss the use of active remote sensing for assessing damages following a major natural disaster. It first explained the basic theories and analytical techniques involved in using active remote sensing data. It also discussed some of the advantages and limitations in using active remote sensing technology. While data derived from active remote sensing may be collected during any part of the day/night and in any weather conditions, they can be costly, complicated to process, and sometimes may be difficult for novice analysts. However, if used and analyzed correctly, they can assess damages for various natural hazards in an efficient and accurate manner.

REFERENCES

Araya, Y. H., & Cabral, P. (2010). Analysis and Modeling of Urban Land Cover Change in Setúbal and Sesimbra, Portugal. *Remote Sensing*, *2*(6), 1549–1563. doi:10.3390/rs2061549

Arciniegas, G. A., Bijker, W., Kerle, N., & Tolpekin, V. A. (2007). Coherence- and amplitude-ased analysis of seismogenic damage in Bam, Iran, using ENVISAT ASAR data. *IEEE Transactions on Geoscience and Remote Sensing*, *45*(6), 1571–1581. doi:10.1109/TGRS.2006.883149

Arp, H., Greisbach, J., & Burns, J. (1982). Mapping in tropical forests: A new aproach using the laser APR. *Photogrammetric Engineering and Remote Sensing*, *48*, 91–100.

Brivio, P. A., Colombo, R., Maggi, M., & Tomasoni, R. (2002). Integration of remote sensing data and GIS for accurate mapping of flooded areas. *International Journal of Remote Sensing*, *23*(3), 429–441. doi:10.1080/01431160010014729

Campbell, J. B., & Wynne, R. H. (2011). Introduction to Remote Sensing (5th ed.). Guilford Press. Retrieved from http://books.google.com.sa/books?id=zgQDZEya6foC

Clark, M. L., Clark, D. B., & Roberts, D. (2004). Small-footprint lidar estimation of sub-canopy elevation and tree height in a tropical rain forest landscape. *Remote Sensing of Environment*, *91*(1), 68–89. doi:10.1016/j.rse.2004.02.008

Creutin, J. D., Borga, M., Gruntfest, E., Lutoff, C., Zoccatelli, D., & Ruin, I. (2013). A space and time framework for analyzing human anticipation of flash floods. *Journal of Hydrology (Amsterdam)*, *482*, 14–24. doi:10.1016/j.jhydrol.2012.11.009

Dash, J., Steinle, E., Singh, R. P., & Bahr, H. P. (2004). Automatic building extraction from laser scanning data: An input tool for disaster management. *Advances in Space Research*, *33*(3), 317–322. doi:10.1016/S0273-1177(03)00482-4

Dolfman, M. L., & Bergman, B. (2007). The effects of Hurricane Katrina on the New Orleans economy. *Monthly Labor Review*, *130*(6), 3–17. Retrieved from http://search.ebscohost.com/login.aspx?direct=true&db=bth&AN=26962930

Dong, Y., Li, Q., Dou, A., & Wang, X. (2011). Extracting damages caused by the 2008 Ms 8.0 Wenchuan earthquake from SAR remote sensing data. *Journal of Asian Earth Sciences*, *40*(4), 907–914. doi:10.1016/j.jseaes.2010.07.009

Fichera, C. R. (2012). Land Cover classification and change-detection analysis using multi-temporal remote sensed imagery and landscape metrics. *European Journal of Remote Sensing*. Doi:<ALIGNMENT.qj></ALIGNMENT>10.5721/EuJRS20124501

Fu, B., Shi, P., Guo, H., Okuayama, S., Ninomiya, Y., & Wright, S. (2011). Surface deformation related to the 2008 Wenchuan earthquake, and mountain building of the Longmen Shan, eastern Tibetan Plateau. *Journal of Asian Earth Sciences*, *40*(4), 805–824. doi:10.1016/j.jseaes.2010.11.011

Gong, J., & Maher, A. (2014). Use of Mobile Lidar Data to Assess Hurricane Damage and Visualize Community Vulnerability. Transportation Research Record. *Journal of the Transportation Research Board*, *2459*(2459), 119–126. doi:10.3141/2459-14

Gruntfest, E., Showalter, P., & Ruin, I. (2008). Flash Flood Research -- Past, Present, and Future. *Natural Hazards Observer*, *33*(1), 11–13.

Hashimoto, M., Enomoto, M., & Fukushima, Y. (2010). Coseismic Deformation from the 2008 Wenchuan, China, Earthquake Derived from ALOS/PALSAR Images. *Tectonophysics*, *491*(1–4), 59–71. doi:10.1016/j.tecto.2009.08.034

Hassan, S., Mahmud-Ul-Islam, S., & Rahman, M. T. (2015). Integration of Remote Sensing and GIS to Assess Vulnerability of Environmental Degradation in North-Western Bangladesh. *Journal of Geographic Information System*, *7*(05), 494–505. doi:10.4236/jgis.2015.75040

Hess, L. L., Melack, J., & Simonett, D. S. (1990). Radar detection of flooding beneath the forest canopy - a review. *International Journal of Remote Sensing*, *11*(7), 1313–1325. doi:10.1080/01431169008955095

Jamali, N. A., & Rahman, M. T. (2016). Utilization of Remote Sensing and GIS to Examine Urban Growth in the City of Riyadh, Saudi Arabia. *Journal of Advances in Information Technology*, *7*(4), 297–301.

Kashani, A. G., & Graettinger, A. J. (2015). Cluster-Based Roof Covering Damage Detection in Ground-Based Lidar Data. *Automation in Construction*, *58*, 19–27. doi:10.1016/j.autcon.2015.07.007

Kornelsen, K. C., & Coulibaly, P. (2013). Advances in soil moisture retrieval from synthetic aperture radar and hydrological applications. *Journal of Hydrology (Amsterdam)*, *476*, 460–489. doi:10.1016/j.jhydrol.2012.10.044

Koukoulas, S., & Blackburn, G. A. (2005). Mapping individual tree location, height and species in broadleaved deciduous forest using airborne LIDAR and multispectral remotely sensed data. *International Journal of Remote Sensing*, *26*(3), 431–455. doi:10.1080/01431160420002982 89

Kundus, P., Karszenbaum, H., Pultz, T., Paramuchi, G., & Bava, J. (2001). Influence of flood conditions and vegetation status on the radar back scatter of wetland ecosystem. *Canadian Journal of Remote Sensing*, *27*(6), 651–662. doi:10.1080/07038992.2001.10854907

Li, M., Cheng, L., Gong, J., Liu, Y., Chen, Z., Li, F., & Song, X. et al. (2008). Post-earthquake assessment of building damage degree using LiDAR data and imagery. *Science in China Series E: Technological Sciences*, *51*(S2), 133–143. doi:10.1007/s11431-008-6014-1

Li, Z., Chen, Q., Zhou, J., & Tian, B. (2009). Analysis of synthetic aperture radar image characteristics for seismic disasters in the Wenchuan earthquake. *Journal of Applied Remote Sensing*, *3*(1), 31685. doi:10.1117/1.3153906

Lim, K., Treitz, P., Baldwin, K., Morrison, I., & Green, J. (2003). Lidar remote sensing of biophysical properties of tolerant northern hardwood forests. *Canadian Journal of Remote Sensing*, *29*(5), 658–678. doi:10.5589/m03-025

Maltamo, M., Mustonen, K., Hyyppä, J., Pitkänen, J., & Yu, X. (2004). The accuracy of estimating individual tree variables with airborne laser scanning in a boreal nature reserve. *Canadian Journal of Forest Research*, *34*(9), 1791–1801. doi:10.1139/x04-055

Martinis, S., Twele, A., & Voigt, S. (2009). Towards operational near real-time flood detection using a split-based automatic thresholding procedure on high resolution TerraSAR-X data. *Natural Hazards and Earth System Sciences*, *9*(9), 303–314. doi:10.5194/nhess-9-303-2009

Massonnet, D., Rossi, M., Carmona, C., Adragna, F., Peltzer, G., Feigl, K., & Rabaute, T. (1993). The displacement field of the Landers earthquake mapped by radar interferometry. *Nature*, *364*(6433), 138–142. doi:10.1038/364138a0

Mata-Lima, H., Alvino-Borba, A., Pinheiro, A., Mata-Lima, A., & Almeida, J. A. (2013). Impacts of natural disasters on environmental and socio-economic systems: What makes the difference? *Ambiente & Sociedade*, *16*(3), 45–64. doi:10.1590/S1414-753X2013000300004

Matgen, P., Schumann, G., Henry, J., Hoffmann, L., & Pfister, L. (2007). Integration of SAR derived river innundation areas, high-precision topographic data and a river flow model toward near real-time flood management. *International Journal of Applied Earth Observation*, *9*(3), 247–263. doi:10.1016/j.jag.2006.03.003

Mileti, D. (1999). *Disasters by Design*. Washington, D.C.: Joseph Henry Press.

Miyagi, Y., Ozawa, T., & Shimada, M. (2009). Crustal deformation associated with an M8.1 earthquake in the Solomon Islands, detected by ALOS/PALSAR. *Earth and Planetary Science Letters*, *287*(3), 385–391. doi:10.1016/j.epsl.2009.08.022

Nahiduzzaman, K. M., Aldosary, A. S., & Rahman, M. T. (2015). Flood induced vulnerability in strategic plan making process of Riyadh city. *Habitat International*, *49*, 375–385. doi:10.1016/j.habitatint.2015.05.034

Oberstadler, R., Honsch, H., & Huth, D. (1997). Assessment of the mapping capabilities of ERS-1SAR data for flood mapping: A case study of Germany. *Hydrological Processes*, *10*(10), 1415–1425. doi:10.1002/(SICI)1099-1085(199708)11:10<1415::AID-HYP532>3.0.CO;2-2

Plank, S. (2014). Rapid Damage Assessment by Means of Multi-Temporal SAR — A Comprehensive Review and Outlook to Sentinel-1. *Remote Sensing*, *6*(6), 4870–4906. doi:10.3390/rs6064870

Popescu, S. C., Wynne, R. H., & Nelson, R. F. (2002). Estimating plot-level tree heights with lidar: Local filtering with a canopy-height based variable window size. *Computers and Electronics in Agriculture*, *37*(1-3), 71–95. doi:10.1016/S0168-1699(02)00121-7

Popescu, S. C., Wynne, R. H., & Nelson, R. F. (2003). Measuring individual tree crown diameter with lidar and assessing its influence on estimating forest volume and biomass. *Canadian Journal of Remote Sensing*, *29*(5), 564–577. doi:10.5589/m03-027

Rahman, M. T. (2016). Detection of Land Use/Land Cover Changes and Urban Sprawl in Al-Khobar, Saudi Arabia: An Analysis of Multi-Temporal Remote Sensing Data. *ISPRS International Journal of Geo-Information*, *5*(2), 15. doi:10.3390/ijgi5020015

Rahman, M. T., Aldosary, A., Nahiduzzaman, K. M., & Reza, I. (2016). Vulnerability of flash flooding in Riyadh, Saudi Arabia. *Natural Hazards*. doi:10.1007/s11069-016-2521-8

Rahman, M. T., & Rashed, T. (2015). Urban tree damage estimation using airborne laser scanner data and geographic information systems: An example from 2007 Oklahoma ice storm. *Urban Forestry & Urban Greening*, *14*(3), 562–572. doi:10.1016/j.ufug.2015.05.008

Ruin, I., Creutin, J.-D., Anquetin, S., Gruntfest, E., & Lutoff, C. (2009). Human vulnerability to flash floods: Addressing physical exposure and behavioral questions. In P. Samuels & S. Huntington (Eds.), *Flood Risk Management: Research and Practice* (pp. 1005–1012). London: Taylor & Francis.

Sanyal, J., & Lu, X. X. (2004). Application of remote sensing in flood management with special reference to monsoon Asia: A review. *Natural Hazards*, *33*(2), 283–301. doi:10.1023/B:NHAZ.0000037035.65105.95

Slattery, M., Willett, K. L., Cobb, G. P., & Benson, W. H. (2010). Multiple facets of environmental impacts from Hurricane Katrina. *Environmental Toxicology and Chemistry*, *29*(7), 1401–1402. doi:10.1002/etc.213 PMID:20821586

Soja, M. J., Persson, H., & Ulander, L. M. H. (2015). Estimation of Forest Height and Canopy Density from a Single InSAR Correlation Coefficient. *IEEE Geoscience and Remote Sensing Letters*, *12*(3), 646–650. doi:10.1109/LGRS.2014.2354551

Solano, E., & Scavuzzo, M. (2013). Remote Sensing to Assessment Damages Post Earthquakes, focusing in Urban Structures. Retrieved from http://aulavirtual.ig.conae.gov.ar/moodle/pluginfile.php/513/mod_page/content/71/PostEarthquake.pdf

Špitalar, M., Gourley, J. J., Lutoff, C., Kirstetter, P.-E., Brilly, M., & Carr, N. (2014). Analysis of flash flood parameters and human impacts in the US from 20062012. *Journal of Hydrology (Amsterdam)*, *519*, 863–870. doi:10.1016/j.jhydrol.2014.07.004

Stramondo, S., Moro, M., Tolomei, C., Cinti, F. R., & Doumaz, F. (2005). InSAR surface displacement field and fault modelling for the 2003 Bam earthquake (southeastern Iran). *Journal of Geodynamics*, *40*(2–3), 347–353. doi:10.1016/j.jog.2005.07.013

Strozzi, T., Wegmuller, U., Tosi, L., Bitelli, G., & Spreckels, V. (2001). Land subsidence monitoring with differential SAR interferometry. *Photogrammetric Engineering and Remote Sensing*, *67*(11), 1261–1270.

Townsend, P. A., & Walsh, S. J. (1998). Modelling flood plain inundation using integrated GIS with radar and optical remote sensing. *Geomorphology*, *21*(98), 295–312. doi:10.1016/S0169-555X(97)00069-X

Zajc, M., Pogačnik, Ž., & Gosar, A. (2014). Ground penetrating radar and structural geological mapping investigation of karst and tectonic features in flyschoid rocks as geological hazard for exploitation. *International Journal of Rock Mechanics and Mining Sciences*, *67*, 78–87. doi:10.1016/j.ijrmms.2014.01.011

Zebker, H. A., Rosen, P. A., Goldstein, R. M., Gabriel, A., & Werner, C. L. (1994). On the derivation of coseismic displacement fields using differential radar interferometry: The Landers earthquake. *Journal of Geophysical Research*, *99*(B10), 19617–19634. doi:10.1029/94JB01179

Zhou, Z., Gong, J., & Guo, M. (2015). Image-Based 3D Reconstruction for Posthurricane Residential Building Damage Assessment. *Journal of Computing in Civil Engineering*. doi:10.1061/(ASCE)CP.1943-5487.0000480

Chapter 11
Site-Suitability Analysis for the Identification of Potential Sites to Construct Rain Water Harvesting Systems

Shreyashi Santra Mitra
Haldia Institute of Technology, India

ABSTRACT

Remote Sensing and Geographic Information Systems are a boon to the modern scientific domain as it provides a strong base for understanding the Earth's behavior in so many ways and help us to understand its response to certain conditions. With the exploitation of these two modules important spatial decisions can be made at ease with quite a satisfactory accuracy level. With the context of global warming water scarcity is being identified as sector demanding strategical implementation of plans for sustainable development. To ensure water supply water conservation is required. Runoff harvesting could be an option to that if the area under plan experience substantial runoff in a year. Lower Kasai Keleghai interfluve situated in West Bengal falls under monsoonal climatic regime. GIS based site-suitability analysis was adopted here based on LULC and Curve Number to identify potential zones for runoff harvesting. The result shows that 18.71% of the interfluve is suitable for providing a base for runoff harvesting.

DOI: 10.4018/978-1-5225-1814-3.ch011

INTRODUCTION

India being a country of climatic variability has a versatile rainfall regimes ranging from highly precipitated to arid (Auffhammer, Ramanathan, & Vincent, 2012). Moreover, across most of the parts of the country precipitation is concentrated over the monsoon months of the year only (Loo, Billa, & Singh, 2015), though if planned properly the amount of water received in a year of normal monsoon precipitation could be sufficient to sustain the regional demand of water to a certain level. However, suitable conservation system is required to store the water of rainy season to meet the demand for the rest of the year. Runoff or groundwater restoration techniques are implemented to serve this purpose. However, the conservation Environmental statistics indicating repetitive flood in Indian subcontinent ensures a possibility of rainfall/runoff harvesting to support water conservation system in some regions in India (Şen, Al Alsheikh, Al-Turbak, Al-Bassam, & Al-Dakheel, 2013). Rainfall or runoff harvesting has now been treated as one of the most dominant methods of water resource conservation that doesn't only control flood, but also controls soil erosion and accelerates the rate of groundwater recharge (Kadam, Kale, Pande, Pawar, & Sankhua, 2012). Runoff is the most dominant parameter capable of predicting potential sites for rainwater harvesting. Runoff potential zone delineation and targeting them to arrest the water in a storing reservoir or any other structure therefore is an excellent solution to the water scarcity (Traboulsi & Traboulsi, 2015). However, identification of the runoff potential of a region considering its geological set-up is a spatial task and needs a thorough analysis of the spatial conditions of various factors in temporal dimensions (Kadam, et al., 2012). Dependence of runoff primarily on soil, landuse landcover, slope, etc. makes the investigation even more complex (Kizza, Rodhe, Xu, & Ntale, 2011; Whigham & Crapper, 2001; Wissinger, Shankar, & Restrepo, 2014). A combined methodological framework of Remote Sensing and GIS could provide a very convincing tool to deal with this problem and has been proved very effective in Spatial Decision Support System of many societies (Ramakrishnan, Bandyopadhyay, & Kusuma, 2009; Rolland & Rangarajan, 2013). Remote Sensing can identify environmental parameters at different spatio temporal scale, while GIS has a proven functionality to analyse the parameters as information to deduce various solutions to a number of problems. A framework based on a combined interplay of RS/GIS helps assess a number of factors responsible from runoff variations.

Runoff predictions are usually executed using runoff models, which in terms of spatial domain can be classed as lumped, distributed and semi-distributed models (Carpenter & Georgakakos, 2006). Among these three models, lumped model is based on an assumption that representation of hydrologic parameters of sub watershed within a watershed can be efficiently done using weighted average method (Seiller,

Hajji, & Anctil, 2015). These models usually avoid complexity and the discharge at the basin outlet is entirely correspondent to the global dynamic of the hydrologic system. Various models/methods have been used so far for this purpose. SCS-CN model is one of the simple and very popular methods to provide efficient outputs especially with an area extension within 15 sq. km. However, periodic modifications were made to make this model applicable to the basins with diverse characteristics all over the world (Mishra, Singh, Sansalone, & Aravamuthan, 2003; Santra Mitra, Wright, Santra, & Ghosh, 2015). Furthermore, ability of this model to fit easily in RS/GIS based approach also has raised its applicability and usability among the researchers. Runoff estimation and its regionalization using SCS-CN method aided by computer oriented modelling provides the hydrologists fairly accurate spatial dynamics of a small watershed and can be used as a basis for water resource planning by the Government at various levels.

Harvesting of runoff as mentioned earlier is considered as one of the most competent approach towards attaining sustainable development. Several studies have been done so far to seek for the suitable sites potentially viable for constructing structures for the purpose of runoff harvesting. Considering the amount of information required for identifying potential sites for runoff harvesting GIS can be considered as a critical tool in hydrological modelling because of its capacity to handle large and complex geospatial datasets (de Winnaar, Jewitt, & Horan, 2007; Mekonnen, Melesse, & Keesstra, 2016). GIS functions, especially related to the spatial analyses helps the researchers in aggregating, comparing and evaluating various hydrologic parameters of a single or multiple spatial units for a simple or complex timesteps. (Coskun & Musaoglu, 2004; Santra Mitra, et al., 2015; Winnar, Jewitt, & Horan, 2007). The main objective of this chapter is to determine the potential runoff sites of lower Kasai-Keleghai interfluves based on SCS-CN runoff modelling processed in GIS platform.

STUDY AREA

The Kasai river basin is the largest drainage basin, covering almost 9658 Sq Km of area in the Western part of West Bengal and Keleghai being one of the most important right hand tributaries of Kasai River, forms its basin having an area of 2015 Sq Km over Medinipur District. The Kasai River originates from the Jabourban hill (642 m altitude) of Purulia District and flows Eastward through the undulating plateau surface of mainly four districts of West Bengal i.e. Purulia, Bankura, Purba and Paschim Medinipur; and unites with Keleghai River to form Haldi River. On the other hand, Keleghai, originating from the plateau surface of Paschim Medinipur, surges to the Easterly direction to form Haldi river after converging with Kasai river. The interfluve, formed by these two rivers is considered as the intermediate common zone of influence the lower Kasai and Keleghai rivers that belong to same drainage system. This interfluve forms the area of the study. The western and eastern limits of this interfluve happens to be the originating point of the Keleghai river and Haldi river respectively, whereas the southern extent are marked by the Kasai and Keleghai river themselves. Geographically the area under the interfluve extends from the 22°0′55″ N to 21°57′15″ N and 87°3′18″ to 87°49′48″E approximately with an area of 2585.28 sq.km (Figure 1).

Methods

In the light of the growing need for the cost-effective and time saving methods of harvesting, it is absolutely necessary to locate and identify areas that are suitable for introducing harvesting technologies. The identification should be completely based on the necessity and feasibility factors. In other words, it is a prior task of the environmentalists to find out the areas where the necessity is there to implement a harvesting system. On the other hand, it becomes inevitable to mark out the

Figure 1. Location of Kasai-Keleghai Interfluve with inset map of West Bengal

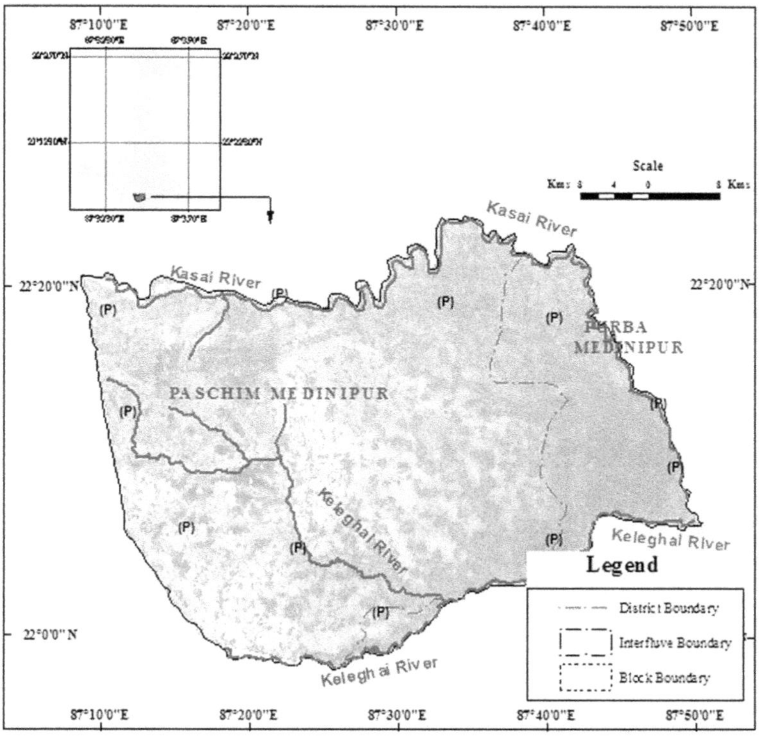

areas that can provide suitable factors to sustain the harvesting system. It is due to the fact that harvesting system implemented in unfavorable physical set-up cannot function properly. Considering these facts, the researcher has tried to identify the favorable sites in the lower Kasai-Keleghai interfluve where harvesting systems may be installed successfully. The entire methodology and outputs for identifying potential sites have been discussed below.

As far as the roof water harvesting is concerned the potential sites have to be built up areas either in form of rural house type or urban house type. So the only criterion for the selection of the type of harvesting has been restricted to settlement sites either the residential or commercial. For the extraction of those areas, landuse/ land cover maps prepared from IRS LISS-III imageries of year 2013have been used.

Methods of Identifying Runoff-Harvesting Zones

A scientific method for identifying potential sites that support runoff harvesting interventions was developed through utilizing GIS. Being site-specific, assessing the suitability of the land for runoff harvesting requires quantitative data and involves integration of specific criteria (Ziadat, Mazahreh, Oweis, & Bruggeman, 2006). GIS provide a useful approach by enabling to develop a frame work for collecting, storing, transforming and analyzing the spatial and non-spatial data to select sites for rain water harvesting (Coskun & Musaoglu, 2004; Padmavathy, Ganesha, Yoga-rajan, & Thangavel, 1993). In this work the author has tried to identify the potential areas for successful runoff harvesting. Runoff harvesting zones were demarcated by identifying potential runoff generating sites with the help of landuse-landcover, Hydrologic Soil group (HSG), slope and the amount of rainfall.

The probabilities of construction of rainwater harvesting systems have been assessed for the entire lower Kasai-Keleghai interfluve. From the detailed analysis of terrain characteristics, slope, landuse land-cover, soil condition etc the best sites have been chosen. The criteria considered for the site selection have been discussed as follows:

CRITARION-I: MODIFIED SCS-CURVE NUMBERS

Modified SCS –Curve number is a composite reflection of interaction between slope, landuse-landcover, hydrologic soil group and its antecedent moisture condition. This variable is an integral part of the model, viz. Modified SCS-CN model The soil conservation services (SCS) method is one of the most widely used techniques for estimating surface runoff as a response to a given rainfall amount from small basin or catchment (Das & Paul, 2006). This method considers slope, landuse-landcover, hydrologic soil group and their interaction. Depending on all these factors CN values have been calculated by the method as discussed in chapter 4 for the entire interfluve. These values help to delineate the areas with various amount of potential runoff. The CN value has been used as one of the criteria for site suitability analysis.

Curve Number (CN) Estimation

As per NRSC, the Curve Number (CN) values are a direct representation of the combined interaction of AMC, Soil and Landuse-Landcover status of the study area. The methods of determining CN values for the study area has been mentioned below.

Extraction of AMC Values

AMC refers to the total moisture content present in the soil just prior to the initiation of the rainfall-runoff event under consideration. Initial Abstraction and Infiltration are governed by this AMC conditions. In this study AMC values have been calculated with the help of the daily rainfall data for 3 years from 2012-2015. For each day AMC value is obtained by the summation of rainfall values (in mm.) of the preceding five days. Depending on the AMC values, all the days have been assigned to the corresponding AMC classes. The range of values for AMC categorization is given in Table 1. According to the US system the AMC class limits should be generated for two seasons separately i.e. for growing season and dormant season. But, in the present study only the growing season limit has been used because of the fact that West Bengal produces different types of crops throughout the year, unlike USA.

Determination of CN Values

CN is a function of landuse/landcover (LULC) and hydrologic characteristics of soils in each spatial unit. The Landuse map has been prepared for each spatial unit using visual interpretation technique applied on IRS-P6 LISS-III imagery. The soil map has been acquired from NATMO. While preparing landuse-landcover map, water bodies and rivers were excluded carefully as their contribution in the expected runoff over this interfluve would be nil. After the landuse/landcover and hydrologic soil maps have been prepared, both of the maps were intersected in GIS environment. The resultant map gave the information about the landuse type for each soil type. Prescribed CN values for each category were assigned with the help of a published CN table. Then the CN value of each class was multiplied with its corresponding area. Then the Sum of these values was normalized by the sum of the area to obtain the weighted CN value. This value is designated as CN-II value

Table 1. Antecedent Moisture Condition (AMC) for determining CN values

AMC Types	Total Rain in Previous Five Days in mm		Remarks
	Dormant Season	Growing Season (Used in this Study)	
I	< 13	<36	Soils are dry but not to wilting point. Satisfactory cultivation has taken place
II	13 – 28	36 - 53	Average Condition
III	>28	>53	Sufficient rainfall has occurred within the immediate 5 days. Saturated soil conditions prevail

Source: US Soil Conservation System

applicable for AMC-II condition. With the help of this value CN-III and CN-I values were calculated using equations written below:

$$CN - I = \left(\frac{CN_{II}}{2.281 - 0.01281\ CN_{II}} \right)$$

Once the CN-I, II and III values have been obtained, these are used to estimate the modified CN-II value. The modified CN-II value includes slope factor also. The slope map was prepared in ARC GIS by using the spot height values as obtained from the Survey of India Toposheet. Then the following relation has been employed to get the final modified CN-II value.

$$ModifiedCN - II = \left(\frac{1}{3} \right) \times \left(CN - III - CN - II \right) \times \left(1 - 2 \times EXP\left(-13.86 \times Slope \right) + CN - II \right)$$

The entire process of Curve Number values retrieval has been stated below:

- Retrieval of CN-II value based on landuse land cover and hydrologic soil classes.
- Retrieval of CN-I and CN-III values based on CN-II value using empirical formulae.
- Retrieval of modified CN-II values dependent on Cn-I, CN-III and Slope values.
- Retrieval of CN-I and CN-III values based on modified CN-II value.

CRITARION-II: LANDUSE LANDCOVER STATUS

Landuse-landcover types can play a dominant role to obstruct or allow runoff dynamics in a given region. In order to find out the dominant LULC crucial for controlling runoff pattern, is identified using regression analyses. From the multivariate analysis it was clear that the built up area and crop-fields have dominant influence on runoff potential. With these findings, the two factors that have been taken into consideration are homesteads and croplands while choosing the potential sites. To extract these two landuse categories, landuse/landcover map was prepared by the author for the identification of crop-fields and settlement areas. From these areas buffers were created based on varying distance intervals taken as a straight line (Euclidean distance). Based on certain logic each buffer class for both cropland and settlement was then classified in suitability ranks to facilitate suitability analysis. Higher the ranking is, more will be the suitability potential for runoff harvesting. The logic applied for ranking are mentioned as below:

Suitability or requirement for runoff or drive-way harvesting decreases with the increasing distance from homesteads (Winnar, et al., 2007). This may be due to the fact that with the increasing distance from the settlement area, the requirement for the water decreases substantially. So it is wise to set up harvesting system near to these landuse classes

In the cropland the runoff potential is minimum. This is due to the obstruction caused by the crop-interception. So in the field where crops are cultivated the runoff harvesting potential is minimal. For this it is difficult to construct the harvesting system in the crop-field itself. But in the immediate surroundings of the cropped area, harvesting should be done to supply water. Again with the increasing distance from the agricultural land, need for the system decreases as demand decreases.

The buffers created by the GIS analysis, the researcher have ranked them according to the necessity classes. The ranking systems are purely based on intense review of literature and field observation. The systems are described below in Table 2:

The buffers created in GIS environment for Settlement area and cropland area are shown in Figure 2

Table 2. Suitability rankings associated with each distance interval class for homesteads and crops, low rankings characterize areas with a high suitability

Interval Class	0 km *	0-250 m	250-500 m	500-1000 m	1500-2000 m	>2000 m
Settlement	1	1	2	3	4	5
Cropland	5	1	2	3	4	5

* 0 km distance means within the particular landuse territory

After classifying the entire interfluve in suitability ranking zones, the author has overlaid it on CN map and depending on some conditional logic, the area has been divided into 5 suitability zones. The logical and spatial operators of ARC GIS 9.2 (ESRI, 2008) have been used for the analysis. The conditions for classifying the interfluve in suitability zones are as follows as given in Table 3:

A conceptual diagram of the entire process of site-suitability analysis has been represented in Figure 3.

Figure 2. Buffers generated in ARC GIS environment from Settlement (a) and Cropland (b)

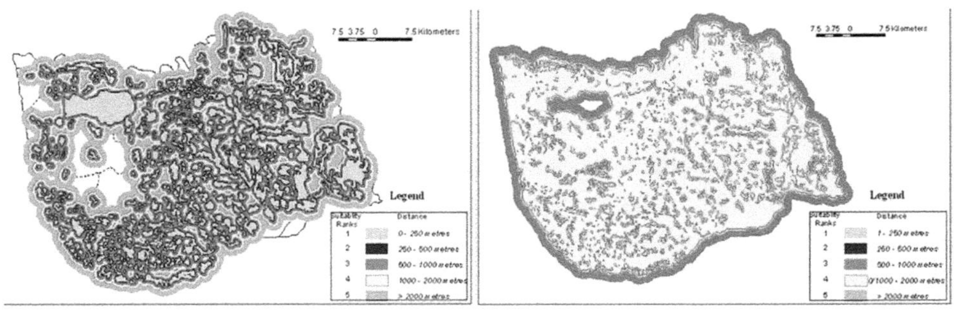

Table 3. Conditions for classifying the interfluve in harvesting-suitability zones

Suitability	Cases	Conditions		
Highest	Case:1	CN	Homestead	Cropland
		>85	1-4	1-4
	Case:2	75-85	1-3	1-3
Moderate	Case:1	75-85	3-4	3-4
	Case:2	65-75	1-3	1-3
	Case:3	50-65	1-2	1-2
Somewhat	Case:1	>50	4-5	4-5
	Case:2	40-50	1-3	1-3
Least	Case:1	<40	1-5	1-5
Not Suitable		<40	5	5

Source: Computed by the Author based on Field survey and Literature review

Site-Suitability Analysis

Figure 3. Conceptual framework of the site-suitability analysis

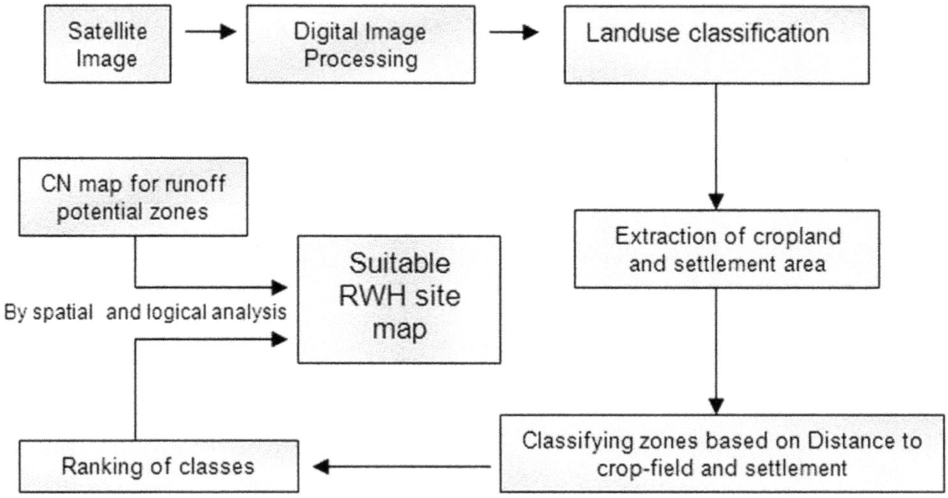

RESULTS AND DISCUSSIONS

Landuse-Landcover of the Interfluve

Landuse-Landcover analysis is indispensable while giving a geographical account of a given areal unit. Activities carried out on a piece land are the foremost criteria for the classification of landuse. On the other hand, the natural distribution of unchanged surface of the Earth can be categorized as landcover features. The study of both of them goes hand in hand.

The landuse-land cover map of the entire interfluve has been prepared using IRS LISS-3 P6 imagery acquired in the year of 2007. The classification has been done by visual interpretation technique in ENVI software with an accuracy level of 81.75%. The major landuse/landcover (LULC) categories are given in Table 4. From the classification it has been observed that the interfluve conforms to the characteristics of any developing economic regions in terms of its landuse features. In order to evaluate the interfluve's potential for runoff trapping two LULC classes were used to set the criteria, i.e. cropland/agriculture and settlement. Almost 76.20% of the total geographical area of the interfluve is used for the agricultural activities. The residential area, including both urban and rural, accounts for 17.31% of the total area (Table 4).

Table 4. Landuse-Landcover characteristics of lower Kasai-Keleghai interfluve

Landuse-Landcover Types	Area in sq.km.	% to total geographical area
Forest	66.00	2.55
Medium Dense Forest	6.02	0.23
Degraded Forest	1.88	0.07
Bush	7.44	0.29
Sand deposit	11.15	0.43
Agriculture	1970.10	76.20
Pastoral	11.83	0.46
Scattered Settlement	447.49	17.31
Water body	63.35	2.45

Curve Number

In order to find out Curve numbers and modified curve numbers the author required precipitation amount of available raingauge stations. The rainfall amount is the controlling factor for estimating Antecedent Moisture Condition (AMC). There were six rain gauge stations within the study are. CN values for each of this station was found by creating theissen polygons treated as a zone of influence of each rain gauge stations. The CN and modified CN is provided with the help of the Table 5. The highest modified CNI, II and III can be found in Tamluk station, whereas lowest is found in Medinipur station signifying high and low runoff potential respectively.

Table 5. Curve number and modified curve numbers for zone of influence of six raingauge stations

Spatial Units	Normal			Modified		
	CNI	CNII	CNIII	CNII	CNI	CNIII
Tamluk	72.28832	85.6119	93.30426	87.7706668	75.88301	94.38458
Keshiary	48.87145	68.55646	83.62293	73.29380096	54.61102	86.53612
Keshpur	50.64676	70.06691	84.5725	73.36933648	54.70674	86.58106
Pingla	54.40024	73.12709	86.43678	76.26138018	58.47854	88.26776
Medinipur	46.61013	66.57023	82.3433	71.62420561	52.5299	85.53097
Jhargram	56.45327	74.72865	87.38199	78.86235683	62.05856	89.73038

Site-Suitability Analysis

As par results obtained from the site-suitability analysis, the various potential harvesting sites have been extracted and shown in Figure 4. The roof water harvesting may be done for about 19.61% of the area (509.94 sq.km) of the entire interfluve. Wherever there is settlement, roof water harvesting may be executed successfully. As far as runoff harvesting is concerned, the entire area has been divided into four suitability zones. The brief description of the zones has been given below:

- **Highly Suitable Zone:** Almost all the blocks within the interfluve has some highly potential sites for setting up runoff harvesting system. But the Eastern part of the interfluve offers more suitable conditions for the harvesting. In Kharagpur-II block in Paschim Medinipur, around the industrial and residential complex, there lies a potential site covering almost 10 sq.km. So far as the Highly Suitable Zone is concerned, the harvesting system can be set up at least in 309.68 sq.km of area accounting for 11.98% of the total area of the entire interfluve.
- **Moderately Suitable Zone:** Moderately suitable sites may be observed mostly in Western parts of the interfluve. These sites account for approximately 174.20 sq.km. and covers 6.73% of the total area.
- **Low Suitable Zone:** The sites which are poorly suitable for harvesting systems may be observed in most parts of the interfluve. This region has varying runoff potential but their distant locations from the required zones have made the sites less suitable. This zone comprises at least 1767.504 sq.km, accounting for 68.37% of the area of the entire interfluve.

Figure 4. Suitability class for runoff harvesting

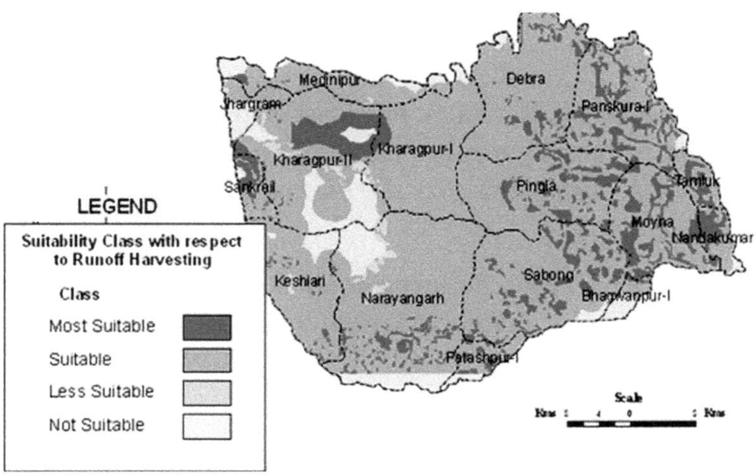

CONCLUSION

Any level of water resource allocation planning requires a concurrent understanding between the availability and requirement status at different spatio-temporal scale. Thereby identifying the scarce and surplus water resource management strategies can be optimized as per requirement. In this study author has attempted to carry out a site suitability analysis so as to identify potential sites suitable for runoff harvesting purely through a Remote Sensing and GIS technique. Remote sending derived products give a synoptic view and information about various physical parameters that can be indispensable for estimating runoff quantity for any region. Using those spatial information as parameters for hydrologic modeling spatio-temporal dynamics of runoff pattern can be assessed. These outputs are the basis for evaluating runoff harvesting potential for an area of any spatial scale. The objective of this study is to identify or rank the area in three suitability classes based on certain criteria thereby enabling the author to understand where surface water conservation is possible. By this study it has been found that 18.71% of the area is suitable for providing a base for runoff harvesting. The rest of the interfluve does not show that much potential for runoff harvesting.

REFERENCES

Auffhammer, M., Ramanathan, V., & Vincent, J. R. (2012). Climate change, the monsoon, and rice yield in India. *Climatic Change*, *111*(2), 411–424. doi:10.1007/s10584-011-0208-4

Carpenter, T. M., & Georgakakos, K. P. (2006). Intercomparison of lumped versus distributed hydrologic model ensemble simulations on operational forecast scales. *Journal of Hydrology (Amsterdam)*, *329*(1–2), 174–185. doi:10.1016/j.jhydrol.2006.02.013

Coskun, M., & Musaoglu, N. (2004). Investigation of rainfall-unoff modelling of Van Lake catchment by using remote sensing and GIS integration. *Paper presented at theTwentieth International Society for Photogrammetry and Remote Sensing (ISPRS) Congress.*

Das, S., & Paul, P. K. (2006). Selection of site for small hydel usinf GIS in the Himalayan region of India. *Journal of Spatial Hydrology*, *6*(1), 18–28.

de Winnaar, G., Jewitt, G. P. W., & Horan, M. (2007). A GIS-based approach for identifying potential runoff harvesting sites in the Thukela River basin, South Africa. *Physics and Chemistry of the Earth Parts A/B/C*, *32*(15–18), 1058–1067. doi:10.1016/j.pce.2007.07.009

ESRI. (2008). *ArcGIS version 9.2*. Environmental System Research Institute.

Kadam, A. K., Kale, S. S., Pande, N. N., Pawar, N. J., & Sankhua, R. N. (2012). Identifying Potential Rainwater Harvesting Sites of a Semi-arid, Basaltic Region of Western India, Using SCS-CN Method. *Water Resources Management*, *26*(9), 2537–2554. doi:10.1007/s11269-012-0031-3

Kizza, M., Rodhe, A., Xu, C.-Y., & Ntale, H. K. (2011). Modelling catchment inflows into Lake Victoria: uncertainties in rainfall–runoff modelling for the Nzoia River. *Hydrological Sciences Journal, 56*(7), 1210-1226. doi:10.1080/02626667.2011.610323

Loo, Y. Y., Billa, L., & Singh, A. (2015). Effect of climate change on seasonal monsoon in Asia and its impact on the variability of monsoon rainfall in Southeast Asia. *Geoscience Frontiers*, *6*(6), 817–823. doi:10.1016/j.gsf.2014.02.009

Mekonnen, M., Melesse, A. M., & Keesstra, S. D. (2016). Spatial Runoff Estimation and Mapping of Potential Water Harvesting Sites: A GIS and Remote Sensing Perspective, Northwest Ethiopia. In M. A. Melesse & W. Abtew (Eds.), *Landscape Dynamics, Soils and Hydrological Processes in Varied Climates* (pp. 565–584). Cham: Springer International Publishing. doi:10.1007/978-3-319-18787-7_26

Mishra, S. K., Singh, V. P., Sansalone, J. J., & Aravamuthan, V. (2003). A Modified SCS-CN Method: Characterization and Testing. *Water Resources Management*, *17*(1), 37–68. doi:10.1023/A:1023099005944

Padmavathy, A. S., Ganesha, R. K., Yogarajan, N., & Thangavel, P. (1993). Checkdam site selection using GIS approach. *Advances in Space Research*, *13*(11), 123–127. doi:10.1016/0273-1177(93)90213-U

Ramakrishnan, D., Bandyopadhyay, A., & Kusuma, K. N. (2009). SCS-CN and GIS-based approach for identifying potential water harvesting sites in the Kali Watershed, Mahi River Basin, India. *Journal of Earth System Science*, *118*(4), 355–368. doi:10.1007/s12040-009-0034-5

Rolland, A., & Rangarajan, R. (2013). Runoff estimation and potential recharge site delineation using analytic hierarchy process. *Geocarto International, 28*(2), 159-170. doi:10.1080/10106049.2012.665499

Santra Mitra, S., Wright, J., Santra, A., & Ghosh, A. R. (2015). An integrated water balance model for assessing water scarcity in a data-sparse interfluve in eastern India. *Hydrological Sciences Journal, 60*(10), 1813-1827. doi:10.1080/02626667.2014.934248

Seiller, G., Hajji, I., & Anctil, F. (2015). Improving the temporal transposability of lumped hydrological models on twenty diversified U.S. watersheds. *Journal of Hydrology: Regional Studies*, *3*, 379–399.

Şen, Z., Al Alsheikh, A., Al-Turbak, A. S., Al-Bassam, A. M., & Al-Dakheel, A. M. (2013). Climate change impact and runoff harvesting in arid regions. *Arabian Journal of Geosciences*, *6*(1), 287–295. doi:10.1007/s12517-011-0354-z

Traboulsi, H., & Traboulsi, M. (2015). Rooftop level rainwater harvesting system. *Applied Water Science*.

Whigham, P. A., & Crapper, P. F. (2001). Modelling rainfall-runoff using genetic programming. *Mathematical and Computer Modelling*, *33*(6), 707–721. doi:10.1016/S0895-7177(00)00274-0

Winnar, G., Jewitt, G. P. W., & Horan, M. (2007). A GIS-based approach for identifying potential runoff harvesting sites in the Thukela River Basin, South Africa. *Physics and Chemistry of the Earth*, *32*(15-18), 1058–1067. doi:10.1016/j.pce.2007.07.009

Wissinger, F., Shankar, R., & Restrepo, J. (2014, March 31-April 3). Hydrologic modeling methodology. *Paper presented at the 2014 8th Annual IEEE Systems Conference (SysCon).*

Ziadat, F. M., Mazahreh, S. S., Oweis, T. Y., & Bruggeman, A. (2006). A GIS based approach fr assessing water harvesting suitability in a Badia bench mark watershed in Jordan. *Paper presented at the Forteenth International Soil COnservatio Organisation Conference: Water Management and Soil Conservation in Semi-Arid Environments.*

Chapter 12
Role of Remote Sensing in Potential Fishing Zone Forecast

Abhisek Santra
Haldia Institute of Technology, India

Debashis Mitra
Indian Institute of Remote Sensing, India

ABSTRACT

Forecasting of Potential Fishing Zone (PFZ) is considered as economically and environmentally significant towards ensuring profitable base of economy and planning for sustenance of existing fishing pool. Changes in environmental conditions affect the distribution, abundance and availability of fish. The traditional sampling approaches for PFZ identification using boats and vessels are not only costly and time consuming but practically absurd considering the vastness of seas and oceans. In this chapter importance of alternative but effective methods of airborne and satellite remote sensing has been given. The chapter elucidates the factors for PFZ identification like thermal condition in sea controlling its thermal circulation, chlorophyll-a concentration estimated from ocean color dynamics, etc. Tools/system to prepare PFZ advisories and also the platforms for dissemination of the same, have been illustrated based on Indian scenario.

DOI: 10.4018/978-1-5225-1814-3.ch012

INTRODUCTION

Coastal environment plays a crucial role in India's economy in terms of its resource base, productive habitat and wide range of biodiversity. India has a long coastline of 8129 Km. and Exclusive Economic Zone (EEZ) of 2.5 million sq. Km. (Figure 1) (Subramanian et al., 2014). Against the harvestable potential of 3.93 MT, India's annual marine fisheries production is about 2.94 MT (Anon, 2008). India has increased its share in world fish production from 3.2% to 4.2% during last decade. The dynamic coastal fisheries sector has undergone considerable changes over the years and support livelihood of millions of people (Nammalwar, Satheesh, & Ramesh, 2013).

As far as fishing is concerned, the seasonal variation of stocks in different areas assumes great importance in developing as well as managing coastal fishery sector. Maximum fishes are harvested by directing fishermen to most profitable fishing grounds through forecasting. Some commercial species are sensitive to environmental changes and undertake migration to the favorable areas (Choudhury et al., 2007; Hela, 1960; Laevastu, 1984; Laevastu & Larkin, 1981; Laevastu & Rosa, 1963). However, nearly 80% of the total fish stocks of the world are now either fully exploited or over exploited (FAO, 2009). Thus, effective monitoring and management of the marine resources are essential for its sustainable use. According to

Figure 1. Coastal stretch of India
(Source: http://www.incois.gov.in/MarineFisheries/TextDataHome?mfid=1&request_locale=en)

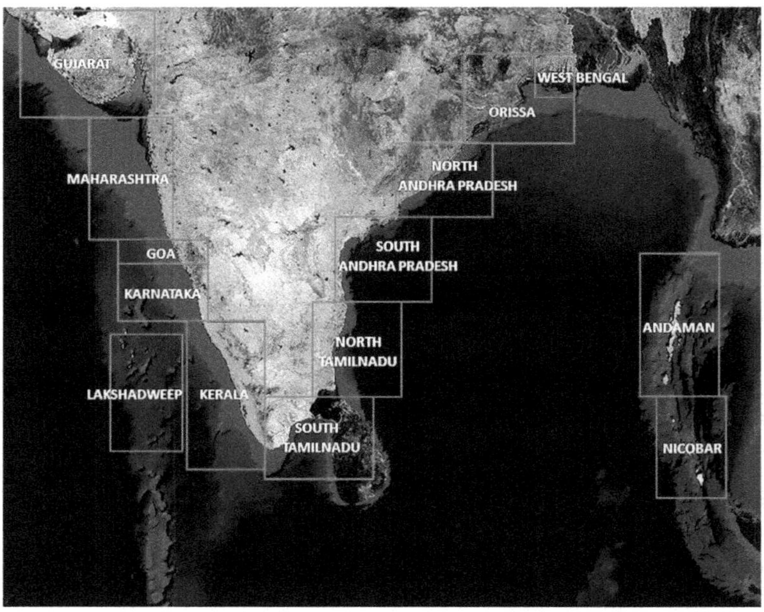

Nammalwar et al. (2013), conventionally potential fishing grounds are located by the way of:

- Fishing in traditionally known fishing areas.
- Locating productive fishing grounds through breaks in color.
- Presence of large number of sea birds over an area.

The traditional sampling research approaches using boats and vessels are both costly and time consuming. Considering the vastness of seas and oceans it is very difficult to study the entire ecosystem. With the advent of remote sensing it has become possible to monitor the global oceans synoptically at regular interval. For example, Sea Surface Temperature (SST) and chlorophyll concentration information derived from the satellite images have been widely used to delineate marine ecological provinces with similar physic-chemical and biological forcing. These time and space variant ecological provinces vary seasonally and inter-annually. These changes affect the survival, condition, distribution pattern and migration of fish species (Chassot et al., 2011; Klemas, 2013; Longhurst, 2010; Moore, Campbell, & Dowell, 2009; Oliver & Irwin, 2008; Stuart, Platt, & Sathyendranath, 2011; Stuart, Platt, Sathyendranath, & Pravin, 2011). The group of fishes can be tracked directly from spotter planes, not from satellite altitudes. However, satellite images are used to locate and predict potential fish aggregation areas based on certain remotely detectable biophysical indicators. These indicators include ocean fronts, different ocean waters based on color and temperature, cooler and more productive greener upwelling areas than background waters, zones of specific temperature preferred by certain fish species, etc. (Klemas, 2013).

ROLE OF SATELLITE REMOTE SENSING

Changes in environmental conditions affect the distribution, abundance and availability of fish. Thus, adequate understanding about these environmental conditions and biological processes are required for Potential Fishing Zone (PFZ) forecasting studies. Now, the question is why satellite remote sensing? Satellite images provide synoptic measurements of different environmental parameters at regular interval, starting from local to global scale (Figure 2). Remote sensing can provide wide range of biophysical information in form of indices. Environmental parameters that are measured frequently using remote sensing include ocean color; yellow substance; chlorophyll pigments; diffuse attenuation coefficients, total suspended matter; salinity; sea surface temperature; wind speed and wind direction; vertical and horizontal circulations, including fronts and gyres; pollution, etc. (Klemas, 2013). PFZ forecast

Figure 2. Components of remote sensing in PFZ forecasting
(Source: http://community.telecentre.org/m/blogpost?id=2086278%3ABlogPost%3A48571)

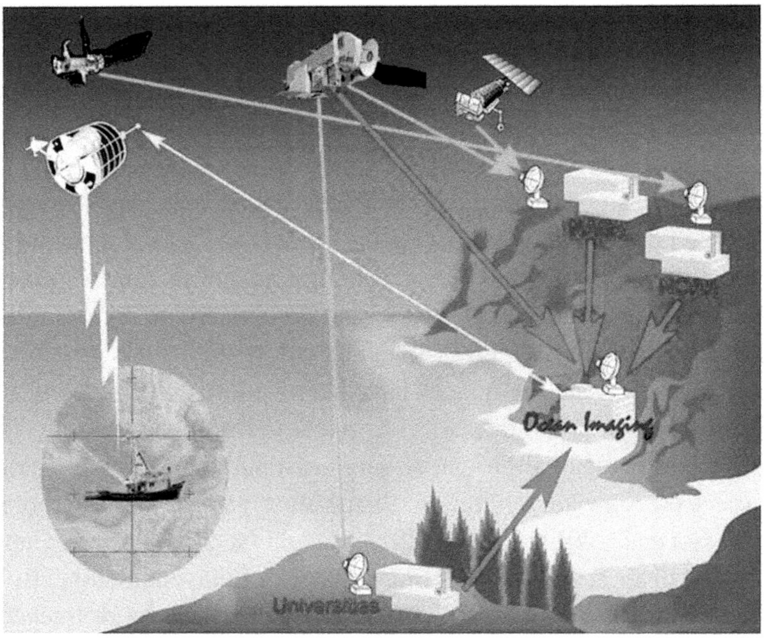

models mostly dependent on the satellite and/or airborne images along with some important ancillary information. Another important advantage of satellite remote sensing is that the oceanic primary production variability and magnitude are poorly known on global scale, mostly because of high spatial and temporal variability of marine phytoplankton concentrations.

The widely used common optical sensors to detect ocean color are NIMBUS-7 Coastal Zone Color Scanner (CZCS), Orbview-2 Sea-Viewing Wide Field of View Sensor (SeaWiFS), Aqua Moderate Resolution Imaging Spectroradiometer (MODIS), Envisat-1 MEdium Resolution Imaging Spectrometer (MERIS) etc. However, due to their limited spatial resolution, these sensor data are more suitable for monitoring open ocean. Landsat, SPOT and IRS series satellite sensors possess medium to fine spatial resolution and are frequently used to observe near-shore coastal waters. However, these optical images are not useful to track rapidly varying coastal processes due to their inability to penetrate clouds. Among the above mentioned coarse resolution images, hyperspectral images MODIS data is considered most useful for coastal upwelling area mapping and for discriminating coastal water complexity. In this regard, Hyperspectral Images for the Coastal Ocean (HICO) data is used. HICO, the first space borne imaging spectrometer with 90 m spatial resolution is designed exclusively to study coastal ocean. The data possesses full spectral coverage (380-

960 nm) and a very high signal-to-noise ratio to resolve complexity of coastal waters (Klemas, 2013).

Sea-surface temperature (SST) can be estimated from thermal infrared (TIR) sensors. The widely used small scale TIR sensors are NOAA – Advanced Very High Resolution Radiometer (AVHRR), ERS2 – Along –Track Scanning Radiometer (ATSR), Terra and Aqua – MODIS and Geostationary Operational Environmental Satellite (GOES). Landsat 5 TM and 7 ETM+ data are widely used to monitor coastal waters. Landsat 8 TIRS data joined subsequently for the same purpose with dual thermal channel capability (bands 10 and 11). Advanced Spaceborne Thermal Emission and Reflection Radiometer (ASTER) on the other hand, got multi-channel capability (bands 10-14). The TIR radiance depends on the temperature and emissivity of water which is known and nearly 98% constant. This paves the success of SST measurement from TIR data (Ikeda & Dobson, 1995). The accuracy of measurement is +/- 0.5°C with certain atmospheric correction (Barton, 1995; Martin, 2004). TIR sensors like SeaWiFS and AVHRR accurately measure the coastal upwelling areas where rising cold water brings nutrients to surface water and attract fish consequently (Martin, 2004; Schofield et al., 2004; Yan, Ho, Zheng, & Klemas, 1993).

Satellite remote sensors can detect the change in color and temperature of the sea/ocean surface due to high concentration of chlorophyll pigments and phytoplanktons as a result of wind induced upwelling of nutrients to the surface. The long term upwelling events off Peru and California coasts influence the abundance of fish species over periods of months (Klemas, 2013). Solanki, Dwivedi & Nayak (2001) adopted an integrated approach for fishery resources exploration based on IRS P4 OCM derived chlorophyll concentration and NOAA AVHRR derived SST data. Solanki, Mankodi, Nayak & Somvanshi (2005) later developed a remote sensing based methodology to forecast PFZ. Zainuddin, Kiyofuji, Saitoh & Saitoh (2006) used satellite sensor based SST and ocean color data to examine fishing grounds and demarcated the high probability areas of albacore prey and tuna abundance corresponding to the frontal zone and anti-cyclonic eddy locations. Zagaglia, Lorenzzetti & Stech (2004) investigated the relationship between yellowfin tuna off the Brazilian coast in the tropical Atlantic with the NOAA – AVHRR SST; SeaWiFS – SeaStar chlorophyll concentration; TOPEX – Poseidon sea surface height anomaly; and Scatterometer – ERS-1 and -2 wind velocity information. They applied statistical methods of Generalized Additive Models (GAMs) to analyze the relationship between the satellite data and catch data and found SST, chlorophyll-a concentration and sea-surface height anomaly information were statistically significant. Valavanish et al. (2008) in EnviEFH project produced essential fish-habitat maps for small, large pelagic and demersal fishes in Mediterranean and adjacent seas based on the synergetic use of remote sensing, GIS and spatial data analysis. Wedding, Friedlander, McGranaghan, Yost & Monaco (2008) scrutinized the relation between the complexity of habitat and Hawaiian reef fish assembly characteristics.

Purkis, Graham & Riegl (2008) extracted the bathymetric and benthic information of a reef system in Diego Garcia from high resolution IKONOS images. Brock & Purkis (2009) commented that the airborne Lidar could provide valuable information about the seascape structure and can be used to prioritize areas for conservation and management. Satellite image derived chlorophyll concentration and SST data are useful for detecting toxic dinoflagellate blooms in relation to habitat mapping (Kumari, Raman, & Mali, 2009; Sanchez, Demestre, Recasens, Maynou, & Martin, 2008). Daqamseh, Mansor, Pradhan, Billa & Mahmood (2013) applied MODIS derived sea surface salinity (SSS), SST and chlorophyll-a data to map potential fish habitat in South China Sea coastal areas of Malaysia. SST and chlorophyll-a data proved to be very useful in identifying potential fishing grounds of Indian Mackerel (*Rastrellinger kanagurta*) (Nurdin, Mustapha, Lihan, & Ghaffar, 2015). Tanveer, Mukherjee & Vinu Chandran (2015) applied artificial neural network (ANN) technique to predict chlorophyll-a and SST from MODIS images in demarcation of fishing zone and successfully compared the results with the NASA derived data. However, meticulous calibration and validation are required to generate global ocean chlorophyll concentration and water constituent data from that wide range of optical images. Instrumented ships, buoys and ocean gliders are used for this purpose (Bailey & Werdell, 2006; Klemas, 2013; Robinson, 2004; Schofield et al., 2004).

ROLE OF AIRBORNE REMOTE SENSING

Airborne remote sensing is advantageous because user can control its deployment, flight planning, swath, spatial resolution and focal length. User can also opt for desired atmospheric, illumination and tidal conditions (Clark, Ripley, Green, Edwards, & Mumby, 1997; Myers & Miller, 2005). Above all, for near-shore coastal fish habitat studies the small single-engine planes are proved to be cost-effective (Klemas, 2013). Lo, Jacobsen & Squire (1992) and Santos (2000) have used commercial aerial fish-spotter data, SST and mixed layer depth information and delta-lognormal models to develop cost effective index for estimating relative abundance of pelagic fish species like northern anchovy.

Aerial photography has been successfully adopted in fishery researches since 1970s. The airborne sensors include low-light-level television, digital cameras, TIR radiometers, Radar and Lidar (Borstad, Hill, Kerr, & Nakashima, 1992; Santos, 2000). O'Brien, Woodworth & Wright (1974) systematically measured SST to forecast potential availability areas of Coho Salmon off the Oregon coast. Low-light-level television is equipped with an image intensifier tube which amplifies light thousands of times producing a brighter image that can be displayed on a television screen. This enables to detect fish shoals that cannot been observed in naked eye at night (Klemas, 2013). This becomes essentially effective, whenever planktons are stimulated into bioluminescence by fish shoal movements at night (Roithmayr, 1970).

To overcome the limitations poised by passive observation systems, active Lidar and Radar systems are applied. Powerful Lidar pulses can penetrate surface water up to three times the Secchi depth of water column. Fish shoals' dimension and depth can be determined from the return pulses. Besides, the reflectance of suspended particles helps to discriminate fish shoals (Klemas, 2013). To fly small aircrafts mounted with Lidar system requires low power consumption and becomes cost effective (Churnside & Hunter, 1996).

High resolution airborne images and Lidar data are used to map complex habitats of coral reef fish and other marine life (Andréfouët et al., 2005; Hamel & Andréfouët, 2010; Purkis, Graham, & Riegl, 2008). Pittman, Costa & Battista (2009) applied the 4m Lidar bathymetric grid data to generate benthic habitat maps and faunal distribution maps to support ecosystem-based management. These Lidar generated orthophotographs and relief maps were used to develop side channel maps and fish habitat suitability maps in upstream water.

The Side-Looking Airborne Radar (SLAR) follows the similar active image acquisition mechanism to that of the Lidar system. Unlike Lidar, SLAR emits radar pulses and receives backscattered signals from ocean surface. Small scales wave, (2-20 cm), generated by fishes swimming close to the surface can be detected by airborne microwave sensors (Santos, 2000). The intensity and size of waves help to determine the size and behavior of fish shoals (Klemas, 2013). In comparison with the visual spotters, the airborne SLAR systems can search wider areas even at higher wind speeds. Griffiths, Lyne, Harris & Parslow (1989) observed the better detection capability of microwave sensors in case of identifying shoals of jack mackerel, skipjack tuna, southern blue-fin tuna and dolphins. These microwave radiometers also add valuable information about the salinity condition of the seas and fresh water plumes from rivers in connection to the potential fishing habitat zone studies.

INDIAN SCENARIO

PFZ is a dependable and short-term forecast on the fish aggregation zone in the open sea. In India, PFZ is identified through the use of medium and large sized fishing vessels owned by different government organizations such as Fishery Survey of India (FSI), Central Marine Fisheries Research Institute (CMFRI), National Institute of Ocean Technology (NIOT), National Institute of Oceanography (NIO) etc. and some private agencies equipped with acoustic, navigational and fishing aids, e.g.

Echosounders, Sonars and Nansen Sampler, Rosette Sampler, Bathythermograph, etc. Satellite derived information about SST, SSS, chlorophyll, nutrients, dissolved oxygen, winds and currents are used by scientific communities to study the feeding and breeding, migratory habits of fishes. These are converted to scientific indicators of PFZs. Indian National Centre for Ocean Information Services (INCOIS), Hyderabad disseminates PFZ advisories on Mondays, Wednesdays and Fridays to around 200 stations covering the entire Indian coastline (Nammalwar et al., 2013). Satellite image based chlorophyll and SST information over the Arabian Sea and Bay of Bengal are the basic inputs for generating such information. TIR channels of NOAA – AVHRR and optical bands of IRS – P4 OCM and Aqua – MODIS data are used in this regard (Subramanian et al., 2014). A uniform ban on marine fishing has been imposed by the Indian Government along the east and west coasts from April 15 to May 31 and June 10 to August 15 respectively and during monsoon season every year to maintain fish stocks at the sustainable level and to conserve fishery wealth (Nammalwar et al., 2013).

To improve coverage, Electronic Display Board (EDB) has been developed and installed at various fish landing centers along the coast (Figure 3). PFZ forecasts are directly transmitted to the EDBs from INCOIS using telephone and other modern dedicated telecommunication technologies. EDBs are 32 inches LCD panel designed to facilitate dissemination of satellite images, animations, short-films, ocean state information in forms of PFZ maps and tables, information about wind speed & direction and wave height. These are also equipped with disaster information and disaster warning such as tsunami. Secured siren system with audibility up to 1 Km is installed in each EDB in this respect to warn fishermen and local coastal people about the possible danger. The fishermen can get their required information about the PFZ advisories directly from EDBs before venturing into the sea to catch fishes without waiting for fax message (Subramanian et al., 2014). INCOIS developed WebGIS application for major fish landing centers of India (Figure 4).

The PFZ advisory consists of two components – table and map. The table provides information on name of landing centre, location, direction, depth and size of fish shoals. The map on the other hand (Figure 5), highlights the darker PFZ lines where possible fish congregation is available. The dotted lines give bathymetric information. The shifting and original features are shown by dark squares and circles respectively on wind based maps. The table also validates the map. This means if fishing is carried out on the first day after receiving of the map, the probability of getting fish is higher than the remaining days (Subramanian et al., 2014). Thus, PFZ is helpful to locate fish shoals, save fuel and reduce search time. The chlorophyll and SST based PFZ forecast is quite useful to pelagic gears and helpful to traditional fishermen.

Figure 3. Location of fish landing centers in India
(Source: http://iomenvis.nic.in/index3.aspx?sslid=1137&subsublinkid=127&langid=1&mid=1)

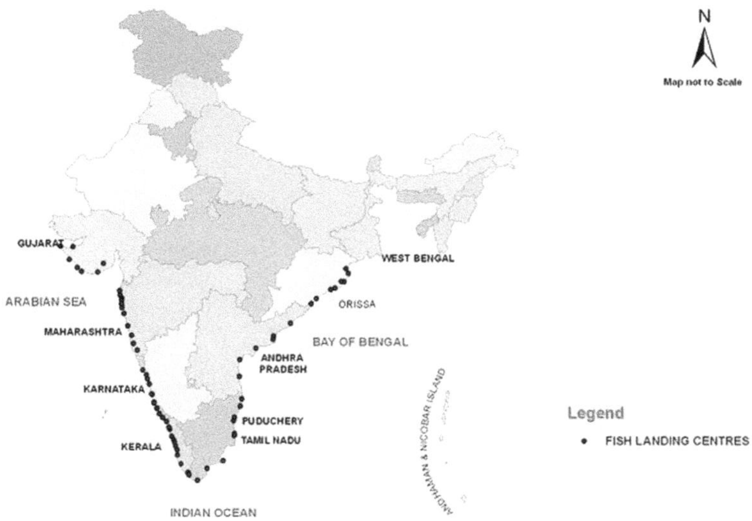

Figure 4. WebGIS for major fish landing centers of part of east coast of India
(Source: http://www.incois.gov.in/MarineFisheries/MfsWebGis)

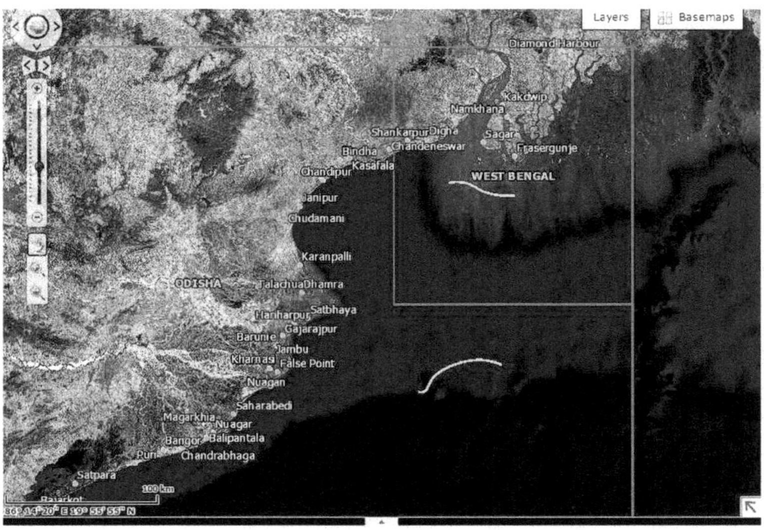

Figure 5. PFZ advisory map for tuna fish
(Source: http://www.incois.gov.in/MarineFisheries/TunaAdvisory)

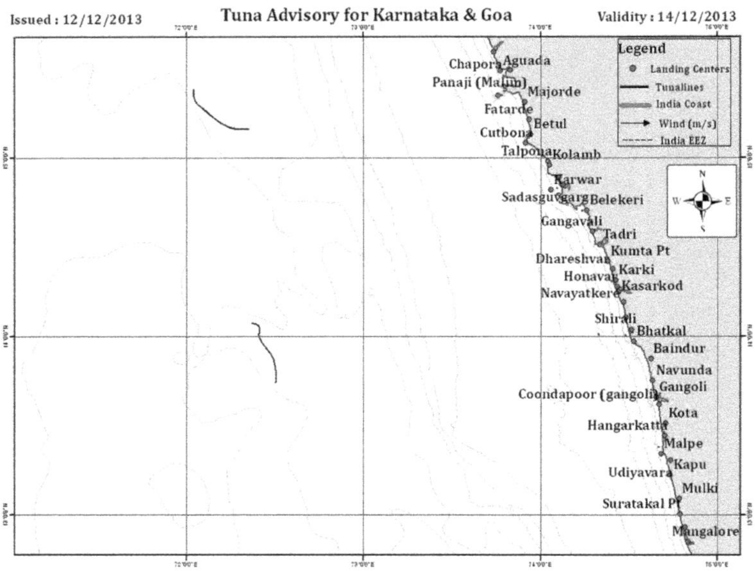

CONCLUSION

Productivity of fish across the globe has been declining over the past 40 years due to overfishing, pollution, habitat and climate change. Most of the fish stocks are now mostly exploited. Thus, sustainable use of marine resources demands effective monitoring and proper management. Traditional approaches of sampling using boats and vessels exposed their limitations in spatio-temporal dimensions. These samples are too weak to represent the entire marine ecosystems. The capabilities of remote sensing techniques in combination with the conventional sampling methods provide a successful tool for the efficient and cost effective exploitation and management of coastal resources. Satellite and airborne sensors are producing series of good quality images at regular interval starting from local to global scale since past few decades. These synoptic images provide valuable information about SST, SSS, ocean color, wind speed & direction, waves, bathymetry, etc. These remotely sensed datasets offer great advantage in forecasting the PFZs and are being applied successfully by different countries. Satellite data based PFZ delineation aid fishermen with real time information that reduce search time for productive fishing areas with less effort. The process is also cost effective. However, to increase the success rate, long time series climatic, oceanographic and fish population data act as important pre-requisites for forecasting potential fishing zones.

REFERENCES

Andréfouët, S., Gilbert, A., Yan, L., Remoissenet, G., Payri, C., & Chancerelle, Y. (2005). The remarkable population size of the endangered clam *Tridacna maxima* assessed in Fangatau atoll (Eastern Tuamotu, French Polynesia) using in situ and remote sensing data. *ICES Journal of Marine Science*, *62*, 1037–1048.

Anon., (2008). *Annual Report*. Kochi: Central Marine Fisheries Research Institute.

Bailey, S. W., & Werdell, P. J. (2006). A multi-sensor approach for the on-orbit of ocean color satellite data products. *Remote Sensing of Environment*, *102*(1-2), 12–23. doi:10.1016/j.rse.2006.01.015

Barton, I. J. (1995). Satellite - derived sea surface temperatures: Current status. *Journal of Geophysical Research*, *100*(C5), 8777. doi:10.1029/95JC00365

Borstad, G. A., Hill, D. A., Kerr, R. C., & Nakashima, B. S. (1992). Direct digital remote sensing of herring schools. *International Journal of Remote Sensing*, *13*(12), 2191–2198. doi:10.1080/01431169208904262

Brock, J. C., & Purkis, S. J. (2009). The emerging role of LiDAR remote sensing in coastal research and resource management. *Special Issue 53 - Coastal Applications of Airborne LiDAR. Journal of Coastal Research*, *53*, 1–5. doi:10.2112/SI53-001.1

Chassot, E., Bonhommeau, S., Reygondeau, G., Nieto, K., Polovina, J. J., Huret, M., & Demarcq, H. et al. (2011). Satellite remote sensing for an ecosystem approach to fisheries management. *ICES Journal of Marine Science*, *68*(4), 651–666. doi:10.1093/icesjms/fsq195

Choudhury, S. B., Jena, B., Rao, M. V., Rao, K. H., Somvanshi, V. S., Gulati, D. K., & Sahu, S. K. (2007). Validation of integrated potential fishing zone (IPFZ) forecast using satellite based chlorophyll and sea surface temperature along the east coast of India. *International Journal of Remote Sensing*, *28*(12), 2683–2693. doi:10.1080/01431160600987878

Churnside, J. H., & Hunter, J. R. (1996). Laser remote sensing of epipelagic fishes. *Society of Photo - Optical Instrumentation Engineers, 2964*, 38-53.

Clark, C., Ripley, H., Green, E., Edwards, A., & Mumby, P. (1997). Mapping and measurement of tropical coastal environments with hyperspectral and high spatial resolution data. *International Journal of Remote Sensing*, *18*(2), 237–242. doi:10.1080/014311697219033

Daqamseh, S. T., Mansor, S., Pradhan, B., Billa, L., & Mahmud, A. R. (2013). Potential fish habitat mapping using MODIS - derived sea surface salinity, temperature and chlorophyll-a data: South China Sea coastal areas, Malaysia. *Geocarto International*, *28*(6), 546–560. doi:10.1080/10106049.2012.730065

FAO. (2009). *The State of World Fisheries and Aquaculture. 2008*. Rome, Italy: FAO Documentation Group.

Griffiths, F. B., Lyne, V. D., Harris, G. P., & Parslow, J. S. (1989). Tracking fish by airborne radar - fishermen could be the winners. *Australian Fisheries*, *48*, 19–21.

Hamel, M. A., & Andréfouët, S. (2010). Using very high resolution remote sensing for the management of coral reef fisheries: Review and perspectives. *Marine Pollution Bulletin*, *60*(9), 1397–1405. doi:10.1016/j.marpolbul.2010.07.002 PMID:20659750

Hela, I. (1960). The influence of temperature on the behaviour of the fish. *Archaeological Society*, *15*, 1–2.

Ikeda, M., & Dobson, F. W. (1995). *Oceanographic Applications of Remote Sensing*. New York: CRC Press.

Klemas, V. (2013). Fisheries applications of remote sensing: An overview. *Fisheries Research*, *148*, 124–136. doi:10.1016/j.fishres.2012.02.027

Kumari, B., Raman, M., & Mali, K. (2009). Locating tuna forage grounds through satellite remote sensing. *International Journal of Remote Sensing*, *30*(22), 5977–5988. doi:10.1080/01431160902798387

Laevastu, T. (1984). The effect of temperature anolmalies on the fluctuation of stock. *Rapports et Proces-Verbaux des Reunions - Conseil International pour L'Exploration de la Mer*, *185*, 214–215.

Laevastu, T., & Larkin, H. A. (1981). *Marine Fisheries Ecosystem, its Quantitative Evaluation and Management*. Farham, Surrey: Fishing News Books.

Laevastu, T., & Rosa, J. (1963). The distribution and relative abundance of tunas in relation to their environment. *FAO Fisheries Report*, *6*, 1835–1851.

Lo, N. C. H., Jacobsen, L. D., & Squire, J. L. (1992). Indices of relative abundance from fish spotter data based on delta-lognormal models. *Canadian Journal of Fisheries and Aquatic Sciences*, *49*(12), 2526–2551. doi:10.1139/f92-278

Longhurst, A. (2010). *Mismanagement of Marine Fisheries*. Cambridge, UK: Cambridge University Press. doi:10.1017/CBO9781139195928

Martin, S. (2004). *An Introduction to Remote Sensing*. Cambridge, UK: Cambridge University Press.

Moore, T. S., Campbell, J. W., & Dowell, M. D. (2009). A class-based approach to characterizing and mapping the uncertainty of the MODIS ocean chlorophyll product. *Remote Sensing of Environment*, *113*(11), 2424–2430. doi:10.1016/j.rse.2009.07.016

Myers, J. S., & Miller, R. L. (2005). Optical airborne remote sensing. In R. L. Miller, C. E. DelCastillo, & B. A. McKee (Eds.), *Remote Sensing of Coastal Aquatic Environments: Technologies, Techniques and Applications* (pp. 51–68). Dordrecht, The Netherlands: Springer. doi:10.1007/978-1-4020-3100-7_3

Nammalwar, P., Satheesh, S., & Ramesh, R. (2013). Applications of remote sensing in the validations of potential fishing zones (PFZ) along the coast of North Tamil Nadu, India. *Indian Journal of Geo-Marine Sciences*, *42*(3), 283–292.

Nurdin, S., Mustapha, M. A., Lihan, T., & Ghaffar, M. A. (2015). Determination of potential fishing grounds of *Rastrelliger kanagurta* using satellite remote sensing and GIS technique. *Sains Malaysiana*, *44*(2), 225–232. doi:10.17576/jsm-2015-4402-09

O'Brien, J. J., Woodworth, B. M., & Wright, D. J. (1974). The Coho project reports. Living Resources Prediction Feasibility Study.: Vol. I. *III*. Corvallis, OR: School of Oceanography, Oregon State University.

Oliver, M. J., & Irwin, A. J. (2008). Objective global ocean biogeographic provinces. *Geophysical Research Letters*, *35*(15), L15601. doi:10.1029/2008GL034238

Pittman, S. J., Costa, B. M., & Battista, T. A. (2009). Using lidar bathymetry and boosted regression trees to predict the diversity and abundance of fish and corals. *Journal of Coastal Research*, *25*, 27–38. doi:10.2112/SI53-004.1

Purkis, S. J., Graham, N. A. J., & Riegl, B. M. (2008). Predictability of reef fish diversity and abundance using remote sensing data in Diego Garcia (Chagos Archipelago). *Coral Reefs*, *27*(1), 167–178. doi:10.1007/s00338-007-0306-y

Robinson, I. S. (2004). *Measuring the Oceans from Space: The Principles and Methods of Satellite Oceanography*. Berlin: Springer - Praxis.

Roithmayr, C. M. (1970). Airborne low-light sensor detects luminescing fish schools at night. *Commercial Fisheries Review*, *32*, 42–51.

Sanchez, P., Demestre, M., Recasens, L., Maynou, F., & Martin, P. (2008). Combining GIS and GAMS to identify potential habitats of squid Loligo vulgaris in the north-western Mediterranean. *Hydrobiologia*, *612*(1), 91–98. doi:10.1007/s10750-008-9487-9

Santos, A. M. P. (2000). Fisheries oceanography using satellite and airborne remote sensing methods: A review. *Fisheries Research*, *49*(1), 1–20. doi:10.1016/S0165-7836(00)00201-0

Schofield, O., Arnone, R. A., Bissett, W. P., Dickey, T. D., Davis, C. O., Finkel, Z., & Moline, M. et al. (2004). Water colors in the coastal zone: What can we see? *Oceanography (Washington, D.C.)*, *17*(2), 24–31. doi:10.5670/oceanog.2004.44

Solanki, H. U., Dwivedi, R. M., & Nayak, S. R. (2001). Synergistic analysis of SeaWiFS chlorophyll concentration and NOAA-AVHRR SST features for exploring marine living resources. *International Journal of Remote Sensing*, *22*(18), 3877–3882. doi:10.1080/01431160110069845

Solanki, H. U., Mankodi, P. C., Nayak, S. R., & Somvanshi, V. S. (2005). Evaluation of remote-sensing-based potential fishing zones (PFZs) forecast methodology. *Continental Shelf Research*, *25*(18), 2163–2173. doi:10.1016/j.csr.2005.08.025

Stuart, V., Platt, T., & Sathyendranath, S. (2011). The future of fisheries science in management: A remote - sensing perspective. *ICES Journal of Marine Science*, *68*(4), 644–650. doi:10.1093/icesjms/fsq200

Stuart, V., Platt, T., Sathyendranath, S., & Pravin, P. (2011). Remote sensing and fisheries: An introduction. *ICES Journal of Marine Science*, *68*(4), 639–641. doi:10.1093/icesjms/fsq193

Subramanian, S., Manjulekshmi, N., Kolwalkar, J., Fernandes, P. M., Sreekanth, G. B., Singh, N. P., ... (2014). A Manual on The Use of Potential Fishing Zone (PFZ) Forecast. Technical Bulletin No.:40. PFZ Mission Programme: Indian National Centre for Oceanic Information Services, Hyderabad.

Tanveer, M., Mukherjee, C. K., & Vinu Chandran, R. (2015). Application of ANN to predict chlorophyll-a and sea surface temperature in determination of fishing zone. *International Water Technology Journal*, *5*(4), 294–304.

Valavanis, V. D., Pierce, G. J., Zuur, A. F., Palialexis, A., Saveliev, A., Katara, I., & Wang, J. (2008). Modelling of essential fish habitat based on remote sensing, spatial analyis and GIS. *Hydrobiologia*, *612*(1), 5–20. doi:10.1007/s10750-008-9493-y

Wedding, L. M., Friedlander, A. M., McGranaghan, M., Yost, R. S., & Monaco, M. E. (2008). Using bathymetric lidar to define nearshore benthic habitat complexity: Implications for management of reef fish assemblages in Hawaii. *Remote Sensing of Environment*, *112*(11), 4159–4165. doi:10.1016/j.rse.2008.01.025

Yan, X.-H., Ho, C., Zheng, Q., & Klemas, V. (1993). Using satellite IR in studies of the variabilities of the Western Pacific Warm Pool. *Science*, *262*(5132), 440–441. doi:10.1126/science.262.5132.440 PMID:17789951

Zagaglia, C. R., Lorenzzetti, J., & Stech, J. L. (2004). Remote sensing data and longline catches of yellowfin tuna (*Thunnus albacores*) in the equatorial Atlantic. *Remote Sensing of Environment*, *93*(1-2), 267–281. doi:10.1016/j.rse.2004.07.015

Zainuddin, M., Kiyofuji, H., Saitoh, K., & Saitoh, S.-I. (2006). Using multi-sensor satellite remote sensing and catch data to detect ocean hot spots for albacore (*Thunnus alalunga*) in the northwestern North Pacific. *Deep-Sea Research*, *53*(3-4), 419–431. doi:10.1016/j.dsr2.2006.01.007

Chapter 13
Wildlife Habitat Evaluation

Peeyush Gupta
Intergraph SG & I India Pvt. Ltd., India

Swati Goyal
Guru Jambheshwar University of Science & Technology, India

ABSTRACT

Before an individual can evaluate wildlife habitat and make management recommendations, some basic concepts about habitat and its relationships to different wildlife species should be understood. In this chapter, some of the basic concepts will be described; mainly analyzing of habitat alterations, landscape analysis, networking and creation of corridor between protected areas, wildlife habitat suitability analysis using Remote Sensing & GIS. Since most of the contest will be based on these concepts. Like other natural resource fields, wildlife management is both an art and science that deals with complex interactions in the environment. This means that management includes art or judgment based on experience as well as sound factual information based on scientific studies.

DOI: 10.4018/978-1-5225-1814-3.ch013

INTRODUCTION

The worldwide destruction of natural environment is reducing the number of species and the amount of genetic variation within the individual species. There are two principal approaches for protecting and managing nature:

1. To ensure survival of population of a given species abundantly.
2. Conservation of natural community in their original habitats.

During early seventies the 'Protected Area' concept obtained wider acceptability to check the process of depleting biodiversity. In India, this diversity is reflected in the fact that as many as 10 biogeographical regions representing three basic biomes and two natural realms are met (Champion et al., 1968). Major threats to biodiversity include habitat alteration, overharvesting, pollution, climatic change, introduced species and population increase. Of these, habitat alterations due to deforestation, habitat loss, fragmentation and degradation are primarily responsible (McNeely et al., 1990).

ANALYSING HABITAT ALTERATIONS

Tropical forests are facing disturbance of varying magnitudes in the different regions. The disturbances are leading to the following changes in the habitats:

- Loss of particular species and change in cover characteristic.
- Degradation due to reduction in canopy density.
- Changes in understory vegetation and invasion of exotic species.
- Fragmentation of forest areas leading to the loss of corridors, depletion of minimum habitat areas and reduction of home range.
- Habitats are often very susceptible to degradation viz., fire, overgrazing and cultivation. These factors make it difficult for original vegetation to reestablish (Gupta, 1991).

Above situation justifies the need of well- structured information system, which can allow access to the stored data and monitoring at very short time space. The continuous monitoring and habitat inventory should permit detection of changes related to the both, the habitat components and the immediate population for the analysis of their interrelationship and cause of change.

LANDSCAPE ANALYSIS

Evaluation of landscapes has emerged as an important branch to study structure and function and changes in the landscapes. Remote sensing provides information with respect to the homogeneous spatial units. The interrelationships of these units can be best evaluated through landscape analysis. Each landscape is formed of several landscape elements, which appear as patches and vary markedly in size, shape, type, and heterogeneity and boundary characteristics. Analysis of these parameters brings out the disturbance gradient in various habitats (Gupta et al., 2014). The remote sensing derived vegetation maps provide perspective horizontal view and helps in delineating different landscape elements and their spatial characteristics. (Gupta 2014) used satellite derived vegetation map for analyzing landscape elements of TehriGarhwal region. The analysis brings out structural and quantitative landscape perspective in the three management zones of the national park. The patch characteristics of vegetation units viz., size, shape, porosity and patch density has been studied. Medium patch size and porosity have been found to be the most important parameter to discriminate differences in the ecological status of three different zone of the park. The patchiness and shape of different vegetation types also provide valuable information and characteristics of the native, seral and retrogressive vegetation forms (MadhvanUnni, et al., 1990).

NETWORKING AND CREATION OF CORRIDOR BETWEEN PROTECTED AREAS

Forest area managed for wildlife should have a network of patches of old growth forest joined by dispersed corridors of similar structures and composition. Such corridors are multipurpose:

- They ensure that sub-populations are interconnected into demographically functioned unit.
- They offer some protection to surrounding agricultural lands.
- They decrease dispersal morality and release pressure on destination areas.

The demarcation of corridor should be based on an evaluation of land resources and it can be done using GIS using existing land use map and physical set up of the landscape. It is also important to incorporate social and economic data to check land utilization type and description of its key attributes. Such land mapping unit characteristics can be translated into qualities related to networking and corridor requirement using GIS (Jain et al., 1996).

WILDLIFE HABITAT SUITABILITY ANALYSIS

Habitat analysis is considered most important for management and planning of national parks. Wildlife habitat includes wide variety of factors viz., vegetation; cover characterization including human influence on all these, soil, topography and water availability. Since each species required a particular habitat to meet the requirement of space, food, cover and other needs of survival. The database with respect to above parameter can be used to identify the specific needs for particular wildlife for habitat suitability area analysis. Besides, other analysis like home range, tourist route planning and carrying capacity analysis can also be made using GIS (Maslekar, 1974).

AN EXAMPLE

Siwalik Himalayas are considered fragile ecosystem. The proposed Rajaji National Park (UP) located in the foothills of Himalaya represents typical Siwalik landscape. It is known for diverse vegetation types and potential habitat of many wildlife including Elephasmaximus (Asian elephant) and Nemorhaedus goral (goral). The GIS data base consisting of present and historical forest type maps, physiographic map, drainage, road/communications, settlement maps and their attributes have been used for deriving critical habitat management information in the areas of habitat changes, fire risk and habitat suitability.

Dhaulkhand range within proposed Rajaji National Park constitutes representative ecosystem comprising of all the key habitat elements of the Siwalik landscape. The area lies within 29°59’45” to 30°08’48” N latitudes; 70°55’5” to 78°04’42”E longitude. Topography of the area varies with altitude ranging from 320 m above msl in submontane area to 800 m above in Siwalik ranges. In the submontane area the ground is apparently levelled but slopes gently towards south-west. Climate of the area is subtropical. Rainy season commences from late June and continues up to mid-September with the heaviest rainfall recorded in August (range 1200-1500 m). The average temperature in the area varies from 13.1°C in January to 38.9°C in May/June.

Vegetation of the area mainly belongs to northern tropical dry deciduous forest and subtropical forest. Among the tropical dry deciduous forests, main types are dry Siwalik Sal (Shorearobusta) forests, while subtropical forest consists mainly of Chir pine (Pinusroxburhii) forest. Sal forest occurs on the higher slopes of Siwalik as well as on the Bhabhar belt. The area is also rich in wildlife. Elephants are present in good number. There are large carnivores like, tiger and leopards. Herbivorous animals are; Axis axis (cheetal), Axis parcinus (hog deer); Muntiacusverginalis (barking deer). Mountain goat (Nemorhaedus goral) is an endangered species found in the specific habitats of this region.

The following study presents results of GIS applications in conservation area management using vegetation type map, physiographic map, digital elevation model, road/communication network, settlements, fuel value in vegetation types and related socioeconomic data.

HABITAT SUITABILITY ANALYSIS FOR NEMORHAEDUS GORAL

Habitat is a place occupied by a specific population within the community of population and habitat selection is an important part of an organism's life history patterns. The colonization pattern of animals is related to parameters viz., cover, physical set-up and land use, which can be studied using aerospace remote sensing. There is a loss of natural habitat for animals in Rajaji National Park due to the biotic interference in forest ecosystem. The study is an attempt to evaluate habitat of goral (Nemorhaedus goral) using remote sensing derived vegetation map integrated with other parameters like, juxtaposition, interspersion and the restrictive factors in the GIS domain. The habitat of the animal under study requires the very specific habitat like, habitat with good grass cover away from human habitation, high slope areas i.e., $>30^{o}$ as escape terrain and areas with less extensive shrub cover. These habitats are becoming narrower due to deforestation, fragmentation and human influences. It would be of interest to estimate suitable habitat of the endangered mountain goat.

The habitat suitability is derived based on spatial information viz., interspersion, juxtaposition and restrictive factors for goral population. The vegetation type map was derived by visual interpretation of IRS 1B LISS-II data and was used to analyse the spatial pattern of the vegetation. In addition, the terrain features viz., slope, drainage, water bodies, settlements and roads were also used to derive habitat parameters (Gupta et al. 2014).

Interspersion (Is) is defined as a measurement of the spatial intermixing of the vegetation types and was calculated in non-species specific manner. Juxtaposition (Jx) is defined as a measure of the proximity of the vegetation. Its measurement mostly includes relative weightage assigned by the importance of the adjacency of two cover types for the species in question. Therefore, Jx is a species-specific measurement. This was accomplished by defining a grid, which was placed on the forest type map covering 149 sq.km. area. The grid size was 20" * 20" i.e., 533.33 * 533.33 m. In all 511 cells covered the whole area. Grid size selection was based on the size of the 'goral' territory i.e. about 300 m radius circle.

Thus, in present study the Is and Jx were calculated as:

Is = arrangement of land cover types around grid

Jx = the mathematical approximation to juxtaposition is the quality of edge

(1 for diagonal and 2 for horizontal and vertical edge)

The edge effect is defined as the tendency for an increase in variety and density at community junction. The weightage factors were assigned to edges based on the order of preference by 'goral'.

Based on habitat used by goral, three restrictive factors were determined:

1. Availability of water within 1 km (Rfw)
2. Steep slope (Rfs)
3. Area devoid of human habitation (Rfh)

Presence of a restrictive was given a value of one and it has impact on habitat quality, otherwise a factor of zero was assigned which resulted in the habitat quality value being zero. Intermediate values were also assigned based on field conditions and habitat requirement of 'goral'. Habitat quality was determined based on the above parameters with different weightages assigned to each parameters.

All parameters for each grid were determined and maps showing interspersion and juxtaposition were prepared. Restrictive factors were assigned using the slope map, drainage map and human habitation map. The following equation was used for habitat quality calculation:

$$HQ = (0.2 * Is/8 + 0.8 * Jx/12) * Rfw * Rfs * Rfh$$

Thus each cell was assigned certain habitat quality value. A threshold value was set to group the entire area into different habitat quality zones viz., highly suitable, moderately suitable, less suitable and not suitable.

The suitable locations listed down during the field visits were compared by overlaying them on the final map of habitat quality prepared through this model. It was observed that out of total 10 location sited, 4 feel in most suitable areas, 4 fell in moderately suitable areas, 2 fell in less suitable areas whereas none fell in unsuitable areas. The results were also compared with census records of the forest department carried out in May, 1993. During census actual sighting method was used.

REFERENCES

Champion, H.G.& Seth, S.K. (1968). *A Revised Survey of Forest Types of India*. Govt. of India, New Delhi.

Gupta, Peeyush. Uniyal S. &Goyal S. (2014). An Assessment of Phytosociological Diversity of Kuther Watershed, Himachal Pradesh, India. *International Journal of Environment, 3*(4), 74-90.

Gupta, Peeyush & Goyal S. (2014). Analysis of Disturbance Gradient in Tehri Garhwal Region, Uttarakhand, India- A Spatial Landscape Modelling Approach. *International Journal of Applied Forest Ecology, 2*(2), 20-22.

Gupta, P. N. (1991). *Forest Management and Remote Sensing. Mountain Resource Management by Remote Sensing* (pp. 195–207). Dehradun: Surya Publications.

Jain, A., Ravan, S.A., Singh, R.K., Das, K.K. & Roy, P.S. (1996). Forest fire risk modelling using remote sensing and geographic information system. *Current Science*, *70*(10), 928–933.

Madhavan Unni, N. V. (1990). Space and Forest Management in India. Space and Forest Management, Special Current Even Session. *Proceedings of the International Astronautical Federation,31st IAF Congress*, Dresden Germany (pp. 49-69).

Maslekar, A. R. (1974). Remote sensing and its scope in Indian Forestry. *Indian Forester*, *100*(3), 192–201.

McNeely, J.A., Miller, K.R., Ried, W.V., Mittermeir, R.A. & Warner, T.B. (1990). Conserving World's Biological Diversity. IUCN-WRI Report.

Chapter 14

Measuring Dynamics of Ecological Footprint as an Index of Environmental Sustainability at the Regional Level using Geospatial Information Technology:

Measuring Ecological Footprint Using GIS

Laxmikant Sharma
Central University of Rajasthan, India

Suman Sinha
Haldia Institute of Technology, India

ABSTRACT

Ecological Footprint (EF) analysis is the spatial measurement of ecological load exerted by the humans on the earth that arises from the concept of sustainability and sustainable use of Earth's resources. A region-based EF study is conducted for Birla Institute of Technology, Mesra (India) campus to improve its sustainability. Highlight of the study is the explicitness of the methodology for determining the

DOI: 10.4018/978-1-5225-1814-3.ch014

EF that incorporates analysis derived from conversion factors mentioned in the Ecological Footprint consultancy publications along with inputs from GIS domain. Questionnaire-based survey from the respondents regarding resource utilization and geospatial enumeration of land use land cover that harbors the population and their resources are the two integral parts of the analysis. Total EF of the institution campus is calculated to be 0.645 gha/ individual. This analysis provides a strong framework for combining efforts in a manner that can communicate the immediate priorities for improving the sustainability strategy of the campus area.

INTRODUCTION

Ecological Footprint (EF) analysis deals with the procedure to estimate how much of the Earth (or how many planet Earths) it would take to support humanity if everybody followed a given lifestyle. For 2006, humanity's total ecological footprint was estimated at 1.4 planet Earths – in other words, humanity uses ecological services 1.4 times as fast as Earth can renew them. Ecological footprint analysis is now widely used around the globe as an indicator of environmental sustainability. It can be used to measure and manage the use of resources throughout the economy. It can be used to explore the sustainability of individual lifestyles, goods and services, organizations, industry sectors, neighborhoods, cities, regions and nations. Since 2006, a first set of ecological footprint standards exist that detail both communication and calculation procedures.

Ecological footprint analysis compares human demand on nature with the biosphere's ability to regenerate resources and provide services. It does this by assessing the biologically productive land and marine area required to produce the resources a population consumes and absorb the corresponding waste, using prevailing technology. Footprint values at the end of a survey are categorized for Carbon, Food, Housing, and Goods and Services as well as the total footprint number of Earths needed to sustain the world's population at that level of consumption. This approach can also be applied to an activity such as the manufacturing of a product or driving of a car. This resource accounting is similar to life cycle analysis wherein the consumption of energy, biomass (food, fiber), building material, water and other resources are converted into a normalized measure of land area called 'global hectares' (gha).

The current study aims in calculating the EF for Birla Institute of Technology, Mesra campus in order to improve its sustainability[1]. Limitations and assumptions in the analysis are critically dealt with. Highlight of the study is the explicitness of the methodology for determining the EF. Current scenario pinpoints the absence of standardized methodology for calculating EF. The method incorporates the analysis derived from conversion factors mentioned in the Ecological Footprint consultancy publications along with the inputs from GIS domain. Hence, the method reveals the potentiality of Geospatial techniques in the EF analysis. Questionnaire-based survey from the respondents regarding their resource utilization and geospatial enumeration of land use land cover that harbors the population and their resources are the two integral parts of the analysis. The EF calculated in this study can be refined with additional data to remove the assumptions in the calculations, however, this can be achieved with improved data monitoring. This analysis provides a strong framework for combining efforts in a manner that can communicate the immediate priorities for improving the sustainability strategy of BIT.

Background

Ecological Footprint (EF) analysis is the spatial measurement of ecological load exerted by the humans on the earth; i.e., it expresses the measure of human demand on the Earth's ecosystems. The concept of EF arises from the concept of sustainability and sustainable use of available resources on Earth. For a certain population, it is the total area required to produce the resources that the population consumes and to assimilate the wastes that they generate; and compares this to the global bio capacity. It is a standardized measure of demand for natural capital that may be contrasted with the planet's ecological capacity to regenerate. It represents the amount of biologically productive land and sea area necessary to supply the resources a human population consumes, and to mitigate associated waste.

The study undertaken by Athira and Subha (2011) analyses the scope of EFA (Ecological Footprint Analysis) as a sustainable environmental management tool for Kochi, India. According to Zhou et al. (2002), a concept framework and an index system for regional sustainable development evaluation supported by geo-spatial information technologies, were built based on some essential achievements attained domestically and aboard such as Human Development Index (HDI) and Ecological Footprint. Then, an Evaluation Information System for Regional Sustainable Development (EIS/ RSD) was produced based on this concept model, the index system and their algorithms. With the help of Geo-spatial information technologies and the EIS/RSD, dynamic analysis and assessment of progresses toward SD in 32 provinces using time serial statistic data, vector data and remotely sensed images. Many sustainable growth indicators consist of implicit valuations, weighting schemes and policy objectives, which are insufficiently, recognized as such. This contribution tries to show up these issues by means of a review of a recently proposed indicator for ecological-economic analysis, namely the Ecological Footprint that has been developed by Wackernagel and Rees (1996, 1997). According to Haberl et al. (2001), EF differentiates between domestic extraction and import but in sufficient data exist to link each consumption item to its spatially explicit origin. The bio capacity component of EF accounts can, however, be spatially mapped. HANPP can be used for spatial precise analyses and is therefore able to relate a society's metabolism to land use and its effects on ecosystem functioning. It is thus a capable approach for linking analytical tools such as Material Flow Analysis (MFA) or approaches such as industrial ecology to landscape ecology or bio geophysical analyses of environmental change. The analysis reveals that the ecological footprint provides on the basis of waste management practices (Salequzzaman et al., 2006; Solano and Cruz, 2011). Finally, the research recommends that the ecological footprint could be used as a sustainable waste management tool for KCC and other metropolitan cities of Bangladesh environmentally friendly sustainable waste management. Kratena (2004) worked on the derivation of an economic indicator from an extended input–output model with Ecological Footprints, Bio capacity and a technology for additional Biocapacity cultivation by industry. Different measures of the cost of eliminate the Ecological shortfall can be derived from the solution of this extended input–output model. Jinhua and Xinqi (2012) used GIS for constructing land use sustainable evaluation model based on comprehensive ecological footprint. The model applied to Shandong province concluded the defect and prospect of this model. There are large differences between the various land use types in ecological footprint per person; the change rate of the ecological footprint per person is much faster than the change rate of the ecological capacity per person; the ecological surplus or deficit of different types decrease in descending order: building area, forest, water area, arable land grassland. Kuzyk (2011) determined local sustainability; the ecological and carbon footprints are approximated in global hectares based on consumption and income in a geographic information system. Kuzyk (2010) determined the Ecological Footprint and its unit, the global hectare, share a reputation of effectively communicating the connection between local awareness and global impact. Using Geospatial technology in urban planning and decision support the potential role of GIS in ecological footprint studied by Kuzyk (2012). The study undertaken by Klinsky et al. (2010), the local sustainability planning tools that support public engagement with sustainability are essential this research investigates the ability of two geographic information system based tools to promote discussion of sustainability in a suburban context. A local ecological footprint tool and a community environmental atlas were created for seven suburban boroughs of Montreal. Fu et al. (2015) calculated the Resource Productivity by using the EF as an indicator of the natural resource input and gross domestic product (GDP).

MAIN FOCUS OF THE CHAPTER

Problems

The main focus of the chapter is the enumeration of the EF for an institution campus in order to determine and improve its sustainability using integrated geospatial techniques[1]. Based on this, the following research questions are generated that attempts to provide solution for the stated objective:

1. What are the different parameters involved in Ecological Footprint analysis of the campus?
2. How RS and GIS are helpful in assessing the Ecological Footprint?
3. How the Ecological Footprint varies with the different LULC types?

Study Area

The study area selected for performing the present research comprises the campus of the Birla Institute of Technology, Mesra; located within the Ranchi city, the capital of Jharkhand, India situated on the Chotanagpur plateau and is covered in the Survey of India toposheet No. 73 E/7 and is bounded by the spatial extension which is 23° 27' 34 N to 23° 16' 28 N and 85° 18' 14'' E to 85° 26' 54'' E (Figure 1). The study area is predominantly characterized by natural and protected open type sal forest. Main dominant species is *Shorea robusta* (sal). Other than Sal trees the available species are *Accacia* sp., *Eucalyptus* sp. *Bambusa* sp., *Terminalia tomentosa*. Along with the woody trees small grasses, and shrubs like Lantana camera are found here and there. The entire study area is surrounded by Subarnarekha and Jumar River. The vegetation remains green during the rainy and winter season while the summer season is characterized by both leaf shedding and leaf emerging processes. The temperature ranges between as low as 2°C in winters to as high as 38°C in the summer; with the annual rainfall being about 1530 mm.

Materials and Methodology

Inputs

Geospatial tools are used to determine and evaluate Ecological footprint network in the study area. The entire methodology applied in this study includes pre-field work, field work and post-field work is shown in Figure 2. The data obtained are then statistically analyzed accordingly:

Figure 1. Location of the study area

Figure 2. Paradigm for determining ecological footprint of the study area

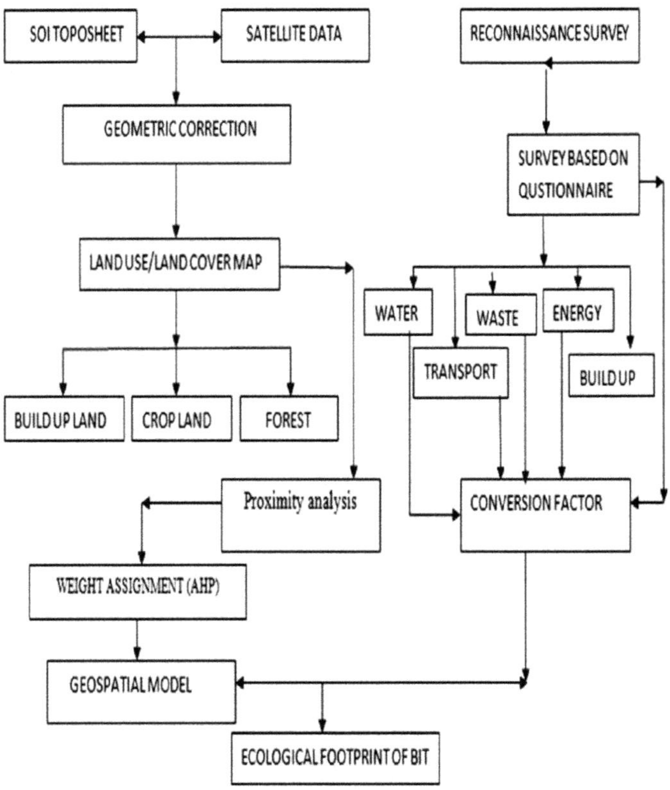

1. **Pre-Field Work:**
    a. Includes collection of data, ancillary information and literature survey.
    b. Base maps are prepared with the help of the SOI toposheet (73E/7) of scale 1: 50,000.
    c. Satellite data of IRS P-6 LISS III (23.5 m resolution) and PAN merged data (path/row-105/55) and Google Earth Imagery was used for improved visualization of ground features.
    d. Management plan secondary data information regarding census data, electricity, maintenance, etc. were procured from BIT authorities.
2. **Field Work:**
    a. Campus survey was a collection of secondary data including gathering of data and information regarding the area under study. LULC distribution data of the campus was provided from BIT.
    b. Expert knowledge was obtained from surveying and discussions with the BIT officials, faculty members, teaching and non-teaching staffs, family members, scientists and researchers.
    c. Certain Information was already gathered during pre-field work while referring to the literatures and similar studies already done. Different land use/ land cover units are identified spectrally, using visual interpretation with interpretation keys.
    d. Signatures are verified with the knowledge of ground truth which provides keys to interpret the satellite data in a meaningful way. GPS is used during field survey for this purpose.
3. **Post-Field Work:** Method for determination of ecological footprint.
4. **Software Used:** ArcGIS 9.3, ERDAS Imagine 9.1, and Microsoft Office 2007.
5. **Questionnaire Survey-Based Method:** Carried out through structured format.
6. **Parameters Considered:**
    a. Build-up land.
    b. Forest.
    c. Agricultural land.
    d. Water.
    e. Waste.
    f. Transportation.
    g. Energy.

Enumeration of Ecological Footprint (EF) and Ecological Capacity (EC)

Ecological capacity is the capacity of ecosystems to produce useful biological materials and to absorb waste materials generated by humans, using current management schemes and extraction technologies (Global Footprint Network, India Report, 2008). For estimation of ecological capacity and ecological footprint of the study area is based on the calculation model (Chang & Xiong, 2005)

$$EF= N*(ef) =N * \sum w_i*a_i \quad (1)$$

where

EF is the total ecological footprint,
N is the number of population,
ef is per capita ecological footprint,
i is the type of consumption,
w_i is equivalence factor, and
A_i is per capita area of land to produce the the i-th consumption for ecological capacity.

$$EC=\sum y_i*w_i*A_i \quad (2)$$

where

EC is the total ecological capacity,
y is the yield factor, and
A is the total area of bio-productive land to produce the i-th consumption.

Geospatial Model

Three parameters or layer are generated from land use/land cover namely, build up land, forest and agricultural land for the analysis. Proximity analysis through buffer is performed for these parameters in GIS (Sinha et al., 2011). Proximity analysis is one way of analyzing locations of features by measuring the distance defined between them and other features in the area. Weights are then assignment through knowledge-based method considering the yield factor and equivalence factor of the parameters using Analytical Hierarchy Process (AHP) (Sinha et al., 2014; Sharma et al., 2012). AHP is a process for developing a numerical score to rank each decision alternative based on how well each alternative meets the decision maker's criteria. The mathematics of the AHP and the calculation techniques are briefly explained in table (Saaty, 1980). Weights are assigned through AHP considering the yield factor and equivalence factor of the parameters. Several studies have used this concept for handling fuzziness in targeting problems (Sinha et al., 2014; Sharma et al., 2012). Weight derived through AHP is fitted to derive geospatial model for Ecological footprint depending upon land use/land cover classification. Integral part of the analysis is reconnaissance survey for collection of resource consumption data which were mainly based on Questionnaire based survey is carried out. Conversion factors adopted from (Chambers et al., 2005) were used for the conversion of data into the international standardized unit of the global hectare (gha). A combined method involving Geospatial and Questionnaire based approach is mentioned in Figure 2.

SOLUTIONS AND RECOMMENDATIONS

This study demonstrates the ability of Remote Sensing and GIS to assess ecological footprint and to make decisions regarding the management strategies for sustainable development of the study area. The necessary factors and constraints for the case study have been determined. The importance of these factors over each other has been analyzed and the comparison ratios have been assigned. GIS software was employed to obtain final maps of the determination of the spatial extent of the study area for ecological footprint. Final maps exhibit the optimum to maximum area for determination of EF with weights obtained from AHP method.

Land Use and Land Cover Classification

In the present study, image of IRS P-6 LISS III has been digitally interpreted. Using ArcGIS software, the image is classified using visual interpretation forming different LULC classes as mentioned in Table 1 and shown spatially in Figure 3. The classification is done based on visual interpretation, along with field verification and ground truth. Classification accuracy of the settlement is high, while other classes gave good results. Area for each land use/ land cover (LULC) units is calculated using ArcGIS software and is summarized in Table 1 and Figure 4. More than 50% of the study area is covered under forest, with open forest covering the maximum area (1091870.76 m^2). Of the total area under investigation, open forest comprises of 35%, agriculture land 17%, build up 3%, dense forest 20%, open space 14%, plantation 1%, water bodies 6% and road 2%. Table 2 documents the complete error matrix in accuracy assessment and Table 3 highlights the accuracy assessment report of the eight classes of land use/land cover with producer accuracy and user accuracy for each LULC classes. Overall classification accuracy of 90% is obtained in the classification (Table 3) which is considered to be highly significant for classification.

Table 1. Area statistics of land use/land cover categories

LU/LC	Area (%)	Area (sq.m.)
Open Forest	35.03	1091870.76
Agriculture land	17.94	559070
Built up	3.41	106162.684
Dense Forest	20.08	625928.65
Open Space	13.91	433549.011
Plantation	1.36	42491.257
Water	6.43	200453
Road	2.00	57026.63

Figure 3. Land use/land cover map of study area

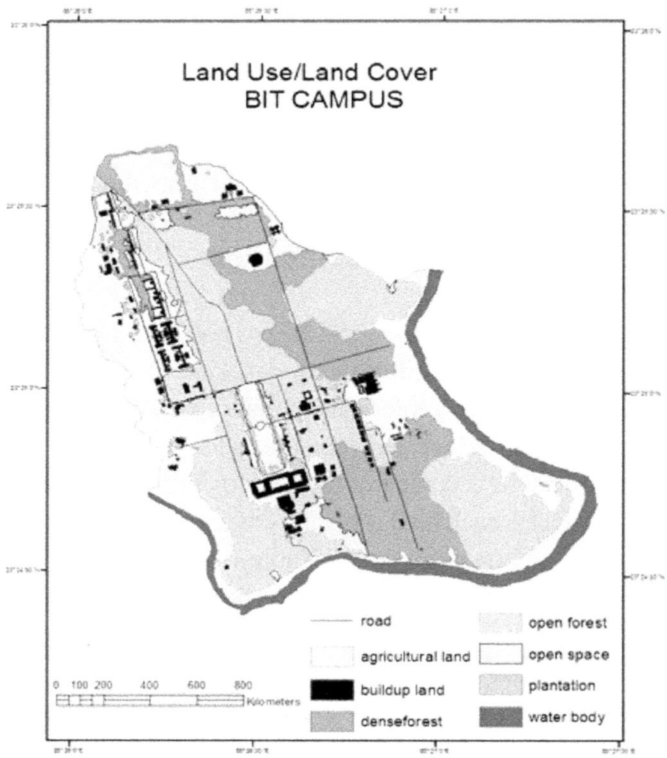

Figure 4. Area statistics of land use/land cover categories

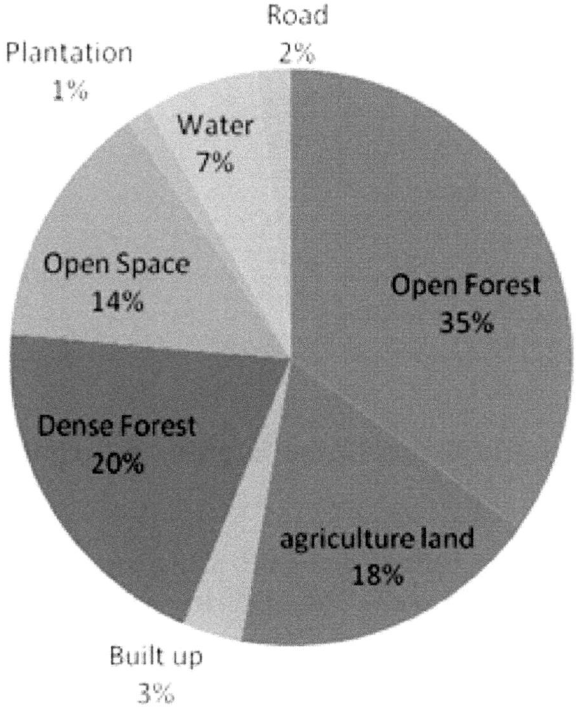

Table 2. Error matrix in accuracy assessment

LULC	Open Space	Dense Forest	Build up Land	Plantation	Open Forest	Agriculture	Water Body	Row Total
Open space	495	10	0	0	0	0	0	505
Dense forest	15	485	0	0	0	0	0	500
Build up land	0	0	72	0	0	10	0	82
Plantation	0	0	4	17	0	0	0	21
Open forest	10	30	4	3	1100	70	0	1217
Agriculture	0	15	0	0	100	300	0	415
Water body	0	0	0	0	0	0	180	180
Column Total	520	540	80	20	1200	380	180	2920

Table 3. Accuracy assessment report

LULC	Producer Accuracy	User Accuracy
Open space	95.00%	98.00%
Dense forest	89.00%	97.00%
Build up land	90.00%	87.00%
Plantation	85.00%	80.00%
Open forest	91.00%	90.00%
Agriculture	78.00%	75.00%
Water body	100.00%	100.00%
Over all classification accuracy	90.00%	

Land Use and Land Cover Classification: Proximity Analysis and AHP

Proximity analysis trough buffer is performed for the parameters namely, build up land, forest and cropland (Figure 5). The weight is assigned through AHP considering yield factor and equivalence factor of the parameters fitted to derived geospatial model for ecological footprint. Weights assigned through AHP are integrated in GIS and the final map showing EF based on LULC classes are derived in GIS domain (Figure 6). Here, EF is reclassified based on the values into four categories depending on the severity in GIS, namely (1) least ecological footprint, (2) low ecological footprint, (3) moderate ecological footprint and (4) high ecological footprint classes. Table 4 documents the area distribution for each EF classes, while Figure 6 shows the spatial distribution of the same. Mostly the area is covered under the moderate EF class covering the forest areas.

Assessment of Ecological Footprint

For ecological footprint assessment on basis of resource use, the questionnaire based data collection for example, transportation, electricity, waste, water and build up land, a resources consumption map of the consumed land of the study area was derived (Figure 7). The figure shows the three classes of consumed land including build up area where all population resides and hence the pressure of population is significant on these particular areas. Figure 7 shows the distribution of land consumption for EF of study area in three categories as mentioned by Rees and Wackernagel (1996) for classifying consumption into one particular measure. The data compiled from questionnaire regarding consumption of resources in BIT are mentioned in Table 5 where the consumption categories are represented in different units. With the help of

Figure 5. Parameter for the EF from LULC category

standardized conversion factors (Chambers et al., 2005), the resource consumption data were converted into global hectares which is categorized in terms of Ecological Footprint of the study area. Table 5 and Figure 8 show that the contribution of energy towards ecological footprint is maximum (2853 gha.), while the minimum is for built up (37.24 gha.). The values for ecological footprints in transportation, waste and water are 1520 gha., 590.4 gha. and 49.896 gha. respectively. The total ecological footprint of the study area is summed up to 5050.53 gha., which is 5 times greater than the area of the study area. The total population (N) considered is approximately 8400 on a rough basis. In terms of per individual, ecological footprint comes out to be 0.6 for resource consumption and 0.045 through GIS based LULC analysis (Table 6); hence totaling to 0.645 gha per person. The total ecological capacity is calculated to 0.058 ha/capita using Equation 2.

Figure 6. EF for LULC categories of BIT defined by AHP

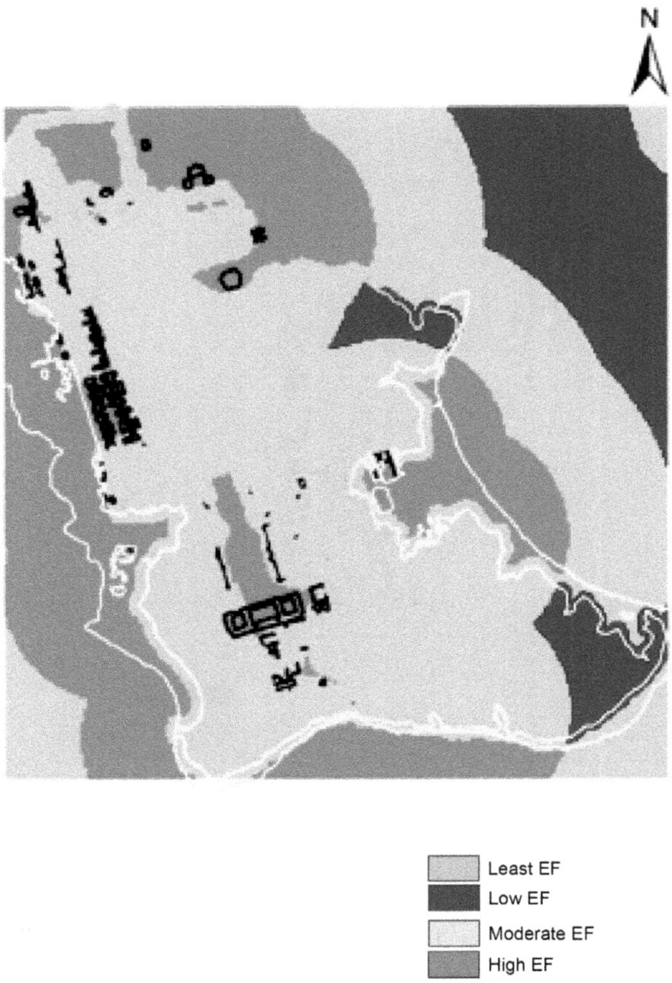

Table 4. EF based on proximity analysis and AHP in GIS

EF Class	Area (sq.km.)
Least ecological footprint	0.021198298
Low ecological footprint	0.660770872
Moderate ecological footprint	3.048387696
High ecological footprint	1.379338813

Figure 7. Distribution of land consumption for EF of study area

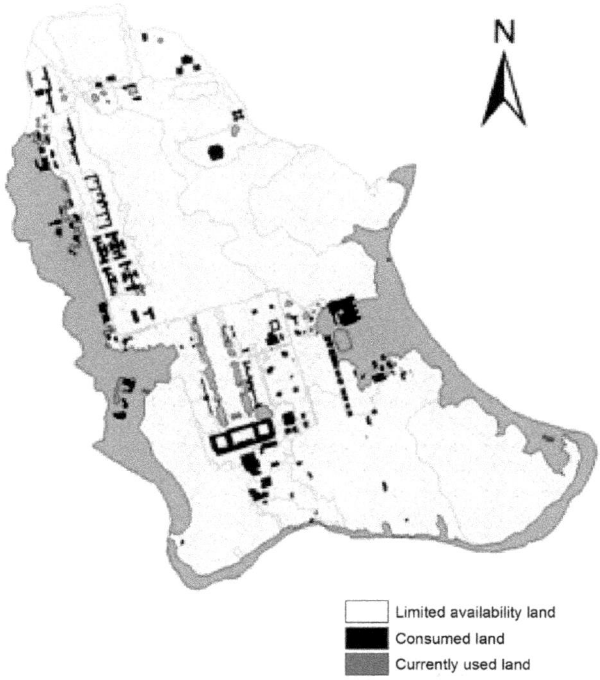

Table 5. EF calculation of BIT campus based on survey generated information

Parameters	Consumption	Conversion Factor	Ecological Footprint
Buildup (ha)	10.61	3.51	37.24
Transportation (10^{3m})	Bus 691.0 Car 829.0	0.09 0.03	1520
Water (10^{6}L)	504	0.099	49.896
Waste (tonne)	144	4.1	590.4
Energy (GWH)	Electricity 18.41 Gas 4.86	143 45	2853

Figure 8. Distribution of EF in the study area

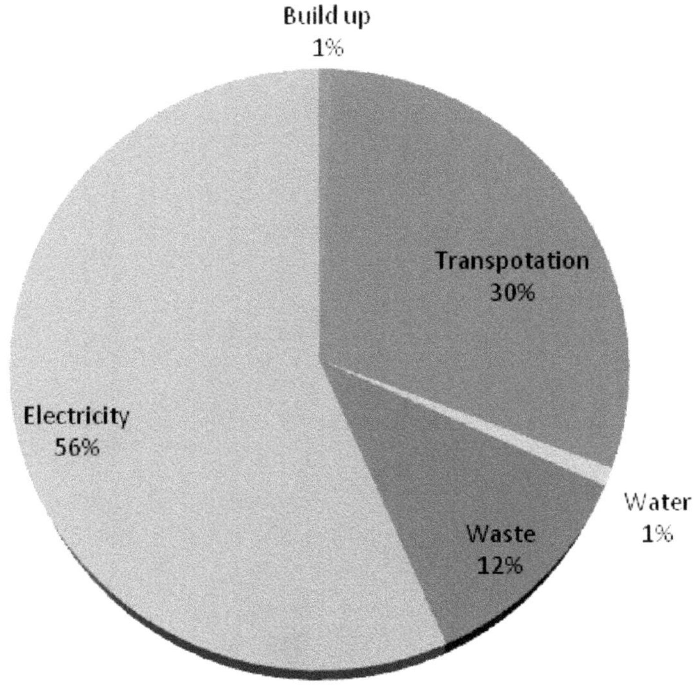

Table 6. Estimation of EF and EC through GIS based LULC analysis

Parameters	Area(ha)	Area per Capita	Equivalence Factor	Yield factor	Ecological Footprint (gha)
Forest	171.7799	0.020761414	1.1	0.91	0.0228375554
Crop land	55.907	0.006756949	2.8	1.66	0.0189194572
Build up	10.616227	0.00128308	2.8	1.66	0.003592624
			Total EF (per person)		0.0453496366
			Total EF		380.93694744
			Total EC		0.058143 ha/capita

FUTURE RESEARCH DIRECTIONS

EF is an integral part of the current environmental issues that evaluates of anthropogenic demand on the Earth's ecosystems and its existing natural resources. The chapter offers a simple yet robust approach for measuring the EF, the method is easily replicable and can be used with numerous parameters at a time that are responsible for EF assessment. However, determination of region-based yield and equivalence factor values is required for effective region-level EF studies. Future research should target in establishing an accurate and effective method for enumerating 'resource and waste flows' that cannot be measured in terms of biologically productive area, leading to a systematic underestimate of the total demand exerted by these flows on ecosystems. There lies a huge potential for advanced GIS and statistics in calculating EF that needs to be explored.

CONCLUSION[1]

The study reveals the utility and efficacy of integrated geospatial information technology in measuring the dynamics of ecological footprint as an index of environmental sustainability at the regional level. A region-based EF study for Birla Institute of Technology, Mesra (India) campus is conducted to improve its sustainability. The total EF of the institution campus is calculated to be 0.645 gha/ individual. The EF calculated in this study can be refined with additional data to remove the assumptions in the calculations, however, this can be achieved with improved data monitoring. This analysis provides a strong framework for combining efforts in a manner that can communicate the immediate priorities for improving the sustainability strategy of the campus area. The same methodology can even be applied on a larger scale for national level assessment.

REFERENCES

Athira, R., & Subha, V. (2011). Ecological foot print analysis-A sustainable environmental management tool for Kochi city. *ACEEE International Journal on Transportation and Urban Development*, *1*(1), 5–7.

Chambers, N., Child, R., Jenkin, N., Lewis, K., Vergoulas, G., & Whiteley, M. (2005). *Stepping forward: A resource flow and ecological footprint analysis of the South West of England. Technical report, The Stepping Forward Report Series*. Oxford, UK: Best Foot Forward Ltd.

Chang, B., & Xiong, L. (2005). Ecological footprint analysis based on RS and GIS in arid land. *Journal of Geographical Sciences*, *15*(1), 44–52. doi:10.1007/BF02873106

Fu, W., Turner, J. C., Zhao, J., & Du, G. (2015). Ecological footprint (EF): An expanded role in calculating resource productivity (RP) using China and the G20 member countries as examples. *Ecological Indicators*, *48*, 464–471. doi:10.1016/j.ecolind.2014.09.023

Global Footprint Network. (2008). *National Footprint Accounts, 2008 Edition.* Retrieved from www.footprintnetwork.org

Haberl, H., Erb, K., & Krausmann, F. (2001). How to calculate and interpret ecological footprints for long periods of time: The case of Austria 19261995. *Ecological Economics*, *38*(1), 25–45. doi:10.1016/S0921-8009(01)00152-5

Jinhua, L., & Xinqi, Z. (2012). A Spatial-EF and Econometrics Model Integrated Approach to explore Land Use Sustainable Forecast Model-in case of Shandong Province. *Procedia Environmental Sciences, 12*(A), 413–420.

Klinsky, S., Sieber, R., & Meredith, T. (2010). Connecting local to global: Geographic information systems and ecological footprints as tools for sustainability. *The Professional Geographer*, *62*(1), 84–102. doi:10.1080/00330120903404892

Kratena, K. (2004). Ecological value added in an integrated ecosystem-economy model: An indicator for sustainability. *Ecological Economics*, *48*(2), 189–200. doi:10.1016/j.ecolecon.2003.09.012

Kuzyk, L. W. (2010). Infill Housing Sustainability Analysis. *AACIP Planning Journal, 2010*(4), 19-22.

Kuzyk, L. W. (2011). Ecological and carbon footprint by consumption and income in GIS: Down to a census village scale. *Local Environment*, *16*(9), 871–886. doi:10.1080/13549839.2011.615303

Kuzyk, L. W. (2012). The ecological footprint housing component: A geographic information system analysis. *Ecological Indicators*, *16*, 31–39. doi:10.1016/j.ecolind.2011.03.009

Rees, W. E., & Wackernagel, M. (1996). Urban ecological footprints: Why cities cannot be sustainable and why they are a key to sustainability. *Environmental Impact Assessment Review*, *16*(4-6), 223–248. doi:10.1016/S0195-9255(96)00022-4

Saaty, L. T. (1980). *The analytic hierarchy process.* New York: McGraw-Hill International.

Salequzzaman, M., Sultana, U. T., Iqbal, A., & Hoque, M. A. (2006). *Ecological footprint of waste generation: A sustainable tool for solid waste management of Khulna City Corporation of Bangladesh.* Retrieved from http://www.necsi.edu

Sharma, L. K., Kanga, S., Nathawat, M. S., Sinha, S., & Pandey, P. C. (2012). Fuzzy AHP for forest fire risk modeling. *Disaster Prevention and Management*, *21*(2), 160–171. doi:10.1108/09653561211219964

Sinha, S., Nathawat, M. S., & Sharma, L. K. (2014). *Geospatial approach for managing relocation and distribution of tigers*. Germany: LAP Lambert Academic Publishing.

Sinha, S., Sharma, L. K., & Nathawat, M. S. (2011). Retrieving tiger habitats: Conserving wildlife geospatially. *Applied Remote Sensing Journal*, *2*(1), 1–5.

Solano, D. M. M., & Cruz, E. (2011, August). Ecological Footprint of Waste Generation in Digos City, Davao Del Sur. *Paper presented at the 7th Mindanao Statistics Congress*, University of Mindanao, Matina Campus, Davao City, Philippines.

Wackernagel, M., & Rees, W. E. (1996). *Our ecological footprint: Reducing human impact on the Earth. Gabriola Island.* New Society Publishers.

Wackernagel, M., & Rees, W. E. (1997). Perceptual and structural barriers to investing in natural capital: Economics from an ecological footprint perspective. *Ecological Economics*, *20*(1), 3–24. doi:10.1016/S0921-8009(96)00077-8

Zhou, Y., Wang, S., Zhou, W., Huang, T., & Yan, S. (2002). Research on assessing the progress towards sustainable development supported by the geo-spatial information technology. *Proceedings of International Society for Photogrammetry and Remote Sensing*, *34*(Part 2), 627–632.

KEY TERMS AND DEFINITIONS

Biocapacity: Biocapacity is the capacity of an area to provide resources and absorb wastes.

Bio-Productivity: Bio-productivity is the capacity to produce biomass.

Ecological Capacity: Ecological capacity is the overall ability of an ecosystem to maintain its natural, original or current condition and to produce goods and services.

Ecological Footprint: The impact of a person or community on the environment, in terms of land required to sustain their use of natural resources.

Geospatial Technology: Geospatial technology refers to equipment used in visualization, measurement, and analysis of earth's features, involving GPS (global positioning systems), GIS (geographical information systems), and RS (remote sensing).

Land Use Land Cover: Land cover indicates the physical land type such as forest or open water whereas land use documents how people are using the land, like agriculture or settlement.

Resource Productivity: Resource productivity is the quantity of good or service that is obtained through the expenditure of unit resource.

Sustainability: Sustainability is the ability or capacity of something that meets the needs of the present without compromising the ability of future generations to meet their own needs.

ENDNOTE

1 The study was performed in the institute campus during 2010-11 by the second author under the supervision of the first author. Please note that the first author was associated as a faculty of the institute during this period; however, currently he serves as a faculty of Central University of Rajasthan

Compilation of References

Agarwal, C. S. (1998). Study of drainage pattern through aerial data in Naugarh area of Varanasi district, U.P. *Journal of the Indian Society of Remote Sensing*, *24*(4), 169–175. doi:10.1007/BF02990795

Aggett, G., & McColl, C. (2006). Evaluating decision support systems for PPGIS applications. (Public Participatory Geographic Information Systems). *Cartography and Geographic Information Science*, *33*(1), 77–92. doi:10.1559/152304006777323163

Agrifoglio, R., & Metallo, C. (2015). Preserving Knowledge Through Community of Practice: A Multiple Case Study. In From Information to Smart Society (pp. 103-111). Springer International Publishing. doi:10.1007/978-3-319-22234-9_3

Al-Ahmadi, F. S., & Hames, A. A. (2009). Comparison of Four Classification Methods to Extract Land Use and Land Cover from Raw Satellite Images forSome Remote Arid Areas, Kingdom of Saudi Arabia. *Earth Science*, *20*(1), 167–191.

Alesheikh, A. A., Helali, H., & Behroz, H. A. (2002). Web GIS: Technologies and Its Applications. *Proceedings of theSymposium on Geospatial Theory, Processing andApplications*, Ottawa.

Anbalagan, R. (1992). Landslide hazard Evaluation and zonation mapping in mountainous terrain. *Engineering Geology*, *32*(4), 269–277. doi:10.1016/0013-7952(92)90053-2

Anderson, J. R., Hardy, E. E., Roach, J. T., & Wirmer, R. E. (1976). A land use land cover classification system for use with remote sensing data. United States Geological Survey (USGS) Professional Paper 964.

Anderson, J. R., Hardy, E. E., Roach, J. T., & Witmer, R. E. (1976). *A land use and land cover classification system for use with remote sensor data*. USGS.

Andréfouët, S., Gilbert, A., Yan, L., Remoissenet, G., Payri, C., & Chancerelle, Y. (2005). The remarkable population size of the endangered clam *Tridacna maxima* assessed in Fangatau atoll (Eastern Tuamotu, French Polynesia) using in situ and remote sensing data. *ICES Journal of Marine Science*, *62*, 1037–1048.

Aniello, C., Morgan, K., Busbey, A., & Newland, L. (1995). Mapping micro urban heat islands using Landsat TM and a GIS. *Computers & Geosciences*, *21*(8), 965–967. doi:10.1016/0098-3004(95)00033-5

Anon., (2008). *Annual Report*. Kochi: Central Marine Fisheries Research Institute.

Araya, Y. H., & Cabral, P. (2010). Analysis and Modeling of Urban Land Cover Change in Setúbal and Sesimbra, Portugal. *Remote Sensing*, *2*(6), 1549–1563. doi:10.3390/rs2061549

Arciniegas, G. A., Bijker, W., Kerle, N., & Tolpekin, V. A. (2007). Coherence- and amplitude-ased analysis of seismogenic damage in Bam, Iran, using ENVISAT ASAR data. *IEEE Transactions on Geoscience and Remote Sensing*, *45*(6), 1571–1581. doi:10.1109/TGRS.2006.883149

Arp, H., Greisbach, J., & Burns, J. (1982). Mapping in tropical forests: A new aproach using the laser APR. *Photogrammetric Engineering and Remote Sensing*, *48*, 91–100.

Arrington, M. (2005). *The Ugly (but necessary?) side of Wikipedia*. Retrieved from http://www.techcrunch.com/2005/09/07/the-ugly-but-necessary-side-of-wikipedia/

Athira, R., & Subha, V. (2011). Ecological foot print analysis-A sustainable environmental management tool for Kochi city. *ACEEE International Journal on Transportation and Urban Development*, *1*(1), 5–7.

Auffhammer, M., Ramanathan, V., & Vincent, J. R. (2012). Climate change, the monsoon, and rice yield in India. *Climatic Change*, *111*(2), 411–424. doi:10.1007/s10584-011-0208-4

Ayalew, L., & Yamagishi, H. (2005). The application of GIS-based logistic regression for landslide susceptibility mapping in the Kakuda-Yahiko Mountains, Central Japan. *Geomorphology*, *65*(1-2), 15–31. doi:10.1016/j.geomorph.2004.06.010

Ayotte, D., & Hunger, O. (2000). Calibration of a runout prediction model for debris flows and avalanches. In G.F. Wieczorek & N.D. Naeser (Eds.), *Proc. of the 2ndInternational Conference on Debris Flows*, Taipei (pp. 505-514). Balkema: Rotterdam.

Ayyangar, R. S. (1978). Experiment to evolve methods for separation and identification of agricultural crops from multispectral information.*Proceedings of 12th International Symposium on Remote Sensing of Environment*, Environmental Research Institute of Michigan, Ann Arbor, Michigan.

Ayyangar, R. S., Krishna Rao, M. V., & Rao, K. R. (1980). Interpretation and analysis of multispectral data for agricultural crop cover types- role of crop spectral response and crop spectral signature. *Journal of the Indian Society of Remote Sensing*, *8*(2), 39–48.

Backes, M., & Jacobi, J. (2006). Classification of weed patches in QUICKBIRD images: verification by ground truth data. In EARSeL ePROCEEDINGS (pp.173-178).

Bailey, S. W., & Werdell, P. J. (2006). A multi-sensor approach for the on-orbit of ocean color satellite data products. *Remote Sensing of Environment*, *102*(1-2), 12–23. doi:10.1016/j.rse.2006.01.015

Barsi, J. A., Barker, J. L., & Schott, J. R. (2003). An atmospheric correction parameter calculator for a single thetmal band Earth-Sensing instrument. *Paper presented at the IGARSS 03, Centre de Congres Pierre Baudis*, Toulouse, France.

Barsi, J. A., Schott, J. R., Palluconi, F. D., & Hook, S. J. (2005, August). Validation of a web-based atmospheric correction tool for single thermal band instruments. Earth Observing Systems X. *Paper presented at theProc. SPIE*, San Diego, CA, USA (*Vol. 5882)*.

Bartolucci, L. A. (1985). Calibration of Landsat-5 TM Thermal IR data. *Paper presented at theProceedings of the First National Symposium on Remote Sensing*, Baghdad, Iraq.

Bartolucci, L. A., & Chang, M. (1988). Look-up tables to convert Landsat TM thermal IR data to water surface temperatures. *Geocarto International*, *3*(3), 61–67. doi:10.1080/10106048809354167

Barton, I. J. (1995). Satellite - derived sea surface temperatures: Current status. *Journal of Geophysical Research*, *100*(C5), 8777. doi:10.1029/95JC00365

Basu, S. N., & Pal, P. K. (1986). Mineral resources of Midnapore District: Scope of some mineral based industries. *Proceedings of the National Seminar on Small Mineral Deposits: their development and industrial possibilities, New Delhi.* The Mining and Metallurgical Institute of India, Kolkata.

Bathrellos, G., Kalivas, D., & Skilodimou, H. (2009). GIS-based landslide susceptibility mapping models applied to natural and urban planning in Trikala, Central Greece. *Estudios Geology,* 65(1), 49-65. Doi:10.3989/egeol.08642.036

Beaujardière, J. L., Mitchell, H., Raskin, R. G., & Rao, A. (2000). *The NASA Digital Earth Testbed* (pp. 47–53). ACM-GIS.

Becker, F., & Li, Z.-L. (1990). Towards alocal split window method over land surface. *International Journal of Remote Sensing*, *11*(3), 369–393. doi:10.1080/01431169008955028

Becker, F., & Li, Z.-L. (1995). Surface temperature and emissivity at different scales: Definition, measurement and related problems. *Remote Sensing Reviews*, *12*(3-4), 225–253. doi:10.1080/02757259509532286

Bhagat, S, V. (. (2012). Use of remote sensing techniques for robust digital change detection of land: A review. *Recent Patents on Space Technology*, *2*(2), 123–144. doi:10.2174/1877611611202020123

Bhattacharya, A. (1983). Geologic and geomorphic controls in land use patterns. *Journal of the Indian Society of Remote Sensing*, *11*(3), 57–66.

Bist, K. S., & Sah, M. P. (1999). The devastating landslide of August 1998 in Ukhimath area, Rudraprayag district, Garhwal Himalaya. *Current Science*, *76*(4), 481–484.

Biswas, S. K. (1992). Tertiary stratigraphy of Kutch. *Journal of the Palaeontological Society of India*, *37*, 1–29.

Biswas, S., Sudhakar, S., & Desai, V. R. (1999). Prioritization of sub-watersheds based on morphometric analysis and drainage basin. a Remote Sensing and GIS approach. *Journal of Indian Society of Remote Sensing*, *27*(3), 155–166. doi:10.1007/BF02991569

Borstad, G. A., Hill, D. A., Kerr, R. C., & Nakashima, B. S. (1992). Direct digital remote sensing of herring schools. *International Journal of Remote Sensing*, *13*(12), 2191–2198. doi:10.1080/01431169208904262

Bossard, M., Feranec, J., & Otahel, J. (2000). *CORINE land cover technical guide - Addendum 2000*. European Environment Agency.

Brahmabhatt, V. S., Dalwadi, G. B., Chhabra, S. B., Ray, S. S., & Dadhwal, V. K. (2000). Land use / land cover change mapping in Mahi Canal Command area, Gujarat, using multi-temporal satellite data. *Journal of the Indian Society of Remote Sensing*, *28*(4), 221–232. doi:10.1007/BF02990813

Brivio, P. A., Colombo, R., Maggi, M., & Tomasoni, R. (2002). Integration of remote sensing data and GIS for accurate mapping of flooded areas. *International Journal of Remote Sensing*, *23*(3), 429–441. doi:10.1080/01431160010014729

Brock, J. C., & Purkis, S. J. (2009). The emerging role of LiDAR remote sensing in coastal research and resource management. *Special Issue 53 - Coastal Applications of Airborne LiDAR. Journal of Coastal Research*, *53*, 1–5. doi:10.2112/SI53-001.1

Buckland, S. T., & Elston, D. A. (1994). Use of groundtruth data to correct land cover area estimates from remotely sensed data. *International Journal of Remote Sensing*, *15*(6), 1273–1282. doi:10.1080/01431169408954160

Burns, G. S., & Joyce, A. T. (1982). Evaluation of land cover change detection techniques using Landsat MSS data. *Proceedings of Conference on Remote Sensing with Special Emphasis on Output to Geographic Information Systems in the 1980s.*, 7th PECORA Symposium, Sioux Falls, South Dakota, USA (pp. 252-257).

Caby-Guillet, L., Crave, S., & Ladame, S. (2006). How ICT can facilitate trust inside networks of SME: the role of Professional Virtual Communities. *Proceedings of theInternational conference on Concurrent Engineering,Milano*. doi:10.1109/ICE.2006.7477102

Callan, R. (1999). The essence of Neural Networks: Prentice Hall Europe.

Calvet, J.-C., & Julien, J.-P. (1996). Land surface temperature retrieval in dry conditions from infrared and microwave satellite radiometry. *Remote Sensing Reviews*, *13*(3-4), 235–255. doi:10.1080/02757259609532306

Campbell, J. B., & Wynne, R. H. (2011). Introduction to Remote Sensing (5th ed.). Guilford Press. Retrieved from http://books.google.com.sa/books?id=zgQDZEya6foC

Campbell, J. B., & Browder, J. O. (1995). Field data collection for remote sensing analysis: SPOT data, Rondonia, Brazil. *International Journal of Remote Sensing*, *16*(2), 333–350. doi:10.1080/01431169508954399

Carpenter, R. A. (1987). What to do While Waiting for an Environmental Ethics. *Environment and Progress*, *9*(4), 327–335.

Carpenter, T. M., & Georgakakos, K. P. (2006). Intercomparison of lumped versus distributed hydrologic model ensemble simulations on operational forecast scales. *Journal of Hydrology (Amsterdam)*, *329*(1–2), 174–185. doi:10.1016/j.jhydrol.2006.02.013

CGWB (Central Ground Water Board). (2006). *Report on Reappraisal Hydrological Survey in Parts of West Midnapur District, West Bengal* (Technical Report No. D-187). CGWB-Eastern Region (ER), Ministry of Water Resources, Govt. of India, Kolkata.

Chambers, N., Child, R., Jenkin, N., Lewis, K., Vergoulas, G., & Whiteley, M. (2005). *Stepping forward: A resource flow and ecological footprint analysis of the South West of England. Technical report, The Stepping Forward Report Series*. Oxford, UK: Best Foot Forward Ltd.

Champati ray, P.K., Chattoraj, S.L. and Kannaujiya, S. (2013b, November 11-12). Uttarakhand Disaster 2013: Response and Mitigation measures using remote sensing and GIS. Pre workshop full publication In: National work shop on Geology and Geoheritage sites of Uttarakhand with special reference to geo-scientific development of the region organized by *Indian geological Congress (IGC), Roorkee* jointly with L.S.M. Govt. PG College (pp. 37-45).

Champati ray, P.K., Chattoraj, S.L., Chand, D.S., & Kannaujiya, S. (2013a). Aftermath of Uttarakhand disaster 2013: an appraisal on risk assessment and remedial measures for Yamunotri shrine using satellite image interpretation. *Indian Landslides*, 6(2), 61-70.

Champati Ray, P. K., Dimri, S., Lakhera, R. C., & Sati, S. (2007). Fuzzy-based method for landslide hazard assessment in active seismic zone of Himalaya. *Landslides*, *4*(2), 101–111. doi:10.1007/s10346-006-0068-6

Champion, H.G.& Seth, S.K. (1968). *A Revised Survey of Forest Types of India*. Govt. of India, New Delhi.

Chang, B., & Xiong, L. (2005). Ecological footprint analysis based on RS and GIS in arid land. *Journal of Geographical Sciences*, *15*(1), 44–52. doi:10.1007/BF02873106

Changon, S. A. (1992). Inadvertent weather modification in urban areas: Lessons for global climate change. *Bulletin of the American Meteorological Society*, *73*, 621–627.

Chassot, E., Bonhommeau, S., Reygondeau, G., Nieto, K., Polovina, J. J., Huret, M., & Demarcq, H. et al. (2011). Satellite remote sensing for an ecosystem approach to fisheries management. *ICES Journal of Marine Science*, *68*(4), 651–666. doi:10.1093/icesjms/fsq195

Chattoraj, S. L. (2011). *Glauconite formation in stratigraphic framework of Palaeogene succession in Kutch* [Unpublished Ph.D. thesis]. Indian Institute of Technology Bombay.

Chattoraj, S. L., Champati ray, P.K., & Bandopadhyay, S. (2014, October 28-30). Debris Flow Simulation and Modeling: A Case Study from Kedarnath Area (abstract). *Proceedings of the Geo-Environmental Hazards and Neo-Tectonic Activities in Himalaya*, HNB Garhwal University Campus Badshahi Thaul, Tehri Garhwal (p. 26).

Chattoraj, S. L., Banerjee, S., & Saraswati, P. K. (2009). Glauconites from the Late Palaeocene EarlyEocene Naredi Formation, western Kutch and their genetic implications. *Journal of the Geological Society of India*, *73*(4), 567–574. doi:10.1007/s12594-009-0040-x

Chauhan, E. S., Porwal, M. C., Sharma, L., & Negi, J. D. S. (2003). Change detection in sal forest in Dehradun forest division using Remote Sensing and Geographical Information System. *Journal of the Indian Society of Remote Sensing*, *31*(3), 211–218. doi:10.1007/BF03030827

Chaurasia, R., Loshali, D. C., Dhaliwal, S. S., Minakshi, , Sharma, P. K., Kudrat, M., & Tiwari, A. K. (1996). Land use change analysis for agricultural management-A case study of Tehsil Talwandi Sabo. Punjab. *Journal of the Indian Society of Remote Sensing*, *24*(2), 115–123. doi:10.1007/BF03016124

Chen, J., He, C., Jiang, J., & Han, G. (2008). Reconciliation of inconsistent perspectives in collaborative GIS. *Cartography and Geographic Information Science*, *35*(2), 77–89. doi:10.1559/152304008784090586

Chen, S., Chen, X., Chen, W., Su, Y., & Li, D. (2011). A simple retrieval method of land surface temperature from AMSR-E passive microwave data - A case study over Sothern China during the strong snow disaster of 2008. *International Journal of Applied Earth Observation and Geo-information*, *13*(1), 140–151. doi:10.1016/j.jag.2010.09.007

Cherem, G. J., & Driver, B. L. (1983). Visitor employed photography: A technique to measure common perceptions of natural environments. *Journal of Leisure Research*, *15*(1), 65–83.

Cheruy, F., Chevallier, F., Morcrette, J.-J., Scott, N. A., & Chedin, A. (1996). Une methode utilisant les techniques neuronales pour le calcul rapide de la distribution verticale du bilan radiatif thermique terreste. *C. R. Acad. Sci., Paris, t. 322. Serie II b*, 665-672.

Cheval, S., & Dumitrescu, A. (2009). The July urban heat island of Bucharest as derived from MODIS images. *Theoretical and Applied Climatology*, *96*(1-2), 145–153. doi:10.1007/s00704-008-0019-3

Cheval, S., Dumitrescu, A., & Bell, A. (2009). The urban heat island of Bucharest during the extreme high temperatures of July 2007. *Theoretical and Applied Climatology*, *97*(3-4), 391–401. doi:10.1007/s00704-008-0088-3

Chitale, D. V. (1987). Natroalunite in a laterite profile over Deccan Trap basalt at Matanumadh Kutch India. *Clays and Clay Minerals*, *35*(3), 196–202. doi:10.1346/CCMN.1987.0350305

Chopada, G., & Sharma, S. (2014). Integrating Remote sensing into Plant Pathology. *Agrobios Newsletter*, *12*(11), 42–43.

Choudhury, S. B., Jena, B., Rao, M. V., Rao, K. H., Somvanshi, V. S., Gulati, D. K., & Sahu, S. K. (2007). Validation of integrated potential fishing zone (IPFZ) forecast using satellite based chlorophyll and sea surface temperature along the east coast of India. *International Journal of Remote Sensing*, *28*(12), 2683–2693. doi:10.1080/01431160600987878

Christen, M. K., Walski, J., & Bartelt, P. (2010). Ramms: Numerical simulation of dense snow avalanches in three-dimensional terrain. *Cold Regions Science and Technology*, *63*(1/2), 1–14. doi:10.1016/j.coldregions.2010.04.005

Christian, C. S., & Stewart, G. A. (1953). General Report on Survey of Katherine Darwin Region. The Commonwealth Scientific and Industrial Research Organization (CSIRO).

Christian, C. S., & Stewart, G. A. (1969). Lands of the Oueandeyan-Shoalhaven area. In R.H. Gunn, R. Story, R.W. Galloway, P.J.B. Duffy, G.A. Yapp, & J.R. McAlpine (Eds.). Lands of the Oueandeyan – Shoalhaven Area, A.C.T. and N.S.W. The Commonwealth Scientific and Industrial Research Organization (CSIRO) Publication.

Chung, C. F., & Fabbri, A. G. (1999). Probabilistic prediction models for landslide hazard mapping. *Photogrammetric Engineering and Remote Sensing*, *65*(12), 1389–1399.

Churnside, J. H., & Hunter, J. R. (1996). Laser remote sensing of epipelagic fishes. *Society of Photo - Optical Instrumentation Engineers, 2964*, 38-53.

Clark, C., Ripley, H., Green, E., Edwards, A., & Mumby, P. (1997). Mapping and measurement of tropical coastal environments with hyperspectral and high spatial resolution data. *International Journal of Remote Sensing*, *18*(2), 237–242. doi:10.1080/014311697219033

Clark, M. L., Clark, D. B., & Roberts, D. (2004). Small-footprint lidar estimation of sub-canopy elevation and tree height in a tropical rain forest landscape. *Remote Sensing of Environment*, *91*(1), 68–89. doi:10.1016/j.rse.2004.02.008

Clarmann, T. V., Dudhia, A., Echle, G., Flaud, J.-M., Harrold, C., & Kerridge, B. (1998). Study on the simulation of atmospheric infrared spectra: Final Report.

CNET. (2005). Wikipedia to tighten editorial rules. Retrieved from http://news.cnet.com/Wikipedia-to-tighten-editorial-rules/2100-1025_3-5820655.html

Coll, C., Caselles, V., Galve, J. M., Valor, E., Niclos, R., Sanchez, J., & Rivas, R. (2005). Ground measurements for the validation of land surface temperatures derived from AATSR and MODIS data. *Remote Sensing of Environment*, *97*(3), 288–300. doi:10.1016/j.rse.2005.05.007

Coll, C., Hook, S. J., & Galve, J. M. (2009). Land surface temperatures from the advanced along-track scanning radio meter: Validation over inland waters and vegetated surfaces. *IEEE Transactions on Geoscience and Remote Sensing*, *47*(1), 350–360. doi:10.1109/TGRS.2008.2002912

Congalton, R., & Green, K. (2008). Assessing the Accuracy of Remotely Sensed Data: Principles and Practices. 2nd ed. CRC/Lewis Press, Boca Raton, FL. doi:10.1201/9781420055139

Congalton, R., Green, K., & Teply, J. (1993). Mapping old growth forests on national forest and park lands in Pacific Northwest from remotely sensed data. *Photogrammetric Engineering and Remote Sensing*, *59*(4), 529–535.

Coskun, M., & Musaoglu, N. (2004). Investigation of rainfall-unoff modelling of Van Lake catchment by using remote sensing and GIS integration. *Paper presented at theTwentieth International Society for Photogrammetry and Remote Sensing (ISPRS) Congress.*

Craig, W., Harris, T., & Weiner, D. (1999). Empowerment, marginalization and public participation GIS. In *Report of Varenius Workshop* (pp. 22–24). Santa Barbara, NCGIA.

Creutin, J. D., Borga, M., Gruntfest, E., Lutoff, C., Zoccatelli, D., & Ruin, I. (2013). A space and time framework for analyzing human anticipation of flash floods. *Journal of Hydrology (Amsterdam)*, *482*, 14–24. doi:10.1016/j.jhydrol.2012.11.009

Crinnion, J. (1991). *Evolutionary Systems Development, a practical guide to the use of prototyping within a structured systems methodology*. New York: Plenum Press.

Cruden, D. M., & Varnes, D. J. (1996). Landslides types and processes, In A.K. Turner & R.L. Schuster (Eds.), Landslides Investigation and Mitigation (pp. 36-71). Transport Research Board Report 247.

Dahdouh-Guebas, F., Verheyden, A., Kairo, J. G., Jayatissa, L. P., & Koedam, N. (2006). Capacity building in tropical coastal resource monitoring in developing countries: A re-appreciation of the oldest remote sensing method. *International Journal of Sustainable Development and World Ecology*, *13*(1), 62–76. doi:10.1080/13504500609469662

Dai, F. C., Lee, C. F., & Ngai, Y. Y. (2002). Landslide risk assessment and management: An overview. *Engineering Geology*, *64*(1), 65–87. doi:10.1016/S0013-7952(01)00093-X

Daqamseh, S. T., Mansor, S., Pradhan, B., Billa, L., & Mahmud, A. R. (2013). Potential fish habitat mapping using MODIS - derived sea surface salinity, temperature and chlorophyll-a data: South China Sea coastal areas, Malaysia. *Geocarto International*, *28*(6), 546–560. doi:10.1080/10106049.2012.730065

Das, D. K. (1972). Report on some of the stone deposit in Jhargram Sub-Division: District Midnapore. Kolkata: Geological Survey of India, Report No. G.P. 30/72.

Dash, J., Steinle, E., Singh, R. P., & Bahr, H. P. (2004). Automatic building extraction from laser scanning data: An input tool for disaster management. *Advances in Space Research*, *33*(3), 317–322. doi:10.1016/S0273-1177(03)00482-4

Dash, P., Gottsche, F. M., Olesen, F. S., & Fischer, H. (2002). Land surface temperature and emissivity estimation from passive sensor data: Theory and practice-current trends. *International Journal of Remote Sensing*, *23*(13), 2563–2594. doi:10.1080/01431160110115041

Das, S., & Paul, P. K. (2006). Selection of site for small hydel usinf GIS in the Himalayan region of India. *Journal of Spatial Hydrology*, *6*(1), 18–28.

Das, T. K., Sharma, A. R., & Pathak, H. (2012). *Crop-Weed Balance under Climate Change. Climate Change Impact, Adaptation and Mitigation in Agriculture: Methodology for Assessment and Application.* New Delhi: Indian Agricultural Research Institute.

De Martonne, E. (1934). Les régions arides du nord argentin et chilien. *Bulletin de lAssociation de Geographes Francais*, *79*(79), 58–62. doi:10.3406/bagf.1934.6686

De Smet, R. (1951). Principles elementaires de morphometrie. *Rev. Cercle des Sciences*, *1*(4), 13–16.

de Winnaar, G., Jewitt, G. P. W., & Horan, M. (2007). A GIS-based approach for identifying potential runoff harvesting sites in the Thukela River basin, South Africa. *Physics and Chemistry of the Earth Parts A/B/C*, *32*(15–18), 1058–1067. doi:10.1016/j.pce.2007.07.009

Dhruvanarayana, V. V., Sastry, G., & Patnaik, U. S. (1990). Watershed Management, New Delhi: Indian Council for Agricultural Research (ICAR).

Di Gregorio, A., & Jansen, L. (2005). *Land Cover Classification System. Classification concepts and user manual. Software version 2*. Rome: Food and Agriculture Organization of the United Nations.

Djepa, V. (2011). Drought prediction using the Along Track Scanning Radiometer (ATSR2) on board ERS2 satellite. *Advances in Space Research*, *48*(1), 56–60. doi:10.1016/j.asr.2011.01.026

Dobhal, D. P., Gupta, A. K., Mehta, M., & Khandelwal, D. D. (2013). Kedarnath disaster: Facts and plausible causes. *Current Science*, *105*(2), 171–174.

Dolfman, M. L., & Bergman, B. (2007). The effects of Hurricane Katrina on the New Orleans economy. *Monthly Labor Review*, *130*(6), 3–17. Retrieved from http://search.ebscohost.com/login.aspx?direct=true&db=bth&AN=26962930

Dong, Y., Li, Q., Dou, A., & Wang, X. (2011). Extracting damages caused by the 2008 Ms 8.0 Wenchuan earthquake from SAR remote sensing data. *Journal of Asian Earth Sciences*, *40*(4), 907–914. doi:10.1016/j.jseaes.2010.07.009

Doornkamp, J. C., & King, C. A. M. (1971). *Numerical Analysis in Geomorphology-An Introduction*. London: Edward Arnold.

Dragicevic, S., & Balram, S. (2004). A Web GIS collaborative framework to structure and manage distributed planning processes. *Journal of Geographical Systems*, *6*(2), 133–153. doi:10.1007/s10109-004-0130-7

Dunkerley, D. L. (1988). LEVPLOT. A BASIC program for processing and plotting survey sections (Technical Memorandum). Dept. of Geography, Monash University, Melbourne.

Duong, N. D. (2012). Water Body Extraction from Multispectral Image by Spectral Pattern Analysis. *Paper presented at the22nd ISPRS Congress.*

Dykes, J., Moore, K., & Wood, J. (1999). Virtual environments for student fieldwork using networked component s. *International Journal of Geographical Information Science*, *13*(4), 397–416. doi:10.1080/136588199241274

Easter, K. W., Dixon, J. A., & Hufschmidt, M. M. (1986). Watershed Resources Management. An integrated framework with studies from Asia and the Pacific. Boulder, Colorado: Westview Press Studies in Water Policy and Management.

ESCAVI. (2003). *Executive Steering Committee for Australian Vegetation Information (ESCAVI). Australian Vegetation Attribute Manual: National Vegetation Information System, Version 6.0.* Canberra: Department of the Environment and Heritage.

Escobar-Munoz, J., Chedin, A., Cheruy, F., & Scott, N. A. (1993). Multi-layer neural networks for the retrieval of atmospheric variables from satellite-borne vertical sounding. *Comptes Rendus de l'Academic des Sciences, Serie II (Macanique, Physique. Chimie Sciences de la Terre et de l'Univers*, *317*, 911–918.

ESRI. (2008). *ArcGIS version 9.2*. Environmental System Research Institute.

Estes, J. E., Stow, D., & Jenson, J. R. (1982). Monitoring land use and land cover changes. In C. J. Johannsen & J. L. Sanders (Eds.), *Remote Sensing for Resources Management* (pp. 100–110). Ankeny, Iowa: Soil Conservation Society of America.

Fang, Q., Sang, J., & Xu, C. (2013). Giant: Geo-informative attributes for location recognition and exploration.*Proceedings of the 21st ACM international conference on Multimedia* (pp. 13-22). ACM. doi:10.1145/2502081.2502088

FAO (Food and Agricultural Organization of the United Nations). (1977). *Guidelines for Watershed Management. 1*. Rome: FAO.

FAO (Food and Agricultural Organization of the United Nations). (1987). *Soil and Water Conservation in Semi-Arid Areas. Soil Bulletin 57*. Rome: FAO.

FAO. (2009). *The State of World Fisheries and Aquaculture. 2008*. Rome, Italy: FAO Documentation Group.

Farrand, W., Timothy, D. G., Rice, J. W., Hurowitz, J. A., & Swayze, E. (2009). Discovery of jarosite within the Mawrth Vallis region of Mars: Implications for the geologic history of the region. *Icarus*, *204*(2), 478–488. doi:10.1016/j.icarus.2009.07.014

Fichera, C. R. (2012). Land Cover classification and change-detection analysis using multi-temporal remote sensed imagery and landscape metrics. *European Journal of Remote Sensing*. Doi:<ALIGNMENT.qj></ALIGNMENT>10.5721/EuJRS20124501

Fily, M. (2003). A simple retrieval method for land surface temperature and fraction of water surface determination from satellite microwave brightness temperatures in sub-arctic areas. *Remote Sensing of Environment*, *85*(3), 328–338. doi:10.1016/S0034-4257(03)00011-7

Fischer, M., Hills, S., Staples, E., Johnson, B., Yaich, M., & Solomon, T. (2008). Japanese encephalitis prevention and control: advances, challenges, and new initiatives. In W. M. Scheld, S. M. Hammer, & J. M. Hughes (Eds.), *Emerging Infections 8* (pp. 93–124). Washington, DC: ASM Press. doi:10.1128/9781555815592.ch6

Flickr. Retrieved June 09, 2015, from http://www.flickr.com

Foody, G. M. (2010). Assessing the accuracy of land cover change with imperfect ground reference data. *Remote Sensing of Environment*, *114*(10), 2271–2285. doi:10.1016/j.rse.2010.05.003

Foolad, M. R. (2004). Recent Advances in Genetics of Salt Tolerance in Tomato.[journal article]. *Plant Cell, Tissue and Organ Culture*, *76*(2), 101–119. doi:10.1023/B:TICU.0000007308.47608.88

Forster, P. M. (2004). Psychological Sense of Community in Groups on the Internet. *Behaviour Change*, *21*(2), 141–146. doi:10.1375/bech.21.2.141.55421

Frank, S., Fürst, C., Koschke, L., Witt, A., & Makeschin, F. (2013). Assessment of landscape aesthetics—validation of a landscape metrics-based assessment by visual estimation of the scenic beauty. *Ecological Indicators*, *32*, 222–231. doi:10.1016/j.ecolind.2013.03.026

Fu, B., Shi, P., Guo, H., Okuayama, S., Ninomiya, Y., & Wright, S. (2011). Surface deformation related to the 2008 Wenchuan earthquake, and mountain building of the Longmen Shan, eastern Tibetan Plateau. *Journal of Asian Earth Sciences*, *40*(4), 805–824. doi:10.1016/j.jseaes.2010.11.011

Fu, W., Turner, J. C., Zhao, J., & Du, G. (2015). Ecological footprint (EF): An expanded role in calculating resource productivity (RP) using China and the G20 member countries as examples. *Ecological Indicators*, *48*, 464–471. doi:10.1016/j.ecolind.2014.09.023

Gardin, S., van Laere, S. M., van Coillie, F. M., Anseel, F., Duyck, W., de Wulf, R. R., & Verbeke, L. P. (2011). Remote sensing meets psychology: A concept for operator performance assessment. *Remote Sensing Letters*, *2*(3), 251–257. doi:10.1080/01431161.2010.516280

Ghassemi, F., Jakeman, A. J., & Nix, H. A. (1995). *Salinisation of Land and Water Resources: Human Causes, Extent Management and Case Studies*. Wallingford, England: CAB International.

Ghosh, J. K., & Suri, S. A. (2005). Knowledge based system for assessment of landslide hazard. *Proceedings of the Indian Geotechnical Conference(Vol. 1,* pp. 393–396). Ahmedabad: IGS.

Gienko, G., Govorov, M., Maguire, B., Khmelevsky Y., & Gienko, A. (2007). A Virtual Professional Community to Support Effective Use of Remotely Sensed Imagery. *The South Pacific Journal of Natural Science*, 25.

Gienko, G., & Chekalin, V. (2004), Neurophysiological features of human visual system in augmented photogrammetric technologies.*Proc. of XXth ISPRS Congress*, Istanbul.

Gillespie, A., Rokugawa, S., Matsunaga, T., Cothern, J. S., Hook, S., & Kahle, A. B. (1998). A temperature and emissivity separation algorithm for Advanced Spaceborne Thermal Emission and Reflection Radiometer (ASTER) images. *IEEE Transactions on Geoscience and Remote Sensing*, *36*(4), 1113–1126. doi:10.1109/36.700995

Givri, J. R. (1997). The extension of the split window technique to passive microwave surface temperature assessment. *International Journal of Remote Sensing*, *18*(2), 335–353. doi:10.1080/014311697219105

Global Footprint Network. (2008). *National Footprint Accounts, 2008 Edition*. Retrieved from www.footprintnetwork.org

Gong, J., & Maher, A. (2014). Use of Mobile Lidar Data to Assess Hurricane Damage and Visualize Community Vulnerability. Transportation Research Record. *Journal of the Transportation Research Board*, *2459*(2459), 119–126. doi:10.3141/2459-14

Goodfellow, R. (2003). *Virtual Learning Communities. A Report for the National College for School Leadership*. Open University.

Govorov, M., Khmelevsky, Y., Ustimenko, V., & Khorev, A. (2005). Security for GIS N-Tier Architecture. In P. Fisher (Ed.), *Developments in Spatial Data Handling* (pp. 71–83). Springer Verlag. doi:10.1007/3-540-26772-7_6

Gregory, K. J., & Walling, D. E. (1978). *Drainage Basin Form and Processes*. London: Edward Arnold.

Griffiths, F. B., Lyne, V. D., Harris, G. P., & Parslow, J. S. (1989). Tracking fish by airborne radar - fishermen could be the winners. *Australian Fisheries*, *48*, 19–21.

Grimm, T. (1998). The Human Condition: A Justification for Rapid Prototyping. *Time Compression Technologies*, 3(3).

Gruntfest, E., Showalter, P., & Ruin, I. (2008). Flash Flood Research -- Past, Present, and Future. *Natural Hazards Observer*, *33*(1), 11–13.

GSI (Geological Survey of India). (1998). *Geological Quadrangle Map - 73J (Jamshedpur Quadrangle)*. Kolkata: GSI.

GSI (Geological Survey of India). (2007). *District Resource Map - Medinipur, West Bengal*. Kolkata: GSI.

Gupta, Peeyush & Goyal S. (2014). Analysis of Disturbance Gradient in Tehri Garhwal Region, Uttarakhand, India- A Spatial Landscape Modelling Approach. *International Journal of Applied Forest Ecology, 2*(2), 20-22.

Gupta, Peeyush. Uniyal S. &Goyal S. (2014). An Assessment of Phytosociological Diversity of Kuther Watershed, Himachal Pradesh, India. *International Journal of Environment, 3*(4), 74-90.

Gupta, A., & Chakraborty, S. (1967). Quantitative analysis of landforms-A case study on the Dulung drainage basin.*Proceedings of Seminar on Geomorphological Studies inIndia*, University of Saugar (pp. 190-199).

Gupta, P. N. (1991). *Forest Management and Remote Sensing. Mountain Resource Management by Remote Sensing* (pp. 195–207). Dehradun: Surya Publications.

Gupta, R., & Joshi, B. (1990). Landslide hazard zoning using the GIS approach-a case study from the Ramganga catchment, Himalayas. *Engineering Geology*, *28*(1-2), 119–131. doi:10.1016/0013-7952(90)90037-2

Gupta, V., & Bist, K. S. (2004). The 23 September 2003 Varunavat Parvat landslide in Uttaranchal Township, Uttaranchal. *Current Science*, *87*, 119–131.

Haberl, H., Erb, K., & Krausmann, F. (2001). How to calculate and interpret ecological footprints for long periods of time: The case of Austria 19261995. *Ecological Economics*, *38*(1), 25–45. doi:10.1016/S0921-8009(01)00152-5

Hamel, M. A., & Andréfouët, S. (2010). Using very high resolution remote sensing for the management of coral reef fisheries: Review and perspectives. *Marine Pollution Bulletin*, *60*(9), 1397–1405. doi:10.1016/j.marpolbul.2010.07.002 PMID:20659750

Harpalani, M., Phumprao, T., Bassi, M., Hart, M., & Johnson, R. (2010). Wiki Vandalysis - Wikipedia Vandalism Analysis. *Proceedings of the 4th International Workshop on Uncovering Plagiarism, Authorship, and Social Software Misuse (PAN-10)*, Padua, Italy.

Hashimoto, M., Enomoto, M., & Fukushima, Y. (2010). Coseismic Deformation from the 2008 Wenchuan, China, Earthquake Derived from ALOS/PALSAR Images. *Tectonophysics*, *491*(1–4), 59–71. doi:10.1016/j.tecto.2009.08.034

Hassan, S., Mahmud-Ul-Islam, S., & Rahman, M. T. (2015). Integration of Remote Sensing and GIS to Assess Vulnerability of Environmental Degradation in North-Western Bangladesh. *Journal of Geographic Information System*, *7*(05), 494–505. doi:10.4236/jgis.2015.75040

Hela, I. (1960). The influence of temperature on the behaviour of the fish. *Archaeological Society*, *15*, 1–2.

Henderson, J. M. (1992). Object identification in context: The visual processing of natural scenes. *Canadian Journal of Psychology*, *46*(3), 319–341. doi:10.1037/h0084325 PMID:1486550

Herzog, T. (1995). A cognitive analysis of preference for urban nature. In *Readings in Environmental Psychology: Landscape Perception* (pp. 65-81).

Herzog, T. R. (1989). A cognitive analysis of preference for field and forest environments. *Landscape Research*, *9*(1), 10–16. doi:10.1080/01426398408706092

Hess, L. L., Melack, J., & Simonett, D. S. (1990). Radar detection of flooding beneath the forest canopy - a review. *International Journal of Remote Sensing*, *11*(7), 1313–1325. doi:10.1080/01431169008955095

Hoffman, R., & Markman, A. (2001). *Interpreting Remote Sensing Imagery: Human Factors*. Lewis Publishers.

Hollinger, J. (1991). *DMSP SSM/I Calibration / Validation. Final Report. Parts 1 and 2*. Washington, D.C.: Naval Research Laboratory.

Holzner, S. (2006). Ajax: Your visual blueprint for creating rich Internet applications. Wiley Visual Imprint.

Honore, G. (1999). Our Land, Ourselves-A Guide to Watershed Management in India. New Delhi: Govt. of India.

Horton, R. E. (1945). Erosional development of streams and their drainage basins: Hydrophysical approach to quantitative morphology. *Bulletin of the Geological Society of America*, *56*(3), 275–370. doi:10.1130/0016-7606(1945)56[275:EDOSAT]2.0.CO;2

Hunt, M. E. (1984). Environmental knowing without being there. *Environment and Behavior*, *16*(3), 307–334. doi:10.1177/0013916584163002

Ikeda, M., & Dobson, F. W. (1995). *Oceanographic Applications of Remote Sensing*. New York: CRC Press.

Imhoff, M., Zhang, P., Wolfe, R., & Bounoua, L. (2010). Remote sensing of the urban heat island effect across biomes in the continental USA. *Remote Sensing of Environment*, *114*(3), 504–513. doi:10.1016/j.rse.2009.10.008

Inamdar, A. K., French, A., Hook, S. J., Vaughan, G., & Luckett, W. (2008). Land surface temperature retrieval at high spatial and temporal resolutions over the southwestern United States. *Journal of Geophysical Research*, *113*(D7), 1–18. doi:10.1029/2007JD009048

Indian Space Research Organisation / Regional Remote Sensing Centre / Indian Society of Remote Sensing. (2009, September 15-16). Pre-Symposium Tutorial of National Symposium on Advances in Geo-Spatial Technologies with special emphasis on Sustainable Rainfed Agriculture, Nagpur (Vol. 1 & 2).

Istomina, L. G., von Hoyningen-Huene, W., Kokhanovsky, A. A., & Burrows, J. P. (2010). The detection of cloud-free snow-covered areas using AATSR measurements. *Atmospheric Measurement Techniques*, *3*(4), 1005–1017. doi:10.5194/amt-3-1005-2010

Iverson, R. M., Reid, M. E., & La Husen, R. G. (1997). Debris-flow mobilization from landslides. *Annual Review of Earth and Planetary Sciences*, *25*(1), 85–138. doi:10.1146/annurev.earth.25.1.85

Jain, N., Bhattacharya, S., Chauhan, P., Ajai. (2011). Hyperspectral study of hydrous sulphate minerals from Deccan Volcanic Province of Kutch, India: Implications for aqueous processes on Mars. *Proceedings of EPSC-DPS '11.*

Jain, A., Ravan, S.A., Singh, R.K., Das, K.K. & Roy, P.S. (1996). Forest fire risk modelling using remote sensing and geographic information system. *Current Science*, *70*(10), 928–933.

Jain, L. K., & Gupta, S. (2013). Sustainable Crop Production in Arid and Semi Arid Areas of Rajasthan (Conservation Agriculture: The New Paradigm). In S. G. N. Hada (Ed.), *Biotech* (pp. 27–51). Books.

Jain, L. K., Hada, N., & Bhaduria, S. S. (2014). Alternate use of Land for Better Livelihood in Western Rajasthan. In S. G. S. S. Tomar (Ed.), *Sustainable Rural Development through agriculture* (pp. 345–349). Biotech Books.

Jain, S. K., Seth, S. M., & Nema, R. K. (1995). Morphometric Analysis of Sei Dam Catchment using GIS.*Proceedings of the national Symposium on Hydrology*, Jaipur (pp. 5-7).

Jaiswal, R. K., Saxena, R., & Mukherjee, S. (1999). Application of Remote Sensing technology for land use / land cover change analysis. *Journal of the Indian Society of Remote Sensing*, *27*(2), 123–128. doi:10.1007/BF02990808

Jamali, N. A., & Rahman, M. T. (2016). Utilization of Remote Sensing and GIS to Examine Urban Growth in the City of Riyadh, Saudi Arabia. *Journal of Advances in Information Technology*, *7*(4), 297–301.

Jankowski, P. (2011). *Designing public participation geographic information system. The SAGE Handbook of GIS and Society* (pp. 347–360). London: Sage Publications Ltd.

Jankowski, P., & Nyerges, T. (2001). *Geographic Information Systems for Group Decision Making: Towards a participatory, geographic information science*. New York: Taylor & Francis.

Jeganathan, C., & Chauniyal, D. D. (2000). An evidential weighted approach for landslide hazard zonation from geo-environmental characterisation: A case study of Kelani area. *Current Science*, *79*(2), 238–243.

Jensen, J. R. (1986). *Introductory Image Processing. A remote sensing perspective*. USA: Prentice-Hall.

Jensen, J. R. (2000). *Remote Sensing of the Environment. An Earth Resource Perspective*. Singapore: Pearson Education, Inc.

Jeyaram, A., Mohabey, N. K., & Krishanamurthy, Y. V. N. (1996). Groundwater potential using Remote Sensing and Geographical Information System. In N.K. Mohabey, L.G. Gwalani, & D.M. Mohabey (Eds.), *Mineral and Groundwater Resources of Vidarbha,* Golden Jubilee Symposium, Nagpur University, Nagpur (pp. 257-267).

Jimenez-Munoz, J. C., & Sobrino, J. A. (2003). A generalized single channel method for retrieving land surface temperature from remote sensing data. *Journal of Geophysical Research, 108*(D22), 4688. doi:10.1029/2003JD003480

Jimenez-Munoz, J. C., & Sobrino, J. A. (2008). Split-window coefficients for land surface temperature retrieval from low-resolution thermal infrared sensors. *IEEE Geoscience and Remote Sensing Letters*, *5*(4), 806–809. doi:10.1109/LGRS.2008.2001636

Jimenez-Munoz, J. C., Sobrino, J. A., Skokovich, D., Matter, C., & Cristobal, J. (2014). Land surface temperature retrieval methods from Landsat-8 thermal infrared sensor data. *IEEE Transactions on Geoscience and Remote Sensing*, *11*(10), 1840–1843. doi:10.1109/LGRS.2014.2312032

Jinhua, L., & Xinqi, Z. (2012). A Spatial-EF and Econometrics Model Integrated Approach to explore Land Use Sustainable Forecast Model-in case of Shandong Province. *Procedia Environmental Sciences, 12*(A), 413–420.

Jin, M. (2004). Analysis of land skin temperature using AVHRR observations. *Bulletin of the American Meteorological Society, 85*(4), 587–600. doi:10.1175/BAMS-85-4-587

Jin, M., Dickinson, R., & Zhang, D. (2005). The footprint of urban areas on global climate as characterized by MODIS. *Journal of Climate, 18*(10), 1551–1565. doi:10.1175/JCLI3334.1

Joan, M., & Cesar, C. (2009). Monthly land surface temperature maps over European zone using advanced along track scanning radiometer data for 2007.*Geoscience and Remote Sensing Symposium, 2009, IEEE International, IGARSS 2009, 4,* 292-295. doi:10.1109/IGARSS.2009.5417316

Johnson, B. C., Bruce, S. S., Early, E. A., Housten, J. M., O'Brian, T. R., Thompson, A. K., Mueller, S. L. (1996). *The Fourth SeaWiFS Intercalibration Round-Robin Experiment (SIRREX-4).* NASA Tech Memo 104566 37.

Johnson, C. P., & Johnson, J. (2001, February 7–9). GIS: A Tool for Monitoring and Management of Infectious.*Proceedings of Map India 2001, the 4th Annual International Conference and Exhibition*, New Delhi, India.

Joshi, P. K., & Gairola, S. (2004). Land cover dynamics in Garhwal Himalayas-A case study of Balkhila sub-watershed. *Journal of the Indian Society of Remote Sensing, 32*(2), 199–208. doi:10.1007/BF03030876

Kachhwaha, T. S. (1985, November 21-26). Temporal monitoring of forest land for change detection and forest cover mapping through satellite remote sensing.*Proceedings of the Sixth Asian Conference on Remote Sensing*, Hyderabad (pp. 77-83).

Kadam, A. K., Kale, S. S., Pande, N. N., Pawar, N. J., & Sankhua, R. N. (2012). Identifying Potential Rainwater Harvesting Sites of a Semi-arid, Basaltic Region of Western India, Using SCS-CN Method. *Water Resources Management, 26*(9), 2537–2554. doi:10.1007/s11269-012-0031-3

Kalluri, S., Gilruth, P., Rogers, D., & Szczur, M. (2007). Surveillance of Arthropod Vector-Borne Infectious Diseases Using Remote Sensing Techniques: A Review. *PLoS Pathog, 3*(10), e116. doi: 10.1371/journal.ppat.0030116

Kamel Boulos, M. N., Roudsari, A. V., & Carson, E. R. (2001). Health Geomantic: An Enabling Suite of Technologies in Health and Healthcare (Mythological Review). *Biomed Inform, 34*(3), 195–219. doi:10.1006/jbin.2001.1015

Kanungo, D., Arora, M., Sarkar, S., Gupta, R. (2009). Landslide Susceptibility Zonation (LSZ) Mapping - A Review. *Journal of South Asia Disaster Studies,* 2(1).

Kashani, A. G., & Graettinger, A. J. (2015). Cluster-Based Roof Covering Damage Detection in Ground-Based Lidar Data. *Automation in Construction, 58*, 19–27. doi:10.1016/j.autcon.2015.07.007

Kasturirangan, K., Aravamudan, R., Deekshatulu, B. L., Joseph, G., & Chandrasekhar, M. G. (1996). Indian Remote Sensing satellite (IRS) IC-The beginning of a new era. *Current Science*, *70*(7), 495–500.

Katzy, B. R., & Ma, X. (2002). Virtual Professional Communities - Definitions and Typology. *CeTIM Publications*. Retrieved from http://www.researchgate.net/publication/228808964_Virtual_Professional_Communities-Definitions_and_Typology

Kazman, R., & Kominek, J. (1993). Information Organization in Multimedia Resources. Association for Computing Machinery. *Proceedings of the 11th Annual International Conference on Systems Documentation SIGDOC '93* (pp. 149-162). New York: ACM Press.

Kennedy, L. S., & Naaman, M. (2008). Generating diverse and representative image search results for landmarks.*Proceedings of the 17th international conference on World Wide Web* (pp. 297-306). ACM. doi:10.1145/1367497.1367539

Kerr, Y. H., Lagouarde, J. P., & Imbernon, J. (1992). Accurate land surface temperature retrieval from AVHRR data with use of an improved split window algorithm. *Remote Sensing of Environment*, *41*(2-3), 197–209. doi:10.1016/0034-4257(92)90078-X

Kerr, Y. H., Lagouarde, J. P., Nerry, F., & Ottle, C. (2004). Land surface temperature retrieval techniques and applications: Case of AVHRR. In D. A. Quattrochi & J. C. Luvall (Eds.), *Thermal Remote Sensing in Land Surface Processes* (pp. 33–109). Boca Raton, Florida: CRC Press. doi:10.1201/9780203502174-c3

Key, J. R., Amano, E., Collins, J., & Schweiger, A. J. (1999). FluxNet User's Guide. Technical Report 96-03: Department of Geography, Boston University.

Khan S. A., Dutta P., Khan A.M., Topno R., Chowdhury P., Borah J., Mahanta J., (2011). Japanese encephalitis epidemiology in Arunachal Pradesh, a hilly state in northeast India. *Asian Pacific Journal of Tropical Disease*, 2011, 119-122.

Kim, J. P. (2009). *Land-use planning and the urban heat island effect* (Ph.D. thesis). The Ohio State University.

Kim, A.-J. (2006). *Community-Building on the Web*. Peachpit Press.

Kimothi, M.M., Garg, J.K., Ajay, & Joshi, V. (2005). Slope Ancient religious Uttarkashi town (Garhwal Himalayas, Uttaranchal, Observation from IRS-P6 (Resourcesat-1) high resolution LISS-IV data. Map India.

Kistemann, T. (2001, September 19–20). GIS for communicable disease control: perspectives and pitfalls.*Proceedings of the First European Conference for Geographic Information Sciences in Public Health*, Sheffield, UK.

Kizza, M., Rodhe, A., Xu, C.-Y., & Ntale, H. K. (2011). Modelling catchment inflows into Lake Victoria: uncertainties in rainfall–runoff modelling for the Nzoia River. *Hydrological Sciences Journal, 56*(7), 1210-1226. doi:10.1080/02626667.2011.610323

Klemas, V. (2013). Fisheries applications of remote sensing: An overview. *Fisheries Research, 148*, 124–136. doi:10.1016/j.fishres.2012.02.027

Klemen, Z., & Schroedter-Homscheidt, M. (2009). Parameterization of air temperature in high temporal and spatial resolution from a combination of the SEVIRI and MODIS instruments. *ISPRS Journal of Photogrammetry and Remote Sensing, 64*(4), 414–421. doi:10.1016/j.isprsjprs.2009.02.006

Klinsky, S., Sieber, R., & Meredith, T. (2010). Connecting local to global: Geographic information systems and ecological footprints as tools for sustainability. *The Professional Geographer, 62*(1), 84–102. doi:10.1080/00330120903404892

Kornelsen, K. C., & Coulibaly, P. (2013). Advances in soil moisture retrieval from synthetic aperture radar and hydrological applications. *Journal of Hydrology (Amsterdam), 476*, 460–489. doi:10.1016/j.jhydrol.2012.10.044

Koukoulas, S., & Blackburn, G. A. (2005). Mapping individual tree location, height and species in broadleaved deciduous forest using airborne LIDAR and multispectral remotely sensed data. *International Journal of Remote Sensing, 26*(3), 431–455. doi:10.1080/0143116042000298289

Kratena, K. (2004). Ecological value added in an integrated ecosystem-economy model: An indicator for sustainability. *Ecological Economics, 48*(2), 189–200. doi:10.1016/j.ecolecon.2003.09.012

Krishna, N. D. R., Westinga, E., & Huizing, H. (1999, March 9-11). Monitoring land cover changes using Geoinformatics in some communal lands of Zimbabwe. *Proceedings of "Geoinformatics. Beyond 2000" International Conference on Geoinformatics for Natural Resource Assessment, Monitoring and Management*, Dehradun (pp. 92-101).

Krishnamurthy, J., Srinivas, G., Jayaram, V., & Chandrasekhar, M. G. (1996). Influence of rock type and structure in the development of drainage networks in typical hard rock terrain. *ITC Journal, 3*(4), 252–259.

Kumar, D. N. (1999). Satellite Remote Sensing & GIS for Irrigation Management. Retrieved from civil.iisc.ernet.in/~nagesh/pubs/B01_RS_Irr_Manak_1999.pdf

Kumar, R., & Silva, L. P. (1974). Statistical separability of agricultural cover types in subsets of one to twelve spectral channels. *Proceedings of the Ninth International Symposium on Remote Sensing of Environment*, Environmental Research Institute of Michigan (ERIM), Ann Arbor, Michigan.

Kumar, A., & Pandey, R. H. (1982). Morphometric properties of the polygenetic drainage basin: A case study of the Burha Nadi. *Geographical Review of India, 37*, 329–338.

Kumari, B., Raman, M., & Mali, K. (2009). Locating tuna forage grounds through satellite remote sensing. *International Journal of Remote Sensing, 30*(22), 5977–5988. doi:10.1080/01431160902798387

Kumar, R., Kumar, S., Lohani, A. K., Nema, R. K., & Singh, R. D. (2000). Evaluation of geomorphological characteristics of a catchment using GIS. *GIS India*, *9*(3), 13–17.

Kumar, S. S., Chennaiah, G. Ch., & Gautam, N. C. (1983). Study of landform control over land utilisation pattern in parts of southern U.P.- A remote sensing approach. *Journal of the Indian Society of Remote Sensing*, *11*(1), 49–53.

Kumar, S., Sahoo, P. K., & Singh, R. P. (1999). Monitoring of brightness temperature over Indian and adjoining regions using SSM/I data. *International Journal of Remote Sensing*, *20*(12), 2305–2307. doi:10.1080/014311699212038

Kumar, V. K., Lakhera, R. C., Martha, T. R., Chatterjee, R. S., & Bhattarcharya, A. (2007). Analysis of the 2003 Varunawat Landslide, Uttarkashi, Indian using Earth Observation data. *Environmentalgeology*, *55*(4), 789–799.

Kumbar, B., Hatti, V., Sanjay, M. T., & Saunshi, S. (2014). Strategies to Enhance Nitrogen Use Efficiency in Agriculture. *Agrobios Newsletter*, *XII*(11), 37–40.

Kundus, P., Karszenbaum, H., Pultz, T., Paramuchi, G., & Bava, J. (2001). Influence of flood conditions and vegetation status on the radar back scatter of wetland ecosystem. *Canadian Journal of Remote Sensing*, *27*(6), 651–662. doi:10.1080/07038992.2001.10854907

Kushwaha, G. S., & Singh, V. (2007). Environmental Psychology in Land. *Society, sustainability, and environment*, 184.

Kuzyk, L. W. (2010). Infill Housing Sustainability Analysis. *AACIP Planning Journal, 2010*(4), 19-22.

Kuzyk, L. W. (2011). Ecological and carbon footprint by consumption and income in GIS: Down to a census village scale. *Local Environment*, *16*(9), 871–886. doi:10.1080/13549839.2011.615303

Kuzyk, L. W. (2012). The ecological footprint housing component: A geographic information system analysis. *Ecological Indicators*, *16*, 31–39. doi:10.1016/j.ecolind.2011.03.009

Kyriakidis, P. C., & Dungan, J. L. (2001). A geostatistical approach for mapping thematic classification accuracy and evaluating the impact of inaccurate spatial data on ecological model predictions. *Environmental and Ecological Statistics*, *8*(4), 311–330. doi:10.1023/A:1012778302005

Laboratory for Application of Remote Sensing (LARS). (1968). *Remote Multi-spectral Sensing in Agriculture, 3-Annual Report*. West Lafayette, Indiana: Purdue University.

Laevastu, T. (1984). The effect of temperature anolmalies on the fluctuation of stock. *Rapports et Proces-Verbaux des Reunions - Conseil International pour L'Exploration de la Mer*, *185*, 214–215.

Laevastu, T., & Larkin, H. A. (1981). *Marine Fisheries Ecosystem, its Quantitative Evaluation and Management*. Farham, Surrey: Fishing News Books.

Laevastu, T., & Rosa, J. (1963). The distribution and relative abundance of tunas in relation to their environment. *FAO Fisheries Report*, *6*, 1835–1851.

Land and Land Reforms Department, Govt. of West Bengal. (2010). Focus on Paschim Medinipur. Kolkata: Dept. of L&LR, Govt. of West Bengal.

Landsberg, H. E. (1981). *The Urban Climate*. New York: Academic Press.

Lee, H. (1993). An application of NOAA thermal data to the study of urban heat islands. *Atmospheric Environment*, *27B*, 1–13.

Levin, N. (1999). *Fundamentals of Remote Sensing* (L. Noam, Ed.).

Likens, W., & Maw, K. (1982). Hierarchical modelling for image classification. *Proceedings of Conference on Remote Sensing with Special Emphasis on Output to Geographic Information Systems in the1980s, 7th PECORA Symposium*, Sioux Falls, South Dakota, USA (pp. 290-300).

Lillesand, T. M., Kiefer, R. W., & Chipman, J. W. (2004). Remote Sensing and Image Interpretation. Singapore: John Wiley & Sons (Asia), Pte. Ltd.

Lillesand, T. M., Kiefer, R. W., & Chipman, J. (2015). *Remote sensing and image interpretation* (7th ed.). New York: John Wiley & Sons.

Li, M., Cheng, L., Gong, J., Liu, Y., Chen, Z., Li, F., & Song, X. et al. (2008). Post-earthquake assessment of building damage degree using LiDAR data and imagery. *Science in China Series E: Technological Sciences*, *51*(S2), 133–143. doi:10.1007/s11431-008-6014-1

Lim, K., Treitz, P., Baldwin, K., Morrison, I., & Green, J. (2003). Lidar remote sensing of biophysical properties of tolerant northern hardwood forests. *Canadian Journal of Remote Sensing*, *29*(5), 658–678. doi:10.5589/m03-025

Liu, H., Lisheng, X., Ding, J., Zhuoma, B., Deng, X., & Liu, Z. (2010, August 28-31). Atmospheric correction and land surface temperature retrieval method for FY-3 IR observations. *Paper presented at theSecond IITA International Conference on Geoscience and Remote Sensing (IITA-GRS)*, Qingdao.

Liu, C., Frazier, P., & Kumar, L. (2007). Comparative assessment of the measures of thematic classification accuracy. *Remote Sensing of Environment*, *107*(4), 606–616. doi:10.1016/j.rse.2006.10.010

Liu, L., & Zhang, Y. (2011). Urban heat island analysis using the Landsat TM data and ASTER data: A case study in Hong Kong. *Remote Sensing*, *3*(12), 1535–1552. doi:10.3390/rs3071535

Liu, R., Wen, J., Wang, X. L. W., Tian, H., & Zhang, T. T. (2010). Actual daily evapotranspiration estimated from MERIS and AATSR data over the Chinese Loess Plateau. *Hydrology and Earth System Sciences*, *14*(1), 47–58. doi:10.5194/hess-14-47-2010

Li, Z., Chen, Q., Zhou, J., & Tian, B. (2009). Analysis of synthetic aperture radar image characteristics for seismic disasters in the Wenchuan earthquake. *Journal of Applied Remote Sensing*, *3*(1), 31685. doi:10.1117/1.3153906

Lo, N. C. H., Jacobsen, L. D., & Squire, J. L. (1992). Indices of relative abundance from fish spotter data based on delta-lognormal models. *Canadian Journal of Fisheries and Aquatic Sciences*, *49*(12), 2526–2551. doi:10.1139/f92-278

Longhurst, A. (2010). *Mismanagement of Marine Fisheries*. Cambridge, UK: Cambridge University Press. doi:10.1017/CBO9781139195928

Loo, Y. Y., Billa, L., & Singh, A. (2015). Effect of climate change on seasonal monsoon in Asia and its impact on the variability of monsoon rainfall in Southeast Asia. *Geoscience Frontiers*, *6*(6), 817–823. doi:10.1016/j.gsf.2014.02.009

Lowe, D. G. (1985). *Perceptual Organization and Visual Recognition*. Boston: Kluwer Academic Publishers. doi:10.1007/978-1-4613-2551-2

LP DAAC. (2016). ASTER data: Unlimited public access. Retrieved from http://lpdaac.usgs.gov/user_resources/outreach_materials/aster_no_charge_promo_

Luo, J., Joshi, D., Yu, J., & Gallagher, A. (2011). Geotagging in multimedia and computer vision—a survey. *Multimedia Tools and Applications*, *51*(1), 187–211. doi:10.1007/s11042-010-0623-y

Luong, P. T. (1993). The detection of land use / land cover changes using Remote Sensing and GIS in Vietnam. *Asian-Pacific Remote Sensing Journal*, *5*(2), 63–66.

Mackenzie S John., Gubler J Duane., Petersen R Lyle. (2004). Emerging flaviviruses: the spread and resurgence of Japanese encephalitis, West Nile and dengue viruses. *Nature Medicine Supplement*, 10(12).

Madden, M. E., Bodnar, R. J., & Rimstidt, J. D. (2004). Jarosite as an indicator of water-limited chemical weathering on Mars. *Nature*, *431*(7010), 821–823. doi:10.1038/nature02971 PMID:15483605

Madhavan Unni, N. V. (1990). Space and Forest Management in India. Space and Forest Management, Special Current Even Session. *Proceedings of the International Astronautical Federation, 31st IAF Congress*, Dresden Germany (pp. 49-69).

Mahajan, S., Panwar, P., & Kaundal, D. (2001). Land use changes in Ashwani Khad watershed using GIS techniques. *Journal of the Indian Society of Remote Sensing*, *33*(2), 243–248. doi:10.1007/BF02995730

Maltamo, M., Mustonen, K., Hyyppä, J., Pitkänen, J., & Yu, X. (2004). The accuracy of estimating individual tree variables with airborne laser scanning in a boreal nature reserve. *Canadian Journal of Forest Research*, *34*(9), 1791–1801. doi:10.1139/x04-055

Marc, F.L.N., & Staples, E.J. (2010). Morbidity and Mortality Weekly Report. Recommendations and Reports, 59(RR-1).

Marshall, R. E., Thomson, F., & Kriegler, F. (1969). Use of multispectral recognition techniques for conducting rapid, wide wheat surveys. *Proceedings of the Sixth International Symposium on Remote Sensing of Environment*, Environmental Research Institute of Michigan (ERIM), Ann Arbor, Michigan.

Martha, T., & Kumar, V. K. (2013). September, 2012 landslide events in Okhimath, India-an assessment of landslide consequences using very high resolution satelliete data. *Landslides*, *10*(4), 469–479. doi:10.1007/s10346-013-0420-6

Martinis, S., Twele, A., & Voigt, S. (2009). Towards operational near real-time flood detection using a split-based automatic thresholding procedure on high resolution TerraSAR-X data. *Natural Hazards and Earth System Sciences*, *9*(9), 303–314. doi:10.5194/nhess-9-303-2009

Martin, S. (2004). *An Introduction to Remote Sensing*. Cambridge, UK: Cambridge University Press.

Maslekar, A. R. (1974). Remote sensing and its scope in Indian Forestry. *Indian Forester*, *100*(3), 192–201.

Massonnet, D., Rossi, M., Carmona, C., Adragna, F., Peltzer, G., Feigl, K., & Rabaute, T. (1993). The displacement field of the Landers earthquake mapped by radar interferometry. *Nature*, *364*(6433), 138–142. doi:10.1038/364138a0

Mata-Lima, H., Alvino-Borba, A., Pinheiro, A., Mata-Lima, A., & Almeida, J. A. (2013). Impacts of natural disasters on environmental and socio-economic systems: What makes the difference? *Ambiente & Sociedade*, *16*(3), 45–64. doi:10.1590/S1414-753X2013000300004

Matgen, P., Schumann, G., Henry, J., Hoffmann, L., & Pfister, L. (2007). Integration of SAR derived river innundation areas, high-precision topographic data and a river flow model toward near real-time flood management. *International Journal of Applied Earth Observation*, *9*(3), 247–263. doi:10.1016/j.jag.2006.03.003

Mather, P., & Koch, M. (2011). *Computer processing of remotely-sensed images: an introduction*. John Wiley & Sons. doi:10.1002/9780470666517

Maune, D. F. (Ed.). (2001). *Digital Elevation Model Technologies and Applications. The DEM Users Manual*. Bethesda, Maryland: American Society of Photogrammetry and Remote Sensing.

McCoy, R. M. (2005). *Field Methods in Remote Sensing*. New York: Guilford Press.

McFarland, M. J., Miller, R. L., & Neale, C. M. U. (1990). Land surface temperature derived from SSM/I passive microwave brightness temperatures. *IEEE Transactions on Geoscience and Remote Sensing*, *28*(5), 839–845. doi:10.1109/36.58971

McFeeters, S. K. (1996). The use of the Normalized Difference Water Index (NDWI) in the delineation of open water features. *International Journal of Remote Sensing*, 17(7), 1425-1432. doi: 10.1080/01431169608948714

McNeely, J.A., Miller, K.R., Ried, W.V., Mittermeir, R.A. & Warner, T.B. (1990). Conserving World's Biological Diversity. IUCN-WRI Report.

Mekonnen, M., Melesse, A. M., & Keesstra, S. D. (2016). Spatial Runoff Estimation and Mapping of Potential Water Harvesting Sites: A GIS and Remote Sensing Perspective, Northwest Ethiopia. In M. A. Melesse & W. Abtew (Eds.), *Landscape Dynamics, Soils and Hydrological Processes in Varied Climates* (pp. 565–584). Cham: Springer International Publishing. doi:10.1007/978-3-319-18787-7_26

Messinger, P. R., Stroulia, E., Lyons, K., Bone, M., Niu, R. H., Smirnov, K., & Perelgut, S. (2009). Virtual worlds—past, present, and future: New directions in social computing. *Decision Support Systems*, *47*(3), 204–228. doi:10.1016/j.dss.2009.02.014

Metternicht, G. I., & Zinck, J. A. (2003). Remote sensing of soil salinity: Potentials and constraints. *Remote Sensing of Environment*, *85*(1), 1–20. doi:10.1016/S0034-4257(02)00188-8

Mileti, D. (1999). *Disasters by Design*. Washington, D.C.: Joseph Henry Press.

Mishra, S. K., Singh, V. P., Sansalone, J. J., & Aravamuthan, V. (2003). A Modified SCS-CN Method: Characterization and Testing. *Water Resources Management*, *17*(1), 37–68. doi:10.1023/A:1023099005944

Mittal, S. C. (2013). Role of Information Technology in Agriculture and its Scope in India. Retrieved from http://125.19.12.220/applications/Brihaspat.nsf/6dca49b7264f71ce65256a81003ad1cb/82f2c15ccd4dd9a065256b37001af3fe/$FILE/it_fai.pdf

Miyagi, Y., Ozawa, T., & Shimada, M. (2009). Crustal deformation associated with an M8.1 earthquake in the Solomon Islands, detected by ALOS/PALSAR. *Earth and Planetary Science Letters*, *287*(3), 385–391. doi:10.1016/j.epsl.2009.08.022

Moore, I. D., Grayson, R. B., & Ladson, A. R. (1991). Digital terrain modelling. A review of hydrological, geomorphological and biological applications. *Hydrological Processes*, *5*(1), 3–30. doi:10.1002/hyp.3360050103

Moore, T. S., Campbell, J. W., & Dowell, M. D. (2009). A class-based approach to characterizing and mapping the uncertainty of the MODIS ocean chlorophyll product. *Remote Sensing of Environment*, *113*(11), 2424–2430. doi:10.1016/j.rse.2009.07.016

Moran, M. S., Inoue, Y., & Barnes, E. M. (1997). Opportunities and limitations for image-based remote sensing in precision crop management. *Remote Sensing of Environment*, *61*(3), 319–346. doi:10.1016/S0034-4257(97)00045-X

Morris, K. (2013). How vandals are destroying Wikipedia from the inside. Retrieved from http://www.dailydot.com/society/wikipedia-hoax-vandalism-legolas2186/

Mukhopadhyay, S. C. (1980). *Geomorphology of the Subarnarekha Basin*. Burdwan: University of Burdwan Publication.

Myers, J. S., & Miller, R. L. (2005). Optical airborne remote sensing. In R. L. Miller, C. E. DelCastillo, & B. A. McKee (Eds.), *Remote Sensing of Coastal Aquatic Environments: Technologies, Techniques and Applications* (pp. 51–68). Dordrecht, The Netherlands: Springer. doi:10.1007/978-1-4020-3100-7_3

Naaman, M., Song, Y. J., Paepcke, A., & Garcia-Molina, H. (2004). Automatic organization for digital photographs with geographic coordinates. *Proceedings of the 2004 Joint ACM/IEEE Conference onDigital Libraries* (pp. 53-62). IEEE. doi:10.1145/996350.996366

NABARD (National Bank for Agricultural and Rural Development). (2007). *District Agricultural Plan-East Singhbhum (2008-2009 to 20011-2012)*. Ranchi: NABARD Consultancy Services, Jharkhand Regional Office.

Nag, S. K., & Chakraborty, S. (2003). Influence of rock types and structures in the development of drainage network in hard rock area. *Journal of the Indian Society of Remote Sensing*, *31*(1), 25–35. doi:10.1007/BF03030749

Nahiduzzaman, K. M., Aldosary, A. S., & Rahman, M. T. (2015). Flood induced vulnerability in strategic plan making process of Riyadh city. *Habitat International*, *49*, 375–385. doi:10.1016/j.habitatint.2015.05.034

Naithani, A. K. (2002). The August, 1998 Okhimath tragedy in Rudraprayag district of Garhwal Himalya, Uttranchal, India. *Gaia*, *16*, 145–156.

Najafabadi, A.T., & Pourhassan, M. (2009). Web GIS and Public Health. *Online Journal of Health and Allied Sciences*, 8(4).

Nalepka, R. (1970). Investigation of multispectral discrimination techniques (Final Report 2264-12-1). Willow Run Laboratories, University of Michigan, Michigan, USA.

Nammalwar, P., Satheesh, S., & Ramesh, R. (2013). Applications of remote sensing in the validations of potential fishing zones (PFZ) along the coast of North Tamil Nadu, India. *Indian Journal of Geo-Marine Sciences*, *42*(3), 283–292.

NASA. (2016). Landsat Science. Retrieved from http://www.landsat.gsfc.nasa.gov/?page_id=5377

Natarajan, S., Gajbe, M. Y., & Manchanda, M. L. (1986). Landuse study in Mewat area, Haryana using Remote Sensing techniques. *Journal of the Indian Society of Remote Sensing*, *14*(1), 17–24. doi:10.1007/BF03007219

NATMO (National Atlas and Thematic Mapping Organisation). (2000). *District Planning Map Series - Purbi Singhbhum, Jharkhand*. Kolkata: NATMO.

NATMO (National Atlas and Thematic Mapping Organisation). (2006). *District Planning Map Series - Paschim Medinipur, West Bengal*. Kolkata: NATMO.

Navalgaud, R. R., Jayraman, V., & Roy, P. S. (2007). Remote sensing applications: An overview. *Current Science*, *93*(12), 1747–1766.

Nieto, H., Sandholt, I., Aguado, I., Chuvieco, E., & Stisen, S. (2011). Air temperature estimation with MSG-SEVIRI data: Calibration and validation of the TVX algorithm for the Iberian Peninsula. *Remote Sensing of Environment*, *115*(1), 107–116. doi:10.1016/j.rse.2010.08.010

Nishii, R., & Tanaka, S. (1999). Accuracy and Inaccuracy Assessments in Land-Cover Classification. *IEEE Transactions on Geoscience and Remote Sensing*, *37*(1), 491–498. doi:10.1109/36.739098

Nooka Ratnam, K., Srivastava, Y. K., Venkateswara Rao, V., Amminedu, E., & Murthy, K. S. R. (2005). Check dam positioning by prioritization of microwatersheds using SYI model and morphometric analysis - Remote Sensing and GIS perspective. *Journal of the Indian Society of Remote Sensing*, *33*(1), 25–38. doi:10.1007/BF02989988

Noyes, E., Soria, G., Sobrino, J. A., Remedios, J., Llewellyn-Jones, D. T., & Corlett, G. K. (2007). AATSR land surface temperature product algorithm verification over a WATERMED site. *Advances in Space Research*, *39*(1), 171–178. doi:10.1016/j.asr.2006.02.024

Nurdin, S., Mustapha, M. A., Lihan, T., & Ghaffar, M. A. (2015). Determination of potential fishing grounds of *Rastrelliger kanagurta* using satellite remote sensing and GIS technique. *Sains Malaysiana*, *44*(2), 225–232. doi:10.17576/jsm-2015-4402-09

Oberstadler, R., Honsch, H., & Huth, D. (1997). Assessment of the mapping capabilities of ERS-1SAR data for flood mapping: A case study of Germany. *Hydrological Processes*, *10*(10), 1415–1425. doi:10.1002/(SICI)1099-1085(199708)11:10<1415::AID-HYP532>3.0.CO;2-2

O'Brien, J. J., Woodworth, B. M., & Wright, D. J. (1974). The Coho project reports. Living Resources Prediction Feasibility Study.: Vol. I. *III*. Corvallis, OR: School of Oceanography, Oregon State University.

Ochi, S., & Takagi, M. (1996). *Ground Truth Database with Portable GPS for Education*. Retrieved from http://www.a-a-r-s.org/aars/proceeding/ACRS1996/Papers/EDC96-6.htm

Oke, T. R. (1982). The energetic basis of the urban heat island. *Quarterly Journal of the Royal Meteorological Society*, *108*, 1–24.

Oke, T. R., & Cleugh, H. A. (1987). Urban heat storage derived as energy balance residuals. *Boundary-Layer Meteorology*, *39*(3), 233–245. doi:10.1007/BF00116120

Oke, T. R., Johnson, G. T., Steyn, D. G., & Watson, I. D. (1991). Simulation of surface urban heat islands under heat islands under ideal conditions at night part 2: Diagnosis of causation. *Boundary-Layer Meteorology*, *56*(4), 339–358. doi:10.1007/BF00119211

Oliver, M. J., & Irwin, A. J. (2008). Objective global ocean biogeographic provinces. *Geophysical Research Letters*, *35*(15), L15601. doi:10.1029/2008GL034238

Open Source Initiative (OSI). (n. d.). Retrieved from http://www.opensource.org

Ottle, C., & Vidal-Madjar, D. (1992). Estimation of land surface temperature with NOAA 9 data. *Remote Sensing of Environment*, *40*(1), 27–41. doi:10.1016/0034-4257(92)90124-3

Pachauri, A. K., & Pant, M. (1992). Landslide Hazard mapping based on Geological attributed. *Engineering Geology*, *36*, 71–98.

Padmavathy, A. S., Ganesha, R. K., Yogarajan, N., & Thangavel, P. (1993). Checkdam site selection using GIS approach. *Advances in Space Research*, *13*(11), 123–127. doi:10.1016/0273-1177(93)90213-U

Paine, D., & Kiser, J. (2012). *Aerial Photography and image interpretation* (3rd ed.). John Wiley & Sons, Inc.

Palloff, R., & Pratt, K. (2007). *Building learning communities in cyberspace: effective strategies for the online classroom* (2nd ed.). San Francisco: Jossey-Bass.

Palmer, J. F. (2004). Using spatial metrics to predict scenic perception in a changing landscape: Dennis, Massachusetts. *Landscape and Urban Planning*, *69*(2), 201–218. doi:10.1016/j.landurbplan.2003.08.010

Palmer, J. F., & Hoffman, R. E. (2001). Rating reliability and representation validity in scenic landscape assessments. *Landscape and Urban Planning*, *54*(1), 149–161. doi:10.1016/S0169-2046(01)00133-5

Pandey, A. C. (2001). Geology and Morphotectonic studies in a part of lesser Garhwal Himalaya using Remote Sensing Techniques [Unpublished PhD thesis]. University of Delhi.

Pant, D. N., & Roy, P. S. (1990). Vegetation and land use analysis of Aglar watershed using satellite remote sensing technique. *Journal of the Indian Society of Remote Sensing*, *18*(4), 1–14. doi:10.1007/BF02997068

Parida, M., Dash, P.K., Tripathi, N.K., Ambuj., Sannaragaiah, S., ... Sekhar, K. (2006, September). Japanese Encephalitis Outbreak, India, 2005. *Emerging Infectious Diseases*, *12*(9), 1427-1430.

Patel, A. N., & Singh, S. (2015). Remote Sensing Principles and Applications (2nd rev. ed.). Scientific Publishers: India.

Patel, P. P., & Dasgupta, R. (2009). Flood induced land use change in the Dulung River valley, West Bengal. In Geoinformatics for Monitoring and Modelling Land-Use, Bio-diversity and Climate Change (pp. 103-123). NMCC Publication.

Patel, P. P., & Sarkar, A. (2009). Application of SRTM data in evaluating the morphometric attributes: A Case Study of the Dulung River Basin. *Practising Geographer*, *13*(2), 249–265.

Patel, P. P., & Sarkar, A. (2010). Terrain Characterization Using SRTM Data. *Journal of the Indian Society of Remote Sensing*, *38*(1), 11–24. doi:10.1007/s12524-010-0008-8

Peca, I., Zhi, H., Vrotsou, K., Andrienko, N., & Andrienko, G. (2011). Kd-photomap: Exploring photographs in space and time. *Proceedings of the2011 IEEE Conference onVisual Analytics Science and Technology (VAST)* (pp. 291-292). IEEE.

Peeria, S., McLeoda, A., Lavoiea, P., Ackermanb, S., Gardnera, J., & Parrishc, C. (2013). Field calibration and validation of remote-sensing surveys. *International Journal of Remote Sensing*, *34*(18), 6423–6436. doi:10.1080/01431161.2013.800655

Peng, Z., & Tsou, M. (2003). *Internet GIS: Distributed Geographic Information Services for the Internet and Wireless Network*. John Wiley & Sons, Inc. Photosynth. Retrieved from https://photosynth.net/about.aspx

Peterson, T., Basist, A., Williams, C., & Grody, N. C. (2000). A blended satellite in-situ near-global surface temperaure dataset. *Bulletin of the American Meteorological Society*, *81*(9), 2157–2164. doi:10.1175/1520-0477(2000)081<2157:ABSSNS>2.3.CO;2

Petley, D. (2013). Global pattern of loss of life from landslide. *Geology*, *40*(10), 927–930. doi:10.1130/G33217.1

Pittman, S. J., Costa, B. M., & Battista, T. A. (2009). Using lidar bathymetry and boosted regression trees to predict the diversity and abundance of fish and corals. *Journal of Coastal Research*, *25*, 27–38. doi:10.2112/SI53-004.1

Plank, S. (2014). Rapid Damage Assessment by Means of Multi-Temporal SAR — A Comprehensive Review and Outlook to Sentinel-1. *Remote Sensing*, *6*(6), 4870–4906. doi:10.3390/rs6064870

Pongracz, R., Bartholy, J., & Dezso, Z. (2006). Remotely sensed thermal information applied to urban climate analysis. *Advances in Space Research*, *37*(12), 2191–2196. doi:10.1016/j.asr.2005.06.069

Pongracz, R., Bartholy, J., & Dezso, Z. (2010). Application of remotely sensed thermal information to urban climatology of Central European cities. *Physics and Chemistry of the Earth*, *35*(1-2), 95–99. doi:10.1016/j.pce.2010.03.004

Popescu, S. C., Wynne, R. H., & Nelson, R. F. (2002). Estimating plot-level tree heights with lidar: Local filtering with a canopy-height based variable window size. *Computers and Electronics in Agriculture*, *37*(1-3), 71–95. doi:10.1016/S0168-1699(02)00121-7

Popescu, S. C., Wynne, R. H., & Nelson, R. F. (2003). Measuring individual tree crown diameter with lidar and assessing its influence on estimating forest volume and biomass. *Canadian Journal of Remote Sensing*, *29*(5), 564–577. doi:10.5589/m03-027

Porter, C. E. (2004). A Typology of Virtual Communities: A Multi-Disciplinary Foundation for Future Research. *Journal of Computer-Mediated Communication*, *10*(1).

Porter, C. E., & Donthu, N. (2008). Cultivating trust and harvesting value in virtual communities. *Management Science*, *54*(1), 113–128. doi:10.1287/mnsc.1070.0765

Prata, A. (2002). Land Surface Temperature Measurement from Space: AATSR Algorithm Theoretical Basis Document (pp. 1-34): Contract Report to ESA. CSIRO Atmospheric Research: Aspendale.

Prata, A. J. (1993). Land surface temperatures derived from the Advanced Very High Resolution Radiometer and the Along-Track Scanning Radiometer 1. Theory. *Journal of Geophysical Research, 98*(D9), 16689–16702. doi:10.1029/93JD01206

Preece, J. (2000). *Online Communities: Designing Usability, Supporting Sociability.* John Wiley & Sons Ltd. Retrieved from http://www.amnh.org/our-research/center-for-biodiversity-conservation/biodiversity-informatics/remote-sensing-guides

Price, J. C. (1983). Estimation of surface temperatures from satellite thermal infrared data - a simple formulation for the atmospheric effect. *Remote Sensing of Environment, 13*(4), 353–361. doi:10.1016/0034-4257(83)90036-6

Pringent, C., Rossow, W. B., & Matthews, E. (1997). Microwave land surface emissivities estimated from SSM/I observations. *Journal of Geophysical Research, 102*(D18), 21867–21890. doi:10.1029/97JD01360

Purkis, S. J., Graham, N. A. J., & Riegl, B. M. (2008). Predictability of reef fish diversity and abundance using remote sensing data in Diego Garcia (Chagos Archipelago). *Coral Reefs, 27*(1), 167–178. doi:10.1007/s00338-007-0306-y

Qian, J., Zhoua, Q., & Houa, Q. (2007). Comparison of pixel-based and object-oriented classification methods for extracting built-up areas in arid zone. *Paper presented at the ISPRS Workshop on Updating Geo-spatial Databases with Imagery & the 5th ISPRS Workshop on DMGISs.*

Qin, Z., Xu, B., Zhang, W., Li, W., Chen, Z., & Zhang, H. (2004). Comparison of split window algorithms for land surface temperature retrieval from NOAA-AVHRR data. *Paper presented at theGeoscience and Remote Sensing Symposium IGARSS'04.*

Qin, Z., Karnieli, A., & Berliner, P. (2001). A mono-window algorithm for retrieving land surface temperature from Landsat TM data and its application to the Israel-Egypt border region. *International Journal of Remote Sensing, 22*(18), 3719–3746. doi:10.1080/01431160010006971

Quan Luna, B., Blahut, J., van Westen, C. J., Sterlacchini, S., van Asch, T. W. J., & Akbas, S. O. (2011). The application of numerical debris flow modelling for the generation of physical vulnerability curves. *Natural Hazards and Earth System Sciences, 11*(7), 2047–2060. doi:10.5194/nhess-11-2047-2011

Raghavswamy, V. (1982). Role of satellite remote sensing in land system mapping, land resources inventory and land use planning: A sample study of Kemang River basin, Arunachal Pradesh. *Journal of the Indian Society of Remote Sensing, 10*(3), 31–40.

Rahman, M. T. (2016). Detection of Land Use/Land Cover Changes and Urban Sprawl in Al-Khobar, Saudi Arabia: An Analysis of Multi-Temporal Remote Sensing Data. *ISPRS International Journal of Geo-Information, 5*(2), 15. doi:10.3390/ijgi5020015

Rahman, M. T., Aldosary, A., Nahiduzzaman, K. M., & Reza, I. (2016). Vulnerability of flash flooding in Riyadh, Saudi Arabia. *Natural Hazards.* doi:10.1007/s11069-016-2521-8

Rahman, M. T., & Rashed, T. (2015). Urban tree damage estimation using airborne laser scanner data and geographic information systems: An example from 2007 Oklahoma ice storm. *Urban Forestry & Urban Greening*, *14*(3), 562–572. doi:10.1016/j.ufug.2015.05.008

Ramakrishnan, D., Bandyopadhyay, A., & Kusuma, K. N. (2009). SCS-CN and GIS-based approach for identifying potential water harvesting sites in the Kali Watershed, Mahi River Basin, India. *Journal of Earth System Science*, *118*(4), 355–368. doi:10.1007/s12040-009-0034-5

Ramesh, H. (2012). Change Detection in Harangi Watershed using Multi-temporal Satellite Imagery. *Paper presented at theInternational Symposium on Mountain resource management in a changing environment (MRMCE '12).*

Rao, D. P., Gautam, N. C., Nagaraja, R., & Mohan, P. (1996). IRS-1C applications in land use mapping and planning. *Current Science*, *7*(3), 390–395.

Rautela, P., & Pande, R. (2006). Non-monsoonal landslides in Uttaranchal Himalaya (India): implications upon disaster mitigation strategy. Disaster Prevention and Management, 15(3).

Reddy, G. P. O., & Maji, A. K. (2003). Delineation and characterzation of geomorphological features in a part of lower Maharashtra metamorphic plateau, using IRS-1D, LISS-III data. *Journal of the Indian Society of Remote Sensing*, *31*(4), 241–250. doi:10.1007/BF03007344

Reddy, G. P. O., Maji, A. K., & Gajbhiye, K. S. (2002). GIS for morphometric analysis of drainage basins. *GIS India*, *11*(4), 9–14.

Reddy, G. P. O., Maji, A. K., & Gajbhiye, K. S. (2004). Drainage morphometry and its influence on landform characteristics in basaltic terrain-A Remote Sensing and GIS approach. *International Journal of Applied Earth Observation and Geoinformatics*, *6*(1), 1–16. doi:10.1016/j.jag.2004.06.003

Rees, W. E., & Wackernagel, M. (1996). Urban ecological footprints: Why cities cannot be sustainable and why they are a key to sustainability. *Environmental Impact Assessment Review*, *16*(4-6), 223–248. doi:10.1016/S0195-9255(96)00022-4

Reis, S., Yalcin, A., Atasoy, M., Nisanci, R., Bayrak, T., Erduran, M., & Ekercin, S. et al. (2012). Remote sensing and GIS-based landslide susceptibility mapping using frequency ratio and analytical hierarchy methods in Rize province (NE Turkey). *Environment and Earth Science*, *66*(7), 2063–2073. doi:10.1007/s12665-011-1432-y

Resenzweig, C., Solecki, W. D., Paschall, L., Chopping, M., Pope, G., & Goldberg, R. (2005). Characterizing the urban heat island in current and future climates in New Jersey. *Global Environmental Change Part B: Environmental Hazards*, *6*(1), 51–62. doi:10.1016/j.hazards.2004.12.001

Retalis, A., Paronis, D., Lagouvardos, K., & Kotroni, V. (2010). The heat wave of June 2007 in Athens, Greece - part 1: Study of satellite derived land surface temperature. *Atmospheric Research*, *98*(2-4), 458–467. doi:10.1016/j.atmosres.2010.08.005

Rheingold, H. (1998). *The Virtual Community: Homesteading on the Electronic Frontier*. New York, Addison-Wesley. Retrieved from http://www.rheingold.com/vc/book

Richards, T. B., Croner, C. M., Rushton, G., Brown, C. K., & Fowler, L. (1999). Geographic Information Systems and Public Health: Mapping the Future. *Pub Health Rep*, *114*(4), 359–373. doi:10.1093/phr/114.4.359 PMID:10501137

Rickenmann, D. (2005). Runout prediction methods. In M. Jakob & O. Hungr (Eds.), *Debris-flow Hazard and Relation Phenomena* (pp. 305–324). Chichester: Springer. doi:10.1007/3-540-27129-5_13

Robinson, I. S. (2004). *Measuring the Oceans from Space: The Principles and Methods of Satellite Oceanography*. Berlin: Springer - Praxis.

Rodrigues, M., Montañés, C., & Fueyo, N. (2010). A method for the assessment of the visual impact caused by the large-scale deployment of renewable-energy facilities. *Environmental Impact Assessment Review*, *30*(4), 240–246. doi:10.1016/j.eiar.2009.10.004

Roithmayr, C. M. (1970). Airborne low-light sensor detects luminescing fish schools at night. *Commercial Fisheries Review*, *32*, 42–51.

Rolland, A., & Rangarajan, R. (2013). Runoff estimation and potential recharge site delineation using analytic hierarchy process. *Geocarto International*, *28*(2), 159-170. doi:10.1080/10106049.2012.665499

Roy, D. (2010). Selection and Design of Appropriate Structure for Watershed Management around Upper Reach of Dulung Nala in the Subarnarekha Basin at Belpahari, West Midnapore, West Bengal (Research Project). School of Water Resources Engineering, Jadavpur University, Kolkata, sponsored by Dept. of Science and Technology, Govt. of West Bengal.

Roy, P. S., Kaul, R. N., Roy, M. R., & Gabryal, S. S. (1985). Forest type stratification and delineation of shifting cultivation in the eastern part of Arunachal Pradesh using Landsat MSS data. *International Journal of Remote Sensing*, *6*(3&4), 411–418. doi:10.1080/01431168508948463

Roy, P. S., Ranganath, B. K., Diwakar, P. G., Vohra, T. P. S., Bhan, S. K., Singh, I. H., & Pandian, V. C. (1991). Tropical forest type mapping and monitoring using remote sensing. *International Journal of Remote Sensing*, *12*(11), 2205–2225. doi:10.1080/01431169108955253

Rozenstein, O., & Karnieli, A. (2011). Comparison of methods for land-use classification incorporating remote sensing and GIS inputs. *Applied Geography (Sevenoaks, England)*, *31*(2), 533–544. doi:10.1016/j.apgeog.2010.11.006

Rubec, C. D., & Thie, J. (1978). Land use monitoring with Landsat digital data in southwestern Manitoba.*Proceedings of the Fifth Canadian Symposium on Remote Sensing of Environment*, Victoria, British Columbia (pp. 136-149).

Ruin, I., Creutin, J.-D., Anquetin, S., Gruntfest, E., & Lutoff, C. (2009). Human vulnerability to flash floods: Addressing physical exposure and behavioral questions. In P. Samuels & S. Huntington (Eds.), *Flood Risk Management: Research and Practice* (pp. 1005–1012). London: Taylor & Francis.

Rundquist, D., Lawson, M., Queen, L., & Cerveny, R. (1987). The Relationship between Summer Season Rainfall and Lake-Surface Area. *Water Resources Bulletin, 23*(3), 493–508. doi:10.1111/j.1752-1688.1987.tb00828.x

Saaty, L. T. (1980). *The analytic hierarchy process*. New York: McGraw-Hill International.

Saaty, T. (1977). A scaling method for priorities in hierarchical structures. *Journal of Mathematical Psychology, 15*(3), 234–281. doi:10.1016/0022-2496(77)90033-5

Saaty, T. (1980). *The Analytic Hierarchy Process*. New York: McGraw-Hill.

Saaty, T. L. (2008). Decision making with the analytic hierarchy process. *International Journal of Services Sciences, 1*(1), 83–98. doi:10.1504/IJSSCI.2008.017590

Saha, S. K., & Barrow, C. J. (1981). *River Basin Planning. Theory and Practice*. Chichester: Wiley.

Salequzzaman, M., Sultana, U. T., Iqbal, A., & Hoque, M. A. (2006). *Ecological footprint of waste generation: A sustainable tool for solid waste management of Khulna City Corporation of Bangladesh*. Retrieved from http://www.necsi.edu

Salm, B., Burkhard, A., Gubler, H.U. (1990). Berechnung von Fliesslawinen: Eine Anleitungfuer Praktiker; mit Beispielen. *Mitteilungen des Eidgenoessischen Instituts fuerSchnee- und Lawinenforschung*, 47, 1–37

Sanchez, P., Demestre, M., Recasens, L., Maynou, F., & Martin, P. (2008). Combining GIS and GAMS to identify potential habitats of squid Loligo vulgaris in the north-western Mediterranean. *Hydrobiologia, 612*(1), 91–98. doi:10.1007/s10750-008-9487-9

Santos, A. M. P. (2000). Fisheries oceanography using satellite and airborne remote sensing methods: A review. *Fisheries Research, 49*(1), 1–20. doi:10.1016/S0165-7836(00)00201-0

Santra Mitra, S., Wright, J., Santra, A., & Ghosh, A. R. (2015). An integrated water balance model for assessing water scarcity in a data-sparse interfluve in eastern India. *Hydrological Sciences Journal, 60*(10), 1813-1827. doi:10.1080/02626667.2014.934248

Sanyal, J., & Lu, X. X. (2004). Application of remote sensing in flood management with special reference to monsoon Asia: A review. *Natural Hazards, 33*(2), 283–301. doi:10.1023/B:NHAZ.0000037035.65105.95

Sarkar, A., & Patel, P. P. (2011). Topographic analysis of the Dulung R. Basin. *eTraverse - The Indian Journal of Spatial Science*, 2(1).

Sarkar, A., & Patel, P. P. (2012). Terrain Classification of the Dulung Drainage Basin. *eTraverse - The Indian Journal of Spatial Science*, 3(1).

Sarkar, A., & Patel, P. P. (2009). Drainage analysis of the Dulung Basin. In H. S. Sharma & V. S. Kale (Eds.), *Geomorphology in India (Prof. Savindra Singh Felicitation Volume)* (pp. 133–154). Allahabad: Prayag Pustak Bhavan.

Sarkar, S., Kanungo, D. P., & Sharma, S. (2015). Landslide hazard assessment in the upper Alaknanda valley of Indian Himalayas. *Geomatics Natural Hazards and Risk*, *6*(4), 308–325. doi:10.1080/19475705.2013.847501

Sarma, V. V. L. N., Krishna, G. M., Malini, B. H., & Rao, K. N. (2001). Land use / land cover change detection through Remote Sensing and its climatic implications in the Godavari delta region. *Journal of the Indian Society of Remote Sensing*, *29*(1&2), 85–91. doi:10.1007/BF02989918

Sattari, F., & Hashim, M. (2014). A brief review of land surface temperature retrieval methods from thermal satellite sensors. *Middle-East Journal of Scientific Research*, *22*(5), 757–768.

Schardt, M. K., Schurek, A., & Winter, R. (1990). Forest mapping using satellite imagery - The Riegenburg map sheet 1.2,000,000 as example. *International Journal for Photogrammetry and Remote Sensing (ISPRS). Journal of Photogrammetry and Remote Sensing*, *45*, 33–46. doi:10.1016/0924-2716(90)90004-U

Schiavone, F., & Agrifoglio, R. (2012). Communities of practice and practice preservation: a case study. In Information Systems: Crossroads for Organization, Management, Accounting and Engineering (pp. 331-338). Physica-Verlag HD. doi:10.1007/978-3-7908-2789-7_37

Schofield, O., Arnone, R. A., Bissett, W. P., Dickey, T. D., Davis, C. O., Finkel, Z., & Moline, M. et al. (2004). Water colors in the coastal zone: What can we see? *Oceanography (Washington, D.C.)*, *17*(2), 24–31. doi:10.5670/oceanog.2004.44

Schroedter, M., Olesen, F. S., & Fischer, H. (2003). Determination of land surface temperature distributions from single channel IR measurements: An effective spatial interpolation method for the use of TOVS, ECMWF and radiosonde profiles in the atmospheric correction scheme. *International Journal of Remote Sensing*, *24*(6), 1189–1196. doi:10.1080/01431160110115087

Schweiger, A. J., & Key, J. R. (1997). Estimating surface radiation fluxes in the Arctic from TOVS brightness temperatures. *International Journal of Remote Sensing*, *18*(4), 955–970. doi:10.1080/014311697218854

Seiller, G., Hajji, I., & Anctil, F. (2015). Improving the temporal transposability of lumped hydrological models on twenty diversified U.S. watersheds. *Journal of Hydrology: Regional Studies*, *3*, 379–399.

Şen, Z., Al Alsheikh, A., Al-Turbak, A. S., Al-Bassam, A. M., & Al-Dakheel, A. M. (2013). Climate change impact and runoff harvesting in arid regions. *Arabian Journal of Geosciences*, *6*(1), 287–295. doi:10.1007/s12517-011-0354-z

Shang, H., & Bishop, I. D. (2000). Visual thresholds for detection, recognition and visual impact in landscape settings. *Journal of Environmental Psychology*, *20*(2), 125–140. doi:10.1006/jevp.1999.0153

Shankarnarayan, K. A., & Sen, A. K. (1977). The methodology of mapping land resources for rational utilization. In P.S. Jaiswal (Ed.), Desertification and its Control (Ch. 9, pp. 77-80). New Delhi: Indian Council for Agricultural Research (ICAR).

Sharma, K. P., Garg, P. K., & Verma, S. K. (1986). Spectral reflectance pattern study of some Indian land use / land cover features. *Journal of the Indian Society of Remote Sensing*, *14*(1), 11–16. doi:10.1007/BF03007218

Sharma, K. P., Jain, S. C., & Garg, P. K. (1984). Monitoring land use and land cover changes using Landsat images. *Journal of the Indian Society of Remote Sensing*, *12*(2), 65–70.

Sharma, L. K., Kanga, S., Nathawat, M. S., Sinha, S., & Pandey, P. C. (2012). Fuzzy AHP for forest fire risk modeling. *Disaster Prevention and Management*, *21*(2), 160–171. doi:10.1108/09653561211219964

Shetty, A., Lakshman, N., Thokchom, S., & Rajesh, M. V. S. (2005). Land use - land cover mapping using satellite data for a forested watershed, Udupi district, Karnataka state, India. *Journal of the Indian Society of Remote Sensing*, *33*(2), 233–238. doi:10.1007/BF02990040

Shipra, V., & Gupta, R.D. (2014, December). Spatial and Temporal Variation of Japanese Encephalitis Disease and Detection of Disease Hotspots: A Case Study of Gorakhpur District, Uttar Pradesh, India. Proceedings of Symposium ISPRS Annals, Commission VIII, WG VIII/2, Hyderabad, India.

Shukla, K. K., Patel, D. J., & Radadiya, B. L. (2014). Role of Information Technology in Improvement of Current Scenario in Agriculture. *Oriental Journal of Computer Science and Technology*, *7*(3), 390–395.

Siddaiah, N. S., & Kumar, K. (2009). Discovery of minamiite from the Deccan Volcanic Province, India: Implications for Martian surface exploration. *Current Science*, *97*(11), 1664–1669.

Singh, A. (1989). Digital change detection techniques using remotely sensed data. *International Journal of Remote Sensing*, *10*(6), 989–1003. doi:10.1080/01431168908903939

Singh, B. M., Manchanda, M. L., Iyer, H. S., Pande, L. M., Prasad, J., & Rao, B. R. M. (1983). Use of remote sensing techniques for agricultural land use survey, classification and planning. *Journal of the Indian Society of Remote Sensing*, *11*(3), 1–13.

Sinha, S., Nathawat, M. S., & Sharma, L. K. (2014). *Geospatial approach for managing relocation and distribution of tigers*. Germany: LAP Lambert Academic Publishing.

Sinha, S., Sharma, L. K., & Nathawat, M. S. (2011). Retrieving tiger habitats: Conserving wildlife geospatially. *Applied Remote Sensing Journal*, *2*(1), 1–5.

Skidmore, A. K., Bijker, W., Schmidt, K., & Kumar, L. (1997). Use of Remote Sensing and GIS for sustainable land management. *ITC Journal*, *3*(4), 302–315.

Slattery, M., Willett, K. L., Cobb, G. P., & Benson, W. H. (2010). Multiple facets of environmental impacts from Hurricane Katrina. *Environmental Toxicology and Chemistry*, *29*(7), 1401–1402. doi:10.1002/etc.213 PMID:20821586

Slotterback, C. S. (2011). Planners perspectives on using technology in participatory processes. *Environment and Planning-Part B*, *38*(3), 468–485. doi:10.1068/b36138

Smith, G. H. (1935). The Relative Relief of Ohio. *Geographical Review*, *25*, 272–284.

Snavely, N., Seitz, S. M., & Szeliski, R. (2006). Photo Tourism: Exploring Photo Collections in 3D.*Proceedings of ACM SIGGRAPH*. doi:10.1145/1179352.1141964

Snavely, N., Seitz, S. M., & Szeliski, R. (2008). Modeling the world from internet photo collections. *International Journal of Computer Vision*, *80*(2), 189–210. doi:10.1007/s11263-007-0107-3

Sobrino, J. A., Cool, C., & Caselles, V. (1991). Atmospheric correction for land surface temperature using NOAA-11 AVHRR channels 4 and 5. *Remote Sensing of Environment*, *38*(1), 19–34. doi:10.1016/0034-4257(91)90069-I

Sobrino, J. A., & Cuenca, J. (1999). Angular variation of thermal infrared emissivity for some natural surfaces from experimental measurements. *Applied Optics*, *38*(18), 3931–3936. doi:10.1364/AO.38.003931 PMID:18320001

Sobrino, J. A., Jimenez-Munoz, J. C., & Paolini, L. (2004). Land surface temperature retrieval from Landsat TM 5. *Remote Sensing of Environment*, *90*(4), 434–440. doi:10.1016/j.rse.2004.02.003

Sobrino, J. A., Li, Z. L., Stoll, M. P., & Becker, F. (1996). Multi-channel multi-angle algorithms for estimating sea and land surface temperature with ATSR data. *International Journal of Remote Sensing*, *17*(11), 2089–2114. doi:10.1080/01431169608948760

Soja, M. J., Persson, H., & Ulander, L. M. H. (2015). Estimation of Forest Height and Canopy Density from a Single InSAR Correlation Coefficient. *IEEE Geoscience and Remote Sensing Letters*, *12*(3), 646–650. doi:10.1109/LGRS.2014.2354551

Solanki, H. U., Dwivedi, R. M., & Nayak, S. R. (2001). Synergistic analysis of SeaWiFS chlorophyll concentration and NOAA-AVHRR SST features for exploring marine living resources. *International Journal of Remote Sensing*, *22*(18), 3877–3882. doi:10.1080/01431160110069845

Solanki, H. U., Mankodi, P. C., Nayak, S. R., & Somvanshi, V. S. (2005). Evaluation of remote-sensing-based potential fishing zones (PFZs) forecast methodology. *Continental Shelf Research*, *25*(18), 2163–2173. doi:10.1016/j.csr.2005.08.025

Solano, D. M. M., & Cruz, E. (2011, August). Ecological Footprint of Waste Generation in Digos City, Davao Del Sur. *Paper presented at the 7th Mindanao Statistics Congress*, University of Mindanao, Matina Campus, Davao City, Philippines.

Solano, E., & Scavuzzo, M. (2013). Remote Sensing to Assessment Damages Post Earthquakes, focusing in Urban Structures. Retrieved from http://aulavirtual.ig.conae.gov.ar/moodle/pluginfile.php/513/mod_page/content/71/PostEarthquake.pdf

Soria, G., & Sobrino, J. A. (2007). ENVISAT/AATSR derived land surface temperature over a heterogeneous region. *Remote Sensing of Environment, 111*(4), 409–422. doi:10.1016/j.rse.2007.03.017

Sosio, R., Crosta, G. B., & Hungr, O. (2008). Complete dynamic modeling calibration for the Thurwieser rock avalanche (Italian Central Alps). *Engineering Geology, 100*(1-2), 11–26. doi:10.1016/j.enggeo.2008.02.012

Špitalar, M., Gourley, J. J., Lutoff, C., Kirstetter, P.-E., Brilly, M., & Carr, N. (2014). Analysis of flash flood parameters and human impacts in the US from 20062012. *Journal of Hydrology (Amsterdam), 519*, 863–870. doi:10.1016/j.jhydrol.2014.07.004

Sreenivas, B., & Chary, B. N. (2011, January 18-21). Processing of Satellite Image using Digital Image Processing. *Paper presented at the Geospatial World Forum*, Hyderabad, India.

Srinivasan, T. N. (1988). Population Growth and Economic Development. *Journal of Policy Modeling, 10*(1), 7–27.

Stamps, A. E. III. (1990). Use of photographs to simulate environments: A meta-analysis. *Perceptual and Motor Skills, 71*(3), 907–913. doi:10.2466/pms.1990.71.3.907

Stehman, S. V., & Czaplewski, R. L. (1998). Design and analysis for thematic map accuracy assessment: Fundamental principles. *Remote Sensing of Environment, 64*(3), 331–344. doi:10.1016/S0034-4257(98)00010-8

Stephen, H., Cummings, M., McCubbrey, D., Pinsonneault, A., & Donovan, R. (2006). *Management Information Systems: For the Information Age* (3rd Canadian ed.). New York: McGraw-Hill Ryerson.

Steven, M., Malthusb, T., & Baretc, F. (2007). Intercalibration of Vegetation Indices - an Update. *Proceedings of the10th International Symposium on Physical Measurements and Signatures in Remote Sensing (ISPMSRS07),* Davos, Switzerland.

Steven, M. D. (1987). Ground truth: An underview. *International Journal of Remote Sensing, 8*(7), 1033–1038. doi:10.1080/01431168708954745

Stewart, T. R., Middleton, P., Downton, M., & Ely, D. (1984). Judgments of photographs vs. field observations in studies of perception and judgment of the visual environment. *Journal of Environmental Psychology, 4*(4), 283–302. doi:10.1016/S0272-4944(84)80001-8

Strahler, A. H., Boschetti, L., Foody, G. M., Friedl, M. A., Hansen, M. C., Herold, M., & Woodcock, C. E. (2006). *Global land cover validation: Recommendations for evaluation and accuracy assessment of global land cover maps* (p. 51). Luxembourg: European Communities.

Strahler, A. N. (1957). Quantitative analysis of watershed geomorphology. *Transactions - American Geophysical Union, 38*(6), 913–920.

Stramondo, S., Moro, M., Tolomei, C., Cinti, F. R., & Doumaz, F. (2005). InSAR surface displacement field and fault modelling for the 2003 Bam earthquake (southeastern Iran). *Journal of Geodynamics*, *40*(2–3), 347–353. doi:10.1016/j.jog.2005.07.013

Strathopoulou, M., & Cartalis, C. (2007). Daytime urban heat islands from Landsat ETM+ and CORINE land cover data: An application to major cities in Greece. *Solar Energy*, *81*(3), 358–368. doi:10.1016/j.solener.2006.06.014

Strathopoulou, M., Cartalis, C., & Kermatisoglou, I. (2004). Mapping micro urban heat islands using NOAA/AVHRR images and CORINE land cover: An application to coastal cities of Greece. *International Journal of Remote Sensing*, *25*(12), 2301–2316. doi:10.1080/01431160 310001618725

Strozzi, T., Wegmuller, U., Tosi, L., Bitelli, G., & Spreckels, V. (2001). Land subsidence monitoring with differential SAR interferometry. *Photogrammetric Engineering and Remote Sensing*, *67*(11), 1261–1270.

Stuart, V., Platt, T., & Sathyendranath, S. (2011). The future of fisheries science in management: A remote - sensing perspective. *ICES Journal of Marine Science*, *68*(4), 644–650. doi:10.1093/icesjms/fsq200

Stuart, V., Platt, T., Sathyendranath, S., & Pravin, P. (2011). Remote sensing and fisheries: An introduction. *ICES Journal of Marine Science*, *68*(4), 639–641. doi:10.1093/icesjms/fsq193

Subramanian, S., Manjulekshmi, N., Kolwalkar, J., Fernandes, P. M., Sreekanth, G. B., Singh, N. P., ... (2014). A Manual on The Use of Potential Fishing Zone (PFZ) Forecast. Technical Bulletin No.:40. PFZ Mission Programme: Indian National Centre for Oceanic Information Services, Hyderabad.

Subramanyan, V. (1978). Land system studies for resource evaluation: A demonstration around Sagar. *Madhya Pradesh. Journal of the Indian Society of Remote Sensing*, *6*(1), 31–42.

Subramanyan, V. (1981). Geomorphology of the Deccan Volcanic Province, India. In K. V. Subbarao & R. N. Sukheswala (Eds.), *Deccan Volcanism and related Basalt Provinces in other parts of the World* (pp. 101–116). Bangalore: Geological Society of India.

Sun, D. (2003). Estimation of land surface temperature from a Geostationary Operational Environmental Satellite (GOES-8). *Journal of Geophysical Research*, *108*(D11), 1–15. doi:10.1029/2002JD002422

Sun, D., Pinker, R., & Basara, J. (2004). Land surface temperature estimation from the next generation of geostationary operational environmental satellites: GOES M-Q. *Journal of Applied Meteorology*, *43*(2), 363–372. doi:10.1175/1520-0450(2004)043<0363:LSTEFT>2.0.CO;2

Susskind, J., Rosenfield, J., Reuter, D., & Chahine, M. T. (1984). Remote sensing of weather and climate parameters from HIRS2 / Msu on TIROS-N. *Journal of Geophysical Research*, *89*(D3), 4677–4697. doi:10.1029/JD089iD03p04677

Swayze, G, A., Desborough, G, A., Smith, K.S., Lowers, H, a., Hammarstrom, J, M, Diehl, S.F, Leinz, R. W., & Driscoll, R.L. (2008). Understanding Jarosite-From Mine Waste to Mars. *USGS Circular*, 1328, 8-13.

System Design Strategies. (2005). *ESRI Technical Reference Document*. ESRI Press.

Tanveer, M., Mukherjee, C. K., & Vinu Chandran, R. (2015). Application of ANN to predict chlorophyll-a and sea surface temperature in determination of fishing zone. *International Water Technology Journal*, *5*(4), 294–304.

Teillet, P. M., Markham, B. L., & Irish, R. R. (2006). Landsat cross-calibration based on near simultaneous imaging of common ground targets. *Remote Sensing of Environment*, *102*(3-4), 264–270. doi:10.1016/j.rse.2006.02.005

Terjung, W. H., & ORourke, P. A. (1980). Influences of physical structures on urban energy budgets. *Boundary-Layer Meteorology*, *19*(4), 421–439. doi:10.1007/BF00122343

The Open Geospatial Consortium, Inc. (n. d.). Retrieved from http://www.opengeospatial.org/

Thu, P. M., & Populus, J. (2007). Status and changes of mangrove forest in Mekong Delta: Case study in Tra Vinh, Vietnam. *Estuarine, Coastal and Shelf Science*, *71*(1-2), 98–109. doi:10.1016/j.ecss.2006.08.007

Tobler, W. (2000). The development of analytical cartography: A personal note. *Cartography and Geographic Information Science*, *27*(2), 189–194.

Tomlinson, C. J., Chapman, L., Thornes, J. E., & Baker, C. (2011). Remote sensing land surface temperature for meteorology and climatology: A review. *Meteorological Applications*, *18*(3), 296–306. doi:10.1002/met.287

Tom, S., Nguyen, D. M., Rachel, K., & Mary, G. (2000). Japanese encephalitis. *Journal of Neurology, Neurosurgery, and Psychiatry*, *68*(4), 405–415. doi:10.1136/jnnp.68.4.405 PMID:10727474

Townsend, P. A., & Walsh, S. J. (1998). Modelling flood plain inundation using integrated GIS with radar and optical remote sensing. *Geomorphology*, *21*(98), 295–312. doi:10.1016/S0169-555X(97)00069-X

Toyama, K., Logan, R., Roseway, A., & Anandan, P. (2003). Geographic Location Tags on Digital Images.*Proceedings of the Eleventh ACM International Conference on Multimedia,*Berkeley, California, USA (pp. 156-166). doi:10.1145/957013.957046

Traboulsi, H., & Traboulsi, M. (2015). Rooftop level rainwater harvesting system. *Applied Water Science.*

Trigo, I. F., Monteiro, I. T., Olesen, F. S., & Kabsch, E. (2008). An assessment of remotely sensed land surface temperature. *Journal of Geophysical Research*, *113*(D17), 1–12. doi:10.1029/2008JD010035

Tsai, M. P., Hsu, Y. C., Li, H. C., Shu, H. M., & Liu, K. F. (2011). Application of simulation technique on debris flow hazard zone delineation: A case study in the Daniao tribe, Eastern Taiwan. *Natural Hazards and Earth System Sciences*, *11*(11), 3053–3062. doi:10.5194/nhess-11-3053-2011

Tsou, M. (2004). Integrating Web-based GIS and image processing tools for environmental monitoring and natural resource management. *Journal of Geographical Systems*, *6*(2), 155–174. doi:10.1007/s10109-004-0131-6

Turner II, B.L. (1995). Linking the Natural and Social Sciences. *The Land Use / Cover Change Core Project of IGBP, IGBP Newsletter*, 22.

Umbarkar, P. S., & Wajire, N. S. (2014). Remote sensing: A novel approach for insect pests management. *Agrobios Newsletter*, *XII*(11), 61–62.

Upadhyayula, S.M., Mutheneni, S.R., Nayanoori, H.K., Natarajan, A., Goswami, P. (2012). Impact of weather variables on mosquitoes infected with Japanese encephalitis virus in Kurnool district, Andhra Pradesh. *Asian Pacific Journal of Tropical Medicine*, 5(5), 337-341.

USGS. (2012). *LANDSAT Data Continuity Mission (LDCM) Level 1 (L1) Data Format Control Book (DFCB)*. South Dakota: USGS.

Uuemaa, E., Antrop, M., Roosaare, J., Marja, R., & Mander, Ü. (2009). Landscape metrics and indices: An overview of their use in landscape research. *Living Reviews in Landscape Research*, *3*(1), 1–28.

Valavanis, V. D., Pierce, G. J., Zuur, A. F., Palialexis, A., Saveliev, A., Katara, I., & Wang, J. (2008). Modelling of essential fish habitat based on remote sensing, spatial analyis and GIS. *Hydrobiologia*, *612*(1), 5–20. doi:10.1007/s10750-008-9493-y

Valor, E., & Caselles, V. (1996). Mapping land surface emissivity from NDVI: Application to European, African and South American areas. *Remote Sensing of Environment*, *57*(3), 167–184. doi:10.1016/0034-4257(96)00039-9

Van Westen, C. J. (1993). Application of Geographic Information Systems to landslide hazard zonation. *ITC Publication*, *15*, 245.

Varnes, D. J. (1978). Slope movement types and process. In Special Report 176: Landslides: Analysis and Control. In R.L. Schuster & R.J. Krizek (Eds.), Transportation and Road Research Board, National Academy of Science (pp. 11-33).

Vasquez, D. P., Reyes, F. J. O., & Arboledas, L. A. (1997). A comparative study of algorithms for estimating land surface temperature from AVHRR data. *Remote Sensing of Environment*, *62*(3), 215–222. doi:10.1016/S0034-4257(97)00091-6

Vijith, H., & Satheesh, R. (2006). GIS based morphometric analysis of two major upland subwatersheds of Meenachil River in Kerala. *Journal of the Indian Society of Remote Sensing*, *34*(2), 181–185.

Voogt, J. A., & Oke, T. R. (2003). Thermal remote sensing of urban climates. *Remote Sensing of Environment*, *86*(3), 370–384. doi:10.1016/S0034-4257(03)00079-8

Wachal, D. J., & Hudak, P. F. (2000). Mapping landslide susceptibility in Travis County, Texas, USA. *GeoJournal*, *51*(3), 245–253. doi:10.1023/A:1017524604463

Wackernagel, M., & Rees, W. E. (1996). *Our ecological footprint: Reducing human impact on the Earth. Gabriola Island.* New Society Publishers.

Wackernagel, M., & Rees, W. E. (1997). Perceptual and structural barriers to investing in natural capital: Economics from an ecological footprint perspective. *Ecological Economics*, *20*(1), 3–24. doi:10.1016/S0921-8009(96)00077-8

Wang, F., Qin, Z., Song, C., Tu, L., Karnieli, A., & Zhao, S. (2015). An improved mono-window algorithm for land surface temperature retrieval from Landsat 8 thermal infrared sensor data. *Remote Sensing*, *7*(4), 4268–4289. doi:10.3390/rs70404268

Wang, K., & Liang, S. (2009). Evaluation of ASTER and MODIS land surface temperature and emissivity products using long-term surface long wave radiation observations at SURFRAD sites. *Remote Sensing of Environment*, *113*(7), 1556–1565. doi:10.1016/j.rse.2009.03.009

Wang, S., Anselin, L., Bhaduri, B., Crosby, C., Goodchild, M. F., Liu, Y., & Nyerges, T. L. (2013). CyberGIS software: A synthetic review and integration roadmap. *International Journal of Geographical Information Science*, *27*(11), 2122–2145. doi:10.1080/13658816.2013.776049

Wan, Z. (2008). New refinements and validation of the MODIS land-surface temperature / emissivity products. *Remote Sensing of Environment*, *112*(1), 59–74. doi:10.1016/j.rse.2006.06.026

Wan, Z., & Dozier, J. (1996). A generalized split-window algorithm for retrieving land-surface temperature from space. *IEEE Transactions on Geoscience and Remote Sensing*, *34*(4), 892–905. doi:10.1109/36.508406

Wan, Z., Zhang, Y., Zhang, Q., & Li, Z. (2002). Validation of the land-surface temperature products retrieved from terra moderate resolution imaging spectroradiometer data. *Remote Sensing of Environment*, *83*(1-2), 163–180. doi:10.1016/S0034-4257(02)00093-7

Wan, Z., Zhang, Y., Zhang, Q., & Li, Z. (2004). Quality assessment and validation of the MODIS global land surface temperature. *International Journal of Remote Sensing*, *25*(1), 261–274. doi:10.1080/0143116031000116417

Watson, C. (2012). *Analysis of urban heat island climates along the I-85 / I-40 corridor in central North Carolina* [Ph.D. thesis]. University of North Carolina at Greensboro.

Wedding, L. M., Friedlander, A. M., McGranaghan, M., Yost, R. S., & Monaco, M. E. (2008). Using bathymetric lidar to define nearshore benthic habitat complexity: Implications for management of reef fish assemblages in Hawaii. *Remote Sensing of Environment*, *112*(11), 4159–4165. doi:10.1016/j.rse.2008.01.025

Weismiller, R. A., & Momin, S. J. (1977). Change detection in coastal zone environments. *Journal of Photogrammetric Engineering and Remote Sensing*, *43*(12), 1533–1539.

Weng, F., & Grody, N. C. (1998). Physical retrieval of land surface temperature using special microwave imager. *Journal of Geophysical Research*, *103*(D8), 8839–8848. doi:10.1029/98JD00275

Weng, Q. (2003). Fractal analysis of satellite-detected urban heat island effect. *Photogrammetric Engineering and Remote Sensing*, *69*(5), 555–566. doi:10.14358/PERS.69.5.555

Weng, Q., Lu, D., & Schubring, J. (2004). Estimation of land surface temperature-vegetation abundance relationship for urban heat island studies. *Remote Sensing of Environment*, *89*(4), 467–483. doi:10.1016/j.rse.2003.11.005

Weng, Q., Rajasekar, U., & Hu, X. (2011). Modeling urban heat islands and their relationship with impervious surface and vegetation abundance by using ASTER images. *IEEE Transactions on Geoscience and Remote Sensing*, *49*(10), 4080–4089. doi:10.1109/TGRS.2011.2128874

Wentworth, C. K. (1930). A Simplified Method of Determining the Average Slope of Land Surfaces. *American Journal of Science*, *20*, 184–194.

Wentz, F. J., Chelle, C., Smith, D., & Chelton, D. (2000). Satellite measurements of sea surface temperature through clouds. *Science*, *288*(5467), 847–850. doi:10.1126/science.288.5467.847 PMID:10797004

Whigham, P. A., & Crapper, P. F. (2001). Modelling rainfall-runoff using genetic programming. *Mathematical and Computer Modelling*, *33*(6), 707–721. doi:10.1016/S0895-7177(00)00274-0

WHO, (2003). Communicable Disease Surveillance and Response: Severe acute respiratory syndrome (SARS): Status of the outbreak and lessons for the immediate future.

WHO. (2006). Preventing disease through healthy environments towards an estimate of the environmental burden of disease. WHO Press.

Wickware, G. M., & Howarth, P. J. (1981). Change detection in the Peace-Athabasca delta using digital Landsat data. *Remote Sensing of Environment*, *11*(1), 1, 9–25.

Williams, C. N., Basist, A., Peterson, T. C., & Grody, N. (2000). Calibration and verification of land surface temperature anomalies derived from the SSM/I. *Bulletin of the American Meteorological Society*, *81*(9), 2141–2156. doi:10.1175/1520-0477(2000)081<2141:CAVOLS>2.3.CO;2

Wilson, J. P., & Gallant, J. C. (Eds.). (2000). *Terrain Analysis: Principles and Applications*. New York: Wiley.

Wissinger, F., Shankar, R., & Restrepo, J. (2014, March 31-April 3). Hydrologic modeling methodology. *Paper presented at the 2014 8th Annual IEEE Systems Conference (SysCon).*

Wright, D. J., & Wang, S. (2011). The emergence of spatial cyberinfrastructure. *Proceedings of the National Academy of Sciences of the United States of America*, *108*(14), 5488–5491. doi:10.1073/pnas.1103051108 PMID:21467227

Wulder, M. A., Franklin, S. E., White, J. C., Linke, J., & Magnussen, S. (2006). An accuracy assessment framework for large-area land cover classification products derived from medium-resolution satellite data. *International Journal of Remote Sensing*, *27*(4), 663–683. doi:10.1080/01431160500185284

Xian, G., & Crane, M. (2006). An analysis of urban thermal characteristics and associated land cover in Tampa Bay and Las Vegas using Landsat satellite data. *Remote Sensing of Environment*, *104*(2), 147–156. doi:10.1016/j.rse.2005.09.023

Xiang, X., & Smith, E. A. (1997). Feasibility of simultaneous surface temperature-emissivity retrieval using SSM/I measurements from HAPEX-Sahel. *Journal of Hydrology (Amsterdam)*, *188-189*, 330–360. doi:10.1016/S0022-1694(96)03165-4

Xiaolin, L. (2009). WebGIS based Information Visualization for Infectious Disease Prevention. *Proceedings of the2009 Third International Symposium on Intelligent Information Technology Application*.

Xiaolin, L. (2005). A Framework of Web GIS Based Unified Public Health Information Visualization Platform. Proceedings of ICCSA 05, *LNCS* (Vol. *3482*, pp. 256–265).

Xu, H. (2002). Spatial Expansion of Urban/Town in Fuquing of China and its Driving Force Analysis. *Remote Sensing Technology and Application*, *17*, 86–92.

Yamaguchi, Y., Kahle, A. B., Tsu, H., Kawakami, T., & Pniel, M. (1998). Overview of advanced spaceborne thermal emission and reflection radiometer (ASTER). *IEEE Transactions on Geoscience and Remote Sensing*, *36*(4), 1062–1071. doi:10.1109/36.700991

Yan, X.-H., Ho, C., Zheng, Q., & Klemas, V. (1993). Using satellite IR in studies of the variabilities of the Western Pacific Warm Pool. *Science*, *262*(5132), 440–441. doi:10.1126/science.262.5132.440 PMID:17789951

Yu, J., Huang, Y., & Feng, X. (2001). Study on water bodies extraction and classification from SPOT image[in Chinese]. *Journal of Remote Sensing*, *5*, 214–219.

Zagaglia, C. R., Lorenzzetti, J., & Stech, J. L. (2004). Remote sensing data and longline catches of yellowfin tuna (*Thunnus albacores*) in the equatorial Atlantic. *Remote Sensing of Environment*, *93*(1-2), 267–281. doi:10.1016/j.rse.2004.07.015

Zainuddin, M., Kiyofuji, H., Saitoh, K., & Saitoh, S.-I. (2006). Using multi-sensor satellite remote sensing and catch data to detect ocean hot spots for albacore (*Thunnus alalunga*) in the northwestern North Pacific. *Deep-Sea Research*, *53*(3-4), 419–431. doi:10.1016/j.dsr2.2006.01.007

Zajc, M., Pogačnik, Ž., & Gosar, A. (2014). Ground penetrating radar and structural geological mapping investigation of karst and tectonic features in flyschoid rocks as geological hazard for exploitation. *International Journal of Rock Mechanics and Mining Sciences*, *67*, 78–87. doi:10.1016/j.ijrmms.2014.01.011

Zebker, H. A., Rosen, P. A., Goldstein, R. M., Gabriel, A., & Werner, C. L. (1994). On the derivation of coseismic displacement fields using differential radar interferometry: The Landers earthquake. *Journal of Geophysical Research, 99*(B10), 19617–19634. doi:10.1029/94JB01179

Zhang, J. H., Li, X., Yao, F. M., & Li, X. H. (2009). The progress in retrieving land surface temperature based on thermal infrared and microwave remote sensing technologies. *Spectroscopy and Spectral Analysis, 29*(8), 2103–2107. PMID:19839318

Zhang, T., & Tsou, M. H. (2009). Developing a grid-enabled spatial Web portal for Internet GIServices and geospatial cyberinfrastructure. *International Journal of Geographical Information Science, 23*(5), 605–630. doi:10.1080/13658810802698571

Zheng, Y. T., Zha, Z. J., & Chua, T. S. (2011). Research and applications on georeferenced multimedia: A survey. *Multimedia Tools and Applications, 51*(1), 77–98. doi:10.1007/s11042-010-0630-z

Zhou, T. (2011). Understanding online community user participation: A social influence perspective. *Internet Research, 21*(1), 67–81. doi:10.1108/10662241111104884

Zhou, Y., Wang, S., Zhou, W., Huang, T., & Yan, S. (2002). Research on assessing the progress towards sustainable development supported by the geo-spatial information technology. *Proceedings of International Society for Photogrammetry and Remote Sensing, 34*(Part 2), 627–632.

Zhou, Z., Gong, J., & Guo, M. (2015). Image-Based 3D Reconstruction for Posthurricane Residential Building Damage Assessment. *Journal of Computing in Civil Engineering*. doi:10.1061/(ASCE)CP.1943-5487.0000480

Ziadat, F. M., Mazahreh, S. S., Oweis, T. Y., & Bruggeman, A. (2006). A GIS based approach fr assessing water harvesting suitability in a Badia bench mark watershed in Jordan. *Paper presented at the Forteenth International Soil COnservatio Organisation Conference: Water Management and Soil Conservation in Semi-Arid Environments.*

Zion National Park. (n. d.). USGS - NPS Vegetation Mapping Program. Retrieved from http://biology.usgs.gov/npsveg/zion/index.html

About the Contributors

Abhisek Santra is an Associate Professor of Remote Sensing in the Department of Civil Engineering, Haldia Institute of Technology. He is the founder coordinator of Geoinformatics Cell in the same institute. He has more than 10 years of teaching and research experience in the field of Remote Sensing and GIS. He has published more than 15 research articles in reputed peer reviewed journals and proceedings of national and international repute. He has also published one book and one book chapter in his field of expertise. Presently, Dr. Santra is in the editorial board member of three international journals and an active evaluator of Dept. of Science & Technology, Govt. of India funded research projects. His research areas of interest are environmental modeling, calibration and validation of earth observation data, thermal remote sensing and urban utility mapping. He is currently engaged in several government funded projects.

***

Subhabrata Barman is an Assistant Professor with the Department of Computer Science & Engineering, Haldia Institute of Technology, West Bengal, India. His research interests are in the field of Wireless Networks, Computational Intelligence, Remote Sensing and Geo-Informatics, Parallel and Grid Computing. He has published research papers at various International and National Journals and Conferences. Currently he is working in the area QoS issues and Energy Management in Wireless Ad-hoc and Sensor Networks. He is a Professional Member of IEEE, IACSIT, IAENG and a reviewer of International Journal of Wireless Networks (Springer).

Shovan Lal Chattoraj hails from Durgapur, West Bengal, is presently working as a scientist in Geosciences and Disaster Management Science Group, IIRS, Dehradun. He has completed his PhD from IIT Bombay, Mumbai in 2011. He has experience in working with NHPC Ltd. as a Geologist. Dr. Chattoraj has authored many research articles published in national and international journals. His current interest lies in modelling of debris flows and spectroscopy of minerals.

Jeganathan Chockalingam is a Professor in the Department of Remote Sensing, BIT, Mesra. He has 23+ years of experience in the field of Geo-Spatial Science and Applications. He is the recipient of National Geospatial Award in 2015 and Klaas Jan Beek Award from ITC iin 2003. His fields of interest are Modelling space-time vegetation dynamics, Geostatistics, Automated Information Extraction, Fuzzy applications, Multi-Critieria Multi-Object Spatial Decision Modelling, etc. He has been part many prestigious National Remote Sensing Application Missions in ISRO, and He has developed many modelling software related to Biodiversity Characterisation, Phenology Modelling and Agriculture Drought Monitoring. He worked at various leading institutions in India and Abroad like RRSSC-D, IIRS, ISRO, University of Southampton, UK. He has been a visiting scholar to various international organisations like Nanjing University, Peking University, University of California Santa Barbara, Boston University, and he was a consultant to ICIMOD, Nepal and IWMI Srilanka.

Gennady Gienko holds a PhD degree from Moscow Institute for Engineers in Geodesy, Aerial Surveying and Cartography. Gennady has extensive international teaching, research and consulting experience in Geospatial Science and technologies, including GIS, Remote Sensing, Digital Image Analysis, Photogrammetry, Aerial LiDAR Surveying, and High Density Spatial Data Analysis.

Michael Govorov has extensive international teaching, research and consulting experience in Geographic Information Systems, Digital Cartography and Remote Sensing. His previous academic position was Senior Lecturer and Director of the GIS Unit at the University of the South Pacific, Fiji Islands where he prepared and presented courses in GIS and Remote Sensing at the Department of Geography. Prior to this, he presented courses in Computer Cartography and GIS at the Department of Surveying and Land Studies, Papua New Guinea under a UNDP program. With the Department of Cartography at the Siberian State Academy of Geodesy, Russia, he was the course coordinator for the Bachelor and Master program in Computer Cartography and GIS. At his current position with the Master Program in GIS/Advanced Diploma Program in GIS of Vancouver Island University, he is preparing and presenting courses in GIS and Remote Sensing for postgraduate and undergraduate students. He has a broad understanding of current global tendencies in geo-information science areas and applies these perspective to his teaching and research.

Swati Goyal received the Master degree in Geoinformatics from TERI University, Delhi, in 2014 and currently pursuing Master of technology in Environmental Sciences & Engineering from Guru Jambheshwar University of Science & Technology, Hisar, Haryana. She is a life member of several professional bodies like ISRS, ISG, ISPRS (I.M.). Her research areas are mainly Natural Resource Management.

Peeyush Gupta M.Sc.-Environmental Science, M. Tech-Remote Sensing & GIS, MBA-Environmental Management. Joined USAC on 25th may 2011. 6 year + experience in field of environment & space technology. Expertise: Environmental Impact Assessment Environmental Management/Disaster Management Remote sensing, Geoinformatics, Image Processing, Digital Photogrammetry, Digital Cartography Inventory in forestry and Water resource Topographical Surveying Membership of Professional Bodies: Life member, Indian Science Congress Association Life member, Indian Society of Geomatics Life member, Indian Society of Remote Sensing Life member, Indian Association of Hydrologist Employment: -Intergraph SG & I India Pvt.Ltd. Gurgaon -Uttarakhand Space Application Centre, Dehradun -Disaster Mitigation & Management Centre, Dehradun -Indian Council of Forestry Research and Education, Dehradun Editorial Board Member (International Journals): -International Journal of Environment (ISSN 2091-2854) -World Research Journal of Geoinformatics (ISSN 2278-7003) -International Journal of Engineering & Technical Research (ISSN 2321-0869) -Global Journals US (ISSN: 0975-4172).

Lokesh Jain passed his post-graduation in Agronomy from SKRAU Bikaner Rajasthan with honours and receipt gold medal. He has a vast experience in research (10 years as SRF in AICRP on Dryland Agriculture), extension (3 years as Subject Matter Specialist at KVK Barmer) and teaching as Assistant Professor since last 2 years to B.Sc. Ag. students of College of Agriculture Sumerpur. He has published a lot of publication like 1 book as co-author, contributed 9 chapters in books, 5 manuals and booklets, 50 research paper along with a lot of technical and popular articles. He is also working as Master Trainer of Agronomy for zone IIb of Rajasthan and a member of seed certification committee.

Debashis Mitra is Scientist – SG and Head of Marine and Atmospheric Sciences Department of Indian Institute of Remote Sensing, ISRO, Govt. of India. He has more than 20 years of teaching experience. His research areas of interest are application of Remote Sensing and GIS in coastal hazards; coastal processes & coastal engineering. He is the reviewer of several esteemed national and international journals and Ph.D. thesis supervisor of several reputed Indian academic institutions, e.g. IIT, Kharagpur. He has published 6 book chapters and more than 40 research articles in peer reviewed journals of national and international status. As a successful researcher, teacher and mentor, Dr. Mitra helped innumerable students to flourish in their career in his domain of expertise. Many of his students are now well established in the field of Remote Sensing and GIS across the world.

A. C. Pandey is Professor and Head, in the Centre for Land Resource Management, Central University of Jharkhand (CUJ), Ranchi. He is also the Dean, School of Natural Resource Management, CUJ. He is Coordinator (CUJ) for ISRO, EDUSAT programme. He previously (2004-2013) served as Associate Professor (Remote Sensing) in Birla Institute of Technology, Mesra, Ranchi, for about a decade and as Scientist in Department of Science & Technology, GoH, Chandigarh, for seven years. He acquired Ph.D. degree in Geology from Department of Geology, University of Delhi in 2001. He has been working in diverse areas of Earth Sciences viz., Water Resources, Glaciology, Natural hazards, Urban Environment etc. He has 04 Ph.D. and 25 M.Tech/M.Sc. thesis completed under his guidance. He has more than 45 publications in refereed international journals and 02 edited books to his credit. He is recipient of NASA-SERVIR Fellowship in 2013 to work as mentor on Himalayan glaciers in Zanskar Valley, J&K.

Priyank Pravin Patel is an Assistant Professor of Geography at Presidency University, Kolkata, India. His PhD focused on the terrain analysis of river basins for sustainable development planning. His interests within Geography reside primarily in fluvial geomorphology and geotectonics, GIS and remote sensing applications, analysis of cultural landscapes and artefacts and the mapping of urban landscapes. Currently he is engaged in mapping the badlands developed on lateritic terrain in the western part of West Bengal, India for proper land use management.

Muhammad Tauhidur Rahman is a faculty member in the Department of City and Regional Planning at King Fahd University of Petroleum and Minerals (KFUPM) in Dhahran, Saudi Arabia. He obtained his PhD in Geography from the University of Oklahoma (USA). His wide-ranging research interests include GIS and urban remote sensing, vulnerability assessment and disaster mapping, mapping and monitoring of vegetation growth, land use and land cover changes, urban growth simulation and modeling, urban heat islands, and lowering energy consumption for sustainable urban growth. He has published over dozen articles in international peer-reviewed journals and conference proceedings. Dr. Rahman has served as a Geospatial Consultant for the U.S. Department of Justice and the U.S. Department of Fish and Wildlife Services. He has received almost $3.0 million in research grants from various organizations including NASA, Natural Hazards Center at UC-Boulder, King Abdulaziz City for Science and Technology (KACST), King Abdullah City for Atomic and Renewable Energy (KACARE), and KFUPM.

Pooja Rana is a humanitarian professional with dual Masters in Disaster Management and Remote sensing & GIS (Geographic Information System). Rana has 2.11 years of experience in project management addressing Disaster Risk Reduction in India's premier humanitarian NGO, Sphere India.

Ashis Sarkar is a distinguished teacher, researcher and academician in several fields of Geography. His professional interests lie in the domain of Fluvial Geomorphology, Uban Geography, Geostatistics & Quantitative Geography, Cartography & Geoinformatics, and Geographical / Development Issues. He has taught extensively in several Colleges and Universities in throughout West Bengal (Haldia Government College, Chandernaore Government College, Presidency College (now University) as Lecturer, Assistant Professor and Professor. He has published more than 100 research papers and articles in noted Journals, authored a number books on various aspects of Geography and supervised numerous doctoral students. He is also the Editor-in-Chief of the Indian Journal of Spatial Science.

Laxmi Kant Sharma received the Ph.D. degree with specialization in Ecology & M.Tech in (RS & GIS) from IIRS (ISRO), Dehradun (India). He was associated as faculty of RS& GIS with Birla Institute of Technology (BIT-Mesra) Ranchi form 2007-2012 and Central University of Jharkhand, Ranchi from 2012-2015. Now he is an Associate Professor & Head, Department of Environmental Science at Central University of Rajasthan. Dr. Sharma is the author of more than 40 peer-reviewed national and International journal articles and other publications and three books and also serves as an editorial board member of two International Journals. Dr. Sharma has worked extensively with optical and Microwave remote sensing data, with research support from the Space Application Centre (ISRO), NRSC (ISRO), Ministry of Environment, Forests & Climate change (GoI). His research focuses on remote sensing and GIS analysis of environmental systems, climate change impacts, human–environment interactions and Sustainable Development.

Suman Sinha is a Post-Doctoral research fellow in Haldia Institute of Technology (India) under SERB N-PDF scheme (SERB, DST, GoI). He received the B.Sc. degree in Zoology in 2004 and the M.Sc. degree in Environmental Science in 2006, both from University of Calcutta, India. In 2010, he received the M.Tech. degree in Remote Sensing from Birla Institute of Technology, Mesra, India. He had been an INSPIRE Fellow (DST, GoI) pursuing Ph.D. from Birla Institute of Technology, Mesra, India, with work in the field of application of optical and microwave remote sensing for assessment of forest biomass. His main research interests include remote sensing and GIS analysis of environmental systems, specifically focusing on aboveground biomass modeling, and application of multi-frequency SAR data in forestry. He has more than 25 publications in refereed international and national journals and book and book chapters to his credit.

Shipra Verma, submitted Ph.D. in Geoinforatics, M.N.N.I.T. Allahabad. I did my M.Sc. in Environmntal Sciences. Verma has done a P.G. Diploma in GIS & Remote Sensing from University of Allahabad. Verma has worked experience from WII, Dehradun and IIT, Roorkee as a JRF in the field of Remote Sensing and GIS.

Index

H

I

J

K

L

M

N

O

P

www.ingramcontent.com/pod-product-compliance
Ingram Content Group UK Ltd.
Pitfield, Milton Keynes, MK11 3LW, UK
UKHW050947290926
15470UKWH00005B/1696

9 781522 518143